Recent Advances in Volatile Organic Compound Analysis as Diagnostic Biomarkers

Recent Advances in Volatile Organic Compound Analysis as Diagnostic Biomarkers

Editors

Natalia Drabińska
Ben de Lacy Costello

MDPI • Basel • Beijing • Wuhan • Barcelona • Belgrade • Manchester • Tokyo • Cluj • Tianjin

Editors
Natalia Drabińska
Faculty of Food Science and
Nutrition
Poznań University of Life
Sciences
Poznań
Poland

Ben de Lacy Costello
Institute of Biosensor
Technology
University of the West of
England
Bristol
United Kingdom

Editorial Office
MDPI
St. Alban-Anlage 66
4052 Basel, Switzerland

This is a reprint of articles from the Special Issue published online in the open access journal *Molecules* (ISSN 1420-3049) (available at: www.mdpi.com/journal/molecules/special_issues/ Volatile_Organic_Compound_Analysis).

For citation purposes, cite each article independently as indicated on the article page online and as indicated below:

LastName, A.A.; LastName, B.B.; LastName, C.C. Article Title. *Journal Name* **Year**, *Volume Number*, Page Range.

ISBN 978-3-0365-5350-4 (Hbk)
ISBN 978-3-0365-5349-8 (PDF)

Contents

About the Editors

Natalia Drabińska

Natalia Drabińska is a Post-doc/Assistant Professor at Poznań Univerisity of Life Sciences in Poland. Her research interest is focused on non-targeted metabolic analysis with emphasis on volatile organic compounds (VOCs) analysis in various matrices, the impact of nutrition and food on human health, and the development of innovative food products using waste products. She along with co-workers has published the most current compendium of VOCs released from the healthy human body - the human volatilome.

Ben de Lacy Costello

Dr Ben de Lacy Costello is an Associate Professor of Biosensing and diagnostics at the University of the West of England, Bristol. He has over 25 years experience in the measurement of volatile organic compounds linked to disease states in humans and crops. His PhD involved measurement of VOCs from common storage pathogens of fruit and vegetables and the fabrication of sensor systems for their early warning. More recently his work has focused on utilising volatile organic compounds to diagnose a range of human diseases including cancer, infections and inflammatory disease. Alongside measuring the VOCs using techniques such as GC-MS his work is focused on the production of inexpensive sensor systems that can be used for near patient testing. He along with co-workers has published the first compendium of VOCs released from the healthy human body - the human volatilome.

Article

Modular Breath Analyzer (MBA): Introduction of a Breath Analyzer Platform Based on an Innovative and Unique, Modular eNose Concept for Breath Diagnostics and Utilization of Calibration Transfer Methods in Breath Analysis Studies

Carsten Jaeschke [1,†], Marta Padilla [2,†], Johannes Glöckler [1,†], Inese Polaka [3], Martins Leja [3], Viktors Veliks [3], Jan Mitrovics [2], Marcis Leja [3] and Boris Mizaikoff [1,*]

1   Institute of Analytical and Bioanalytical Chemistry, University of Ulm, Albert-Einstein-Allee 11, 89081 Ulm, Germany; carsten.jaeschke@gmx.de (C.J.); johannes.gloeckler@uni-ulm.de (J.G.)
2   JLM Innovation GmbH, Vor dem Kreuzberg 17, 72070 Tuebingen, Germany; marta.padilla@jlm-innovation.de (M.P.); jan.mitrovics@jlm-innovation.de (J.M.)
3   Institute of Clinical and Preventive Medicine, University of Latvia, LV-1079 Riga, Latvia; inese.polaka@gmail.com (I.P.); martinsleja11@gmail.com (M.L.); viktors.veliks@lu.lv (V.V.); marcis.leja@lu.lv (M.L.)
*   Correspondence: boris.mizaikoff@uni-ulm.de; Tel.: +49-731-502-2750
†   These authors contributed equally.

Citation: Jaeschke, C.; Padilla, M.; Glöckler, J.; Polaka, I.; Leja, M.; Veliks, V.; Mitrovics, J.; Leja, M.; Mizaikoff, B. Modular Breath Analyzer (MBA): Introduction of a Breath Analyzer Platform Based on an Innovative and Unique, Modular eNose Concept for Breath Diagnostics and Utilization of Calibration Transfer Methods in Breath Analysis Studies. *Molecules* 2021, 26, 3776. https://doi.org/10.3390/molecules26123776

Academic Editors: Natalia Drabińska and Ben de Lacy Costello

Received: 18 May 2021
Accepted: 14 June 2021
Published: 21 June 2021

Publisher's Note: MDPI stays neutral with regard to jurisdictional claims in published maps and institutional affiliations.

Abstract: Exhaled breath analysis for early disease detection may provide a convenient method for painless and non-invasive diagnosis. In this work, a novel, compact and easy-to-use breath analyzer platform with a modular sensing chamber and direct breath sampling unit is presented. The developed analyzer system comprises a compact, low volume, temperature-controlled sensing chamber in three modules that can host any type of resistive gas sensor arrays. Furthermore, in this study three modular breath analyzers are explicitly tested for reproducibility in a real-life breath analysis experiment with several calibration transfer (CT) techniques using transfer samples from the experiment. The experiment consists of classifying breath samples from 15 subjects before and after eating a specific meal using three instruments. We investigate the possibility to transfer calibration models across instruments using transfer samples from the experiment under study, since representative samples of human breath at some conditions are difficult to simulate in a laboratory. For example, exhaled breath from subjects suffering from a disease for which the biomarkers are mostly unknown. Results show that many transfer samples of all the classes under study (in our case meal/no meal) are needed, although some CT methods present reasonably good results with only one class.

Keywords: breath analysis; MOX sensors; low sensing chamber volume; calibration transfer; standard samples; piecewise direct standardization; correlation alignment; breath sampling; eNose; pattern recognition

1. Introduction

The importance of exhaled breath gas analysis is increasing in medical diagnostics for early disease detection and therapy progress monitoring over the last decades [1–4]. Exhaled human breath is composed of nitrogen, oxygen, carbon dioxide, water vapor, inert gases and trace amounts of volatile organic compounds (VOCs) [5]. The ancient physician—Hippocrates (460–370 BC)—noticed that the exhaled breath of an ill patient differs from a healthy one and described fetor oris and fetor hepaticus in his essay on breath aroma and disease. In 1971 modern breath analysis started with the experiments of Pauling et al. [6], he showed that human breath contains several hundred different VOCs in low concentrations. Pauling's observation was confirmed by subsequent studies from other research groups.

Phillips et al. approved Pauling's statement and confirmed that exhaled human breath contains more than a thousand different VOCs in low concentrations [7,8]. Furthermore, they observed, by gas chromatography coupled with mass spectrometry, 3481 different VOCs in the breath of 50 healthy humans. On average each human has approximately 204 VOCs in their breath. Moreover, it was observed that only 27 VOCs were equal in the breath of the 50 healthy humans examined. This makes breath analysis a difficult task because there is only a small common core of VOCs in all humans. These VOCs are probably produced by metabolic pathways common to most humans [8]. Moreover, these VOCs can be from exogenous and endogenous origin [9]. Exogenous VOCs are inhaled or absorbed as contaminants via breath, skin or ingestion [10] while endogenous VOCs are produced in the body via the metabolism [3]. The identification of these VOCs is very important and forms the focus of research, as they act or can act as important "markers" for the early detection of a disease [11]. The identification of breath markers should be qualitative and quantitative to distinguish between a diseased group and a healthy one. The differences in the VOCs content between these two groups must be large enough to reach clinical relevance. In the last 30 years, many of these molecules have been identified and correlated to different diseases. The basis of the emission of VOCs is cell biology. Tumor growth causes metabolic changes which are linked to the production of specific volatile compounds [12–15]. Cancer-related blood chemistry changes lead to changes in breath by exchange through the lung [16]. Therefore, some VOCs can be used as cancer markers in exhaled breath [3].

Currently, the gold standard for detecting VOCs in exhaled air is gas chromatography coupled to mass spectrometry (GC-MS) [1,17–26]. Beside GC-MS, other analytical instrumentations are used like proton-transfer reaction-mass spectrometry (PTR-MS) and ion mobility spectrometry (IMS). These techniques enable separation, identification and quantification of the different VOCs in the exhaled breath gas. The main disadvantages of the analytical instrumentation are the need of high skilled operating personnel, being time-consuming (except for IMS) and the high costs.

To reduce costs, chemical sensors integrated into electronic noses (eNoses) for breath analysis in medical point-of-care diagnosis have become an emerging field. Many research groups are pushing forward the frontier of non-invasive, rapid, portable and potentially low cost medical diagnosis tests for different diseases [3,27]. Electronic noses in breath gas analysis are still a noticeably young research field. Different research laboratories use different internal standardized methods for the breath sample collection, but there is no globally accepted standard procedure. The common procedures are total or alveolar breath gas sampling. In total breath sampling, the complete breath is collected including dead space air, and in the alveolar breath gas sampling only the end-tidal, alveolar part of breath is collected [1,28,29]. The method of total breath gas sampling is simple but has a big disadvantage because of the dilution with the dead space air [1]. In comparison, alveolar sampling reduces the concentration of contaminants [1,29].

Advantages and limitations of eNose sensor techniques are associated with different parameters like specificity, response and recovery time, detection range, sensitivity, operating temperature, temperature as well as humidity effect on sensing technique, portability, cost and complexity of measuring circuitry.

In the review by Röck et al., a list of commercially available eNoses is published [30]. Several preliminary studies were conducted with those commercially available eNose systems like Cyranose 320 [31–36], LibraNose [37] and DiagNose [38] using offline sampling with Tedlar or Mylar bags. Other studies with chemical sensors, surface acoustic wave (SAW) sensors [39], metal oxide semiconductor (MOX) sensors [40], colorimetric sensors [41], quartz microbalance (QMB) sensors [42,43], MOX-SAW sensors [44] and trichloro-(phenethyl)silane-silicon nanowire-field effect transistor (TPS-SiNW FET) sensors [45] were carried out. Furthermore, studies were conducted in which an eNose and additionally gas chromatography coupled to mass spectroscopy were used. Those studies showed that different organic functionalized gold nanoparticle (GNP) sensors are suitable to different diseases [46–51]. Amal et al. investigated the detection of gastric cancer utilizing GNP sensors with offline sampling and demonstrated that the sensor technology determines that the breath of cancer patients is different from healthy ones. [49,50].

Over recent years, the field of breath analysis with MOX-based eNose systems is continuously progressing. Different studies of cancer detection via breath gas analysis with eNoses based on MOX sensors have been conducted by Yu et al. [40], Wang, D. et al. [44] and Wang, X.R. et al. [52] and some of those studies also involved other sensor technologies in combination with a MOX sensor array. De Vries et al. integrated an eNose sensing array into an existing diagnostic spirometer. This system is based on five identical commercially available MOX sensor arrays out of four MOX sensors [53].

Special attention shall be given to the experiment design, most of the published studies on breath analysis by an eNose were conducted with one device. For larger scale studies, given the difficulties of obtaining breath measurements from patients with specific conditions, it would be desirable to extend the study to more devices and more places (like different hospitals or recruitment centers). However, sensor to sensor variability, time degradation (drift), cross-sensitivities to background and environmental conditions, etc., causes data models (calibration models) built for one instrument at a given time and place to not be valid for measurements collected by another instrument, or the same instrument in another place or later in time. In other words, the data calibration models degrade when taking measurements under conditions other than those under which the calibration model was created. The effort in building a calibration model is costly and time consuming, and therefore limits the use of eNoses in many applications such as breath analysis.

To reduce the impact of these limitations, data processing techniques exist that help in reducing the effort of full calibrations by transferring information from the main calibration model (built in a so-called master instrument) to be applied to new measurements obtained under different conditions (from so-called slave instruments). These techniques are called calibration transfer (CT) or instrument standardization in chemometrics, and transfer learning tools in machine learning. CT methods have been largely applied in NIR spectrometry and also in eNoses [54–57]. The objective of a calibration transfer method is to perform its task using as few transfer samples (samples to link instruments) as possible. In this way, the costs of calibration of individual (slave) instruments are reduced to a few measurements of transfer samples, instead of a whole large set for proper calibration.

Calibration transfer methods can be grouped according to different criteria. For example, the following approach [58]: (i) no standardization (feature selection, calibration model extension (CME) by including samples from multiples instruments, special pretreatments like orthogonal signal correction (OSC) [59]); (ii) adjusting the output of the calibration model to be used by other instruments, such as the simple univariate slope and bias correction (SBC) [60]; (iii) transforming measurements from slave instruments so that they resemble measurements from the master instrument using direct standardization (DS) and piecewise direct standardization (PDS) [61,62] and (iv) removing differences between instruments that are orthogonal to the calibration model [63]. Indeed, PDS has shown good results and is considered by many to be a reference for novel techniques [64–69].

Other classification of calibration transfer techniques can be made with regard to the domain of the transfer and the transfer samples. For the first one, we can have (a) transfer to the master space or the slave space, such as DS and PDS, (b) transfer to a common subspace such as OSC. Additionally, for the latter, CT methods regarding transfer samples; (a) methods that need transfer samples such as DS, PDS, OSC [70], Shenk's algorithm [71], spectral space transformation (SST) [72] and canonical correlation analysis (CCA) [73] (b) and methods that do not need transfer samples, such as methods from the IR spectra field: multiplicative signal correction (MSC) [66], finite impulse response (FIR) [68,69], stacked partial least-squares (SPLS) [67] and from the machine learning field, like transfer component analysis (TCA) [74] and transfer sample-based coupled task learning (TCTL) [54] and others that have been applied to eNoses [75]. Reviews and discussions can be found in literature about the use of standard samples [76], about the techniques based on orthogonal projections [77,78] and reviews of different techniques [58,65,79].

An important matter in calibration techniques that need transfer samples is the selection of these transfer samples. Such samples must be representative of the samples under study and the respective instruments, keeping other variables such that they can be linked between the instruments. The required transfer samples would be those measured at the same conditions by different instruments, for example, by having the instruments measuring the same sample at the same time; we call them standard samples. When standard samples are not available, we can use reference (or nonstandard) samples, which are measurements made under exact or similar conditions in all master and slave instruments.

In addition, the fact that transfer samples must be representative of the samples under study may be a limitation in out-of-lab applications, such as breath analysis, because representative samples of given cases might be difficult to obtain. In an ideal situation, samples representative of the cases under study can be artificially made, such that measurements of these samples can be made in a laboratory under controlled conditions and thus be used as transfer samples. This way, all instruments would be referred to the same general samples under specified conditions. However, it is difficult to create synthetic samples representative of complex samples such as human breath samples, especially for patients suffering diseases for which the exhaled VOCs pattern is poorly known or even unknown and which is affected by numerous variables. Therefore, the use of on-site measured samples as transfer samples may be helpful, or alternatively hard-to-obtain sample classes could be excluded from the transfer sample set. In summary, for practical reasons we wonder whether on-site measured sample measurements can be used as transfer samples and whether the quality of the calibration transfer would decrease much if one sample class (necessary for the calibration model) is excluded from the transfer sample set. We can find an example in the literature where breath samples from electronic noses were used, although case breath samples were artificially made by mixing control breath samples with chemicals [57].

In this work, we explore the performance of several CT techniques in a real-life breath analysis study using our recently published [80,81] sensing array and three instruments. The experiment consists of the discrimination of breath samples from people before and after eating a specified meal. The performance of the CT methods is evaluated in regard to the number and type of transfer samples (standard or nonstandard) and class membership (transfer samples belonging to one class or both classes meal/no meal), to explore the possibility of using transfer samples from the on-site experiment in the CT methods.

This study is mainly motivated by the difficulty of calibration of eNoses for breath analysis applications, the differences between instruments, the frequent recalibrations needed due to aging and drift and the environmental and other different conditions present in different hospitals which prevents obtaining a unified dataset for deep statistical studies. Another goal in this work, is to present our modular breath analyzer (MBA) platform (shown in Figure 1), a new updated version of our modular eNose [80–82] specifically designed for breath analysis. The previous version of the modular eNose concept was recently presented [80–82]. It consisted of a new eNose platform based on a novel modular sensing chamber, where different kinds of chemoresistive sensors can be combined [80–82]. In one of these works, we combined analog and digital commercial surface mount devices (SMD) MOX gas sensors and checked its potential in an experiment aiming to detect VOCs under a high humidity background in a future application [82]. The other experiment consisted of testing the instrument for on-line monitoring under dry and moderate humid conditions with six concentrations of two VOCs of interest [80]. The MBA is based on the previously presented innovative iLovEnose concept of a modular eNose system [80,81] and incorporates a direct alveolar breath sampling system, which is explicitly used in the breath analysis experiment.

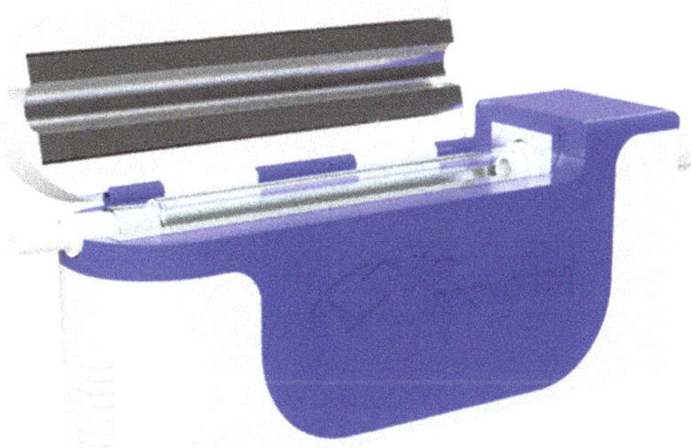

Figure 1. Modular breath analyzer (MBA) platform based on chemoresistive sensors for laboratory and clinical use; external view with cap opened to exchange the glass sampling tube.

The manuscript is organized as follows. Section 2 describes our updated MBA instrument and details both experiments: the breath analysis study. Section 3 explains the analysis methods and followed methodology. Section 4 shows and discusses the results and finally Section 5 derives some conclusions.

2. Material

2.1. Device Description

The presented novel modular breath analyzer (MBA) platform is developed for laboratory and clinical use. The internal components of the MBA platform are shown in Supplementary Materials, Figure S1, while its basic arrangement and the connection of the individual units is shown with a schematic drawing of the MBA platform in Figure 2. It contains a direct breath sampling unit and three modules able to host different types of sensors and technologies. The MBA platform contains three main units: (i) a sampling unit with an internal exhalation monitoring unit (EMU), (ii) a temperature control unit and (iii) a modular sensing chamber unit. The sampling unit is especially designed for breath analysis, based on the buffered-end-tidal (BET) sampling process [83]. Our presented system weights about 2.1 kg and has the dimensions 280 mm × 118 mm × 75 mm.

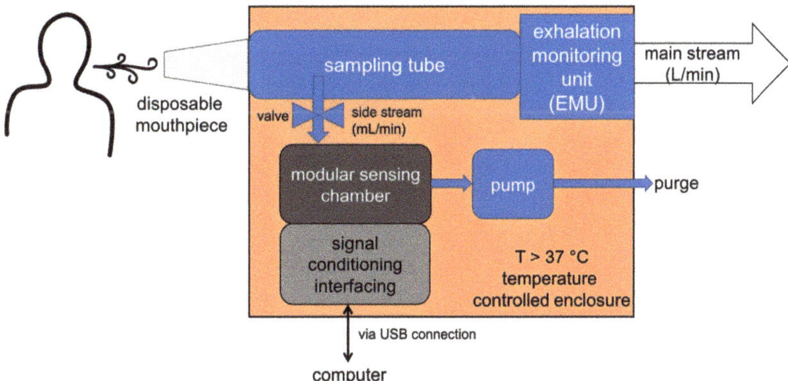

Figure 2. Schematic drawing of the modular breath analyzer platform for laboratory and clinical use showing the linkage of the individual units. Fluidic units are drawn in blue (sampling and exhalation monitoring unit) and dark gray (modular sensing chamber).

The two fluidic units—sampling unit as well as sensing unit—which are described below are integrated into a temperature-controlled aluminum body to ensure that all fluidic paths within the MBA platform are kept above body temperature to avoid temperature effects on the sensors to avoid influences towards the composition of the exhaled breath gas. Without thermostatic control the temperature within the fluidic paths will drop below body temperature and thus vapor will condense and trap water soluble volatile organic compounds (VOCs). For this reason, the temperature inside the sensing chamber of the presented modular breath analyzer platform is 45 °C ± 1 °C.

2.1.1. Buffered-End-Tidal (BET) Sampling and Exhalation Monitoring Unit (EMU)

During an exhalation into the device the volume of the sampling tube is exchanged several times. After the exhalation process has finished, the last 38 mL of the exhaled breath are buffered within the tube and remain there until the sample is transferred into the sensing chamber by a controlled pumping process. The buffered volume of approx. 30 mL allows the system to transfer several times the volume inside the sensor chamber, which can be set by the software. The internal exhalation monitoring unit is coupled to the sampling tube to operate the pump directly when the exhalation stops to draw selected parts of the alveolar air inside the sensor chamber. Sensors within the EMU allow real-time monitoring of the full exhalation process. EMU parameters (like pressure, temperature, humidity) are recorded during the exhalation process of the volunteer or patient to ensure proper repeatable sampling and enable capturing relevant parts of the exhaled breath, which are different portions of the pulmonary volume.

For a reliable breath analyzer platform, it is crucial that the EMU is not only recognizing the end of the exhalation, it is also important to ensure that the volume of the exhaled breath and the profile is within a certain variance. Direct feedback to the patient and the study nurses may help to improve the sampling process. This enables capturing relevant parts of the exhaled breath and allows a more accurate transfer of sample to sensors as well as selective sampling of different portions of the pulmonary volume. To avoid cross-infection between the patients the glass sampling tube can be exchanged and cleaned by sterilization. The disposable mouthpiece is exchanged for every volunteer or patient.

2.1.2. Modular Sensing Chamber Unit

To utilize the buffered end-tidal breath sampling method, we designed a specific valve-controlled inlet for our recently published sensing chamber [80–82] to be able to plug the exchangeable glass sampling tube to the sensing chamber of our eNose system. The current sensing array setup consists of three compartments: one with 8 analog and two with 10 digital sensors each. The low volume of the sensing unit (less than 3 mL) ensures that the volume of the sensor chamber could be flushed several times with the BET air from the sampling tube. The sensing unit consists of a modular sensor array that contains three exchangeable sensor modules with a valve-controlled inlet connected to the sampling tube and one outlet (can be seen in Figure 3). The exchangeable modular design of the sensor unit allows the MBA to host three modules containing sensors of different types, but those three modules can also contain the same type of sensors, which is useful in the study of sensor chip variability.

The current setup contains many of the most relevant analog and digital surface mount devices (SMD) sensors on the market. A list of all integrated sensors, number of obtained signals from each sensor and used heater/supply voltages are summarized in Tables S1 and S2, see Supplementary Materials. The concept of the sensing unit follows a modular structure to allow an easy and simple exchange of the sensors [80].

The cleaning of the sensor chambers is done by a two-step process. First, the pump and the valve shown in Figures 2 and 3 are used to generate a low vacuum for a few seconds to remove the breath gas out of the sensing chamber, and then ambient air is driven through the sensing chamber. The cleaning cycle is programmed in the firmware of the MBA, and it is started after a successful and a cancelled measurement.

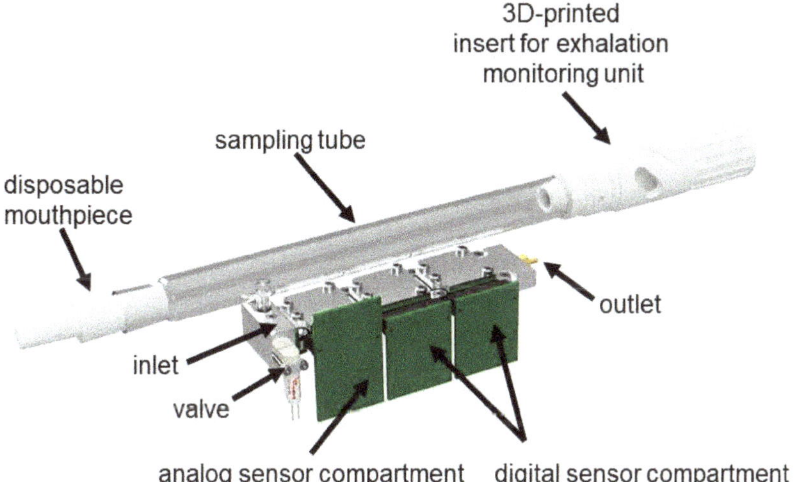

Figure 3. Sensing chamber with three individual compartments, shown here as a combination of analog and two digital metal oxide semiconductor (MOX) sensor compartments and the connection of the sampling tube to the valve-controlled sensing array inlet.

2.2. Experiment–Pilot Study Description

A group of 15 generally healthy 17–18 year old individuals were recruited for an experimental study using three modular breath analyzer prototypes, the first breath sample was obtained following a 12 h fasting period with all three MBA devices; then the participants were given a standardized meal and invited for a follow-up (second) breath sample 4 h thereafter. The test-meal was a hamburger with 0.5 L water. To avoid potential contaminants, on the day of testing the recruited study participants did not use mouthwash, chewing gum, furthermore they did not perform excessive physical activity, did not smoke or consume alcohol for 24 h before the breath test. The same restrictions were applied to the 4-h pause between two measurements. This experimental routine was repeated three times, each measurement day was one week apart from the previous one. The general scheme of the breath sample collection (measurement day) of the pilot study can be seen in Figure 4. Signed consent was obtained from all recruited study participants For study participants under 18 years of age the parents or legal guardians signed this consent.

Figure 4. General scheme of collecting the breath samples for the pilot study.

In this study, the firmware of the used MBA devices was set to 10 s for baseline acquisition and 20 s for breath acquisition.

3. Methods

3.1. Calibration Transfer Algorithms

In this work, we compare three methods for calibration transfer based on different principles: Correlation alignment (CORAL) [84], partial least squares-based calibration transfer (PLSCT) [85], direct standardization (DS) and piecewise direct standardization (PDS) [61,86] and a partial least squares discriminant analysis-based method (PLSDA) [81].

DS, PDS and PLSCT are based on adjusting the slave features to the master by using a set of labeled transfer samples measured in both the master and the slave instruments. These samples can be standard samples, i.e., samples related to both instruments, such as same samples measured at the same time, samples measured at the exact same conditions, etc., or non-standard samples if they are labeled but not related as the standard samples. In turn, CORAL is a simple method that transforms data from the master to the slave space using their covariance structure without the need of labelled samples. Finally, PLSDA finds a common master-slave space by removing components using unlabeled transfer samples, only the information about their membership to the master or the slave instrument is used.

Direct standardization (DS) and piecewise direct standardization (PDS) methods—DS and PDS [61,86,87] methods were created in the field of NIR spectrometry to correct the slave spectra by computing a transfer matrix. This transfer matrix is obtained by relating the master spectra to the slave's spectra by using a small number of labeled transfer samples. The PDS method is in fact an extension of DS by which each wavelength (variable) at the master spectra is related to a sliding window of fixed size in the slave spectra. PDS can deal with having a larger number of variables than samples [64].

DS and PDS are widely used methods which have provided good results in laboratory experiments using a relatively small number of samples [56,79] and are typically employed as a reference for other novel techniques [64].

DS assumes a linear relationship between master and slave instruments such that:

$$X^m_{ct} = X^s_{ct} \, B \tag{1}$$

where B is the transformation matrix and X^m_{ct} and X^s_{ct} are the data measured from the transfer samples at the master and slave instrument, respectively. Therefore, B can be estimated by

$$B = (X^s_{ct})^+ \, X^m_{ct} \tag{2}$$

where $(X^s_{ct})^+$ is the pseudo inverse of X^s_{ct}. The new samples from the slave instrument X^s can be projected onto the master instrument X^m:

$$X^m = X^s \, B \tag{3}$$

In turn, PDS creates local PLS models relating the master instrument j-th variable to a sliding window of size w centered at the j-th variable in the slave instrument. The resulting transformation matrix B_{PDS} has a diagonal structure:

$$B_{PDS} = \text{diag}(b^{1T}, b^{2T}, \dots,) \tag{4}$$

where k is the number of variables on both instruments. Finally, X^s can be projected on the master instrument by:

$$X^{mT} = X^{sT} \, B_{PDS} \tag{5}$$

Partial least squares discriminant analysis-based calibration transfer (PLSDA)—the PLSDA-based method builds a PLSDA model relating transfer unlabeled samples from both master and slave instruments with a dummy vector containing their membership (master or slave) label. Furthermore, the predicted data from this model is removed from the original data set.

If W and P are notations for the resulting PLSDA weight and latent variable matrices, respectively, X^m and X^s denote the original data from the master and the slave instrument, respectively, and $X^{m\prime}$ and $X^{s\prime}$ denote the transformed data from the master and the slave instrument, respectively:

$$X^{m\prime} = X^{om} - X^{om} \, W \, ((P)^T \, W)^{-1} \, P \tag{6}$$

$$X^{s\prime} = X^{os} - X^{os}\, W\, ((P)^T\, W)^{-1}\, P \qquad (7)$$

The number of components or latent variables (LVs) to be removed must be selected. Finally, a classification/regression method can be built on the $X^{m\prime}$ and be used to predict $X^{s\prime}$.

Partial least squares-based calibration transfer (PLSCT)—In PLSCT [85], a PLS model is built in the master instrument and a subset of samples are used to relate master and slave instruments. This operation is made in the PLS low dimensional space between projected spectra from transfer samples in both master and slave instrument.

If X^m and Y^m are the calibration set data and label matrices in the master instrument, respectively, W^m, P^m and β^m are the weight, latent variable and regression coefficient matrices of the PLS model for X^m and Y^m in the master instrument (with selected number of latent variables (LVs)), the projection of the master's transfer samples X^m_{ct} in the PLS model T^m_{ts} and the projection of the slave's transfer samples X^s_{ct} in the PLS model T^s_{ts} are given by:

$$T^m_{ts} = X^m_{ct}\, W^m\, ((P^m)^T\, W^m)^{-1} \qquad (8)$$

$$T^{\prime s}_{ts} = X^s_{ct}\, W^m\, ((P^m)^T\, W^m)^{-1} \qquad (9)$$

Assuming a linear relationship between the projection matrices:

$$T^s_{ts} = T^{\prime s}_{ts}\, M = T^m_{ts} \qquad (10)$$

where M can be obtained by the ordinary least squares method:

$$M = ((T's ts)T\ T's ts) - 1\ (T's ts)T\ Tm ts \qquad (11)$$

when M is obtained, a classification/regression method can be applied on the projected matrices T or from the PLS model already built in the master instrument:

$$Y^s = X^s\, \beta^m \qquad (12)$$

Since in this work we use PLSDA for a classification problem, we will call this method PLSDA-CT instead of PLSCT.

Correlation alignment (CORAL)—CORAL [84] is an unsupervised domain adaptation method coming from the machine learning field that attempts to minimize the differences in the data distributions between two domains (master and slave instruments in our case) by transferring the data structure of the target domain (slave).

Notating X^m and X^s subsets of unlabeled data from master and slave, respectively, having N_m and N_s number of features each (X matrix columns), the proposed transformation is:

$$C_m = (X^m)^T\, X^m + \lambda\, I_{Nm \times Nm} \qquad (13)$$

$$C_s = (X^s)^T\, X^s + \lambda\, I_{Ns \times Ns} \qquad (14)$$

$$X^m = X^m\, C_m^{-1/2}\, C_s^{1/2} \qquad (15)$$

where C^m and C^s are the master's and slave's data covariance matrices, respectively, adapted with a small regularization parameter λ that allows it to be full rank and thus the square root to be computed. Therefore, CORAL uses two steps to align both data distributions: whitening the master data and re-coloring it with the slave covariance.

3.2. Data Analysis Methodology

In this work, we study the performance of several CT algorithms to transfer calibration models between pairs of 3 instruments in an experiment consisting of an on-site real breath analysis study for discrimination of breath samples from subjects before and after eating a specific meal. Thus, data is classified into classes "meal" and "no-meal". We use device 1 as the master device, thus devices 2 and 3 are considered the slave devices. Denoting the master instrument as M and the slave instrument as S, two instrument pairs considered are: M1–S2 and M1–S3.

To illustrate the effect of the different CT algorithms, we followed a procedure consisting of three steps. In the first step, dimensionality reduction and a classifier using data from M is built and evaluated with selection of their optimal parameters for each of the 3 devices. In the second step, a small number of samples (transfer samples) from both M and S are selected and the CT is performed. Finally, in the third step, the CT algorithm is evaluated using the classifier from the first step on S data. The result of the M classifier applied on the M data is considered the reference to be achieved, while the result of the M classifier applied on each S data is considered the threshold to be overcome.

Since only two classes are involved in the classification task, the classification results are given in area under the receiver operating characteristic curve (AUC) and standard errors (SE). Results for each S device are compared before and after the application of the CT algorithms; if the results of the corrected S data (after CT) are similar to the M's reference classification result we consider a successful CT, if the S's result is higher than the one before the CT (threshold) we consider good CT.

Classifier—First, the basic pre-processing step here is given by the ratio of the sensor's conductance (1/R) with the baseline, which gives R0/R, where R0 is the sensor output resistance to room air (baseline) and R is the sensor's response resistance to the breath exposure. The considered R is one value that summarizes the sensor's response to the whole analyte exposure. It corresponds to the mean of the latest measurements before the cleaning step, in this case the last 5 s, when the sensor responses are most stable (steady state). Therefore, the resulting data sets from measurements with our 18 sensors have 18 columns, one per sensor.

As specified above, the results of the classifier built in M and applied on each S is considered the reference for the evaluation of the CTs. The considered classifier is linear discriminant analysis (LDA) with a previous dimensionality reduction task using principal component analysis (PCA). Therefore, the number of PCA components (nPCs) for the PCA + LDA classifier is the parameter to be selected. For this, cross-validation is made on two random subsets of the M's data: a training set containing 48 samples with equally represented classes (24 + 24), and a validation set with the remaining samples (~39). The procedure is repeated 20 times for each parameter value to obtain the optimal nPC.

Transfer samples—Once the optimal nPCs for the classifier in M are selected, the CT task is carried out as follows. For every pair M–S, a sample subset (M-training set) is selected from M with both classes equally represented. The M's transfer samples are selected from this subset. Then, the S's transfer samples are selected according to the CT algorithm as described below, and the CT is performed.

The CT algorithms need a number of samples (transfer samples), labeled or not, from both M and S. To select the transfer samples from M, we use first the Kennard–Stone algorithm (KS) [88], and then for CT methods that use labeled transfer samples, the transfer samples from S are selected according to standard samples or nonstandard samples. Instead, for CT methods that use unlabeled transfer samples, KS is also used to select the S's transfer samples. The standard samples correspond to the M's equivalent samples in S, this is, the samples that were taken close in time (the subjects exhaled on each instrument right one after the other) from the same patients by the 3 instruments. In turn, the non-standard transfer samples are an alternative to the standard samples and correspond to the very likely case where standard samples as defined above are not available. Non-standard transfer samples are selected as follows: once M's transfer set is selected with KS, from a subset of 20 known (labeled) S samples, each selected S transfer sample is the closest to each M's transfer sample.

In addition, for cases with labeled samples the selection of M and S transfer samples is made according to 2 class membership conditions: samples belonging to (a) both classes in the experiment (meal/no-meal) or (b) only one class corresponding to no-meal. As for the healthy class in a disease-control breath analysis experiment, subjects belonging to no-meal are easier to collect and thus, a CT method based on only such sample class would be more practical.

CT evaluation—To assess the CT algorithms' performance, the selected M classifier is built using the M's training samples and the S's transfer samples if they are labeled. The classifier is then applied on the corrected S's samples excluding its transfer samples (Figure 5).

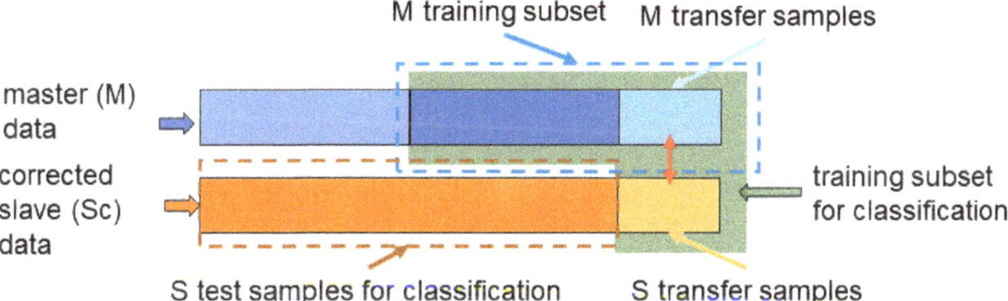

Figure 5. Visual scheme of training set for the calibration transfer evaluation.

An additional reference can be considered for the CT algorithms that need known labeled transfer samples from S; the calibration extension method (CEM) which consists of the LDA classifier built with M's training samples plus original (non-corrected by CT) S's labeled transfer samples, this is S's samples without CT. Then, the CT can be considered good if its AUC overcomes the CEM's AUC. The procedure is repeated 20 times and the AUC and SE are computed.

The number of considered transfer samples is 10, 20, 30 and 40. If the transfer samples are labeled, as for DS, PDS and PLSDA-CT, the classes are equally represented within them.

4. Results and Discussion

4.1. The Dataset

After removing few outliers using PCA, the data set composition of the breath analysis experiment is shown in Table 1.

Table 1. Data set composition.

Device	Meal	No-Meal
1	41	45
2	43	45
3	42	45

Figures 6 and 7 show the PCA scores plots of the complete data set according to the devices and the meal status (classes), respectively. Measurements from the three devices depend strongly on the device, since the breath samples come from the same individuals and the sampling is made on each device one right after the other (Figure 6). On the other hand, a certain degree of overlap between the data classes meal/no-meal can be seen in Figure 7. This happens in every device, as it is shown by the sample symbols shown in Figure 6.

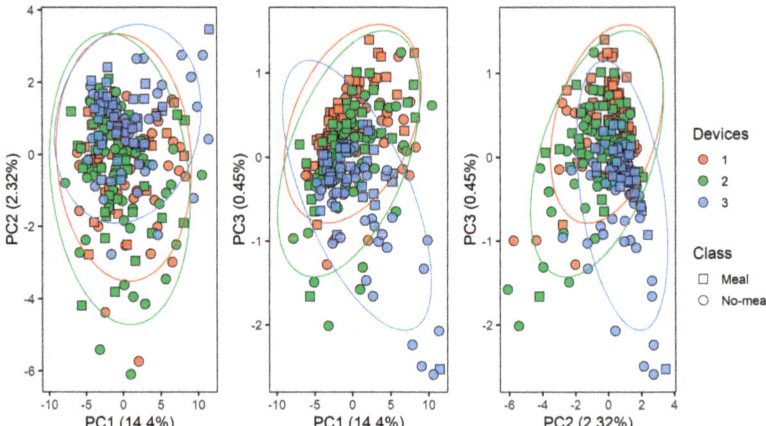

Figure 6. Principal component analysis (PCA) scores plot of the data according to the devices.

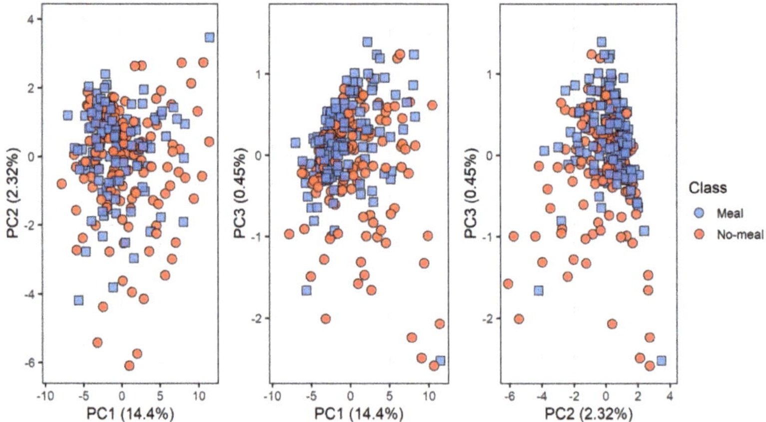

Figure 7. PCA scores plot of the data according to the classes: meal/no-meal.

4.2. Classification

The references to compare the performance of the CTs are the results of the classification of every device's data using a PCA + LDA model built with M's training set data (48 samples) before the CT. Table 2 shows the classification results for the PCA + LDA models giving the best AUC according to device master-slave pairs, which indicates: device to build model-device to test model. For example, for pairs with the same device as M1–M1, it indicates that a training set (~48 samples) from M1 was used to build the model and a test set (~39 samples) from the same device was used to evaluate it. When the pairs are formed by different devices, the number of samples in the training set is ~48 samples but for the test set is ~80 samples.

Table 2. Classification results before calibration transfer (CT).

Pair Train-Test Device	AUC (%)	Accuracy (%)	Sensitivity (%)	Specificity (%)	PCA nPCs
M1–M1	89.26 ± 0.87	80.01 ± 0.14	84.11 ± 0.20	75.75 ± 0.22	13
S2–S2	93.34 ± 0.65	86.41 ± 0.10	85.53 ± 0.17	87.25 ± 0.16	-
S3–S3	91.03 ± 0.12	81.56 ± 0.17	80.63 ± 0.30	82.50 ± 0.21	15
M1–S2	73.15 ± 1.15	64.83 ± 0.09	49.65 ± 0.26	79.66 ± 0.20	13
M1–S3	75.72 ± 2.40	66.88 ± 0.18	79.00 ± 0.38	54.75 ± 0.61	13

We can see in Table 2 that results for pairs of the same devices show good results for discriminating human breath before and after the meal for the individual devices. However, there is a significant performance decrease when the classifier is built with the master device, which is a clear indicator of the fact that the devices differ.

The selected reduced dimension obtained after cross-validation (13 PCs) corresponds to the model with optimal results in M1 (89.26 ± 0.87). The same model applied to devices 2 and 3 gave 73.15 ± 1.15 and 75.72 ± 2.40, respectively. These values are the lower value reference for the evaluation of the methods (shadowed in Table 2).

4.3. Calibration Transfer Using Two-Class Transfer Samples

Figures 8 and 9 show the results of CT methods CORAL and PLSDA for both slave devices. These are methods that do not need labeled samples from S. However, since results for CORAL depend on λ and PLSDA depend on nLV, some known samples in the slave device must be known in order to find an optimal value. Results for CORAL depend on the device but good results are obtained for both devices at high λ with low dependency on the number of transfer samples. AUC results for the low λ increase with the number of transfer samples. The lowest λ give the best results for device 3, while it is the contrary for device 2. In turn, for PLSDA the optimal number of PLSDA components to be removed depends on the slave device and the number of transfer samples. Best results are obtained with the maximum number of transfer samples. Figures 10 and 11 show the results of CT methods DS, PDS and PLSDA-CT for both slave devices, when using standard samples. CEM is a reference which shows whether it is worth applying any of these CT methods or simply building the CEM classifier with labeled samples from each pair of M and S devices together.

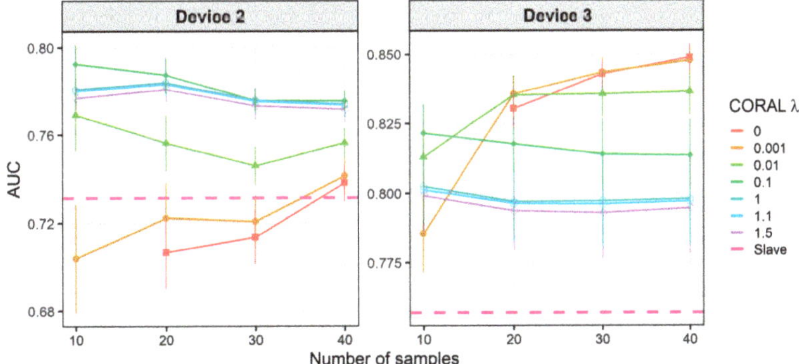

Figure 8. Classification results for correlation alignment (CORAL) using two-class standard transfer samples.

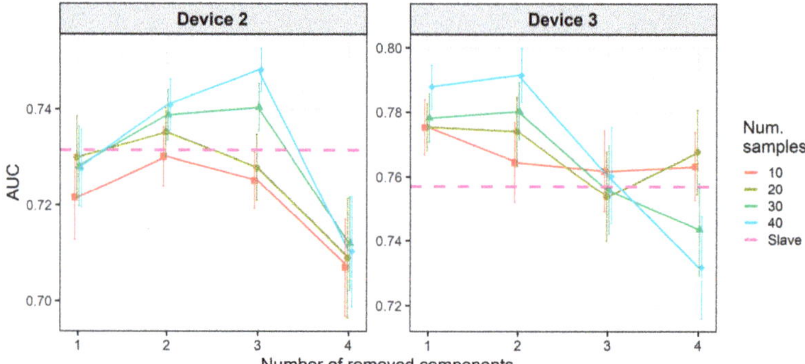

Figure 9. Classification results for partial least squares discriminant analysis (PLSDA) using two-class standard transfer samples.

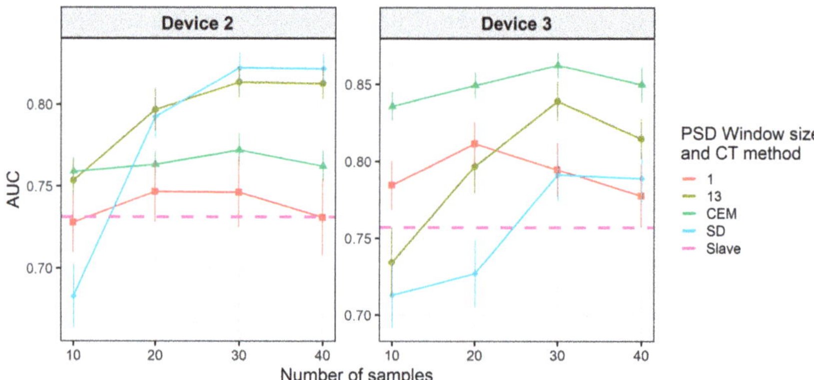

Figure 10. Classification results for CEM, DS and PDS using two-class standard transfer samples.

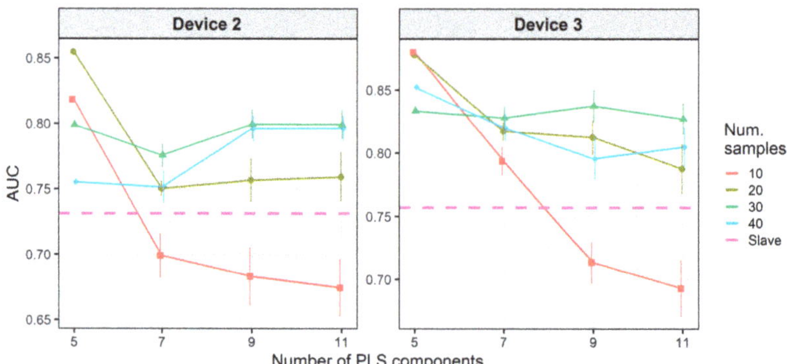

Figure 11. Classification results for using two-class standard transfer samples.

Figure 10 shows that when using standard transfer samples, PDS can give good results in on-field experiments, although with many more transfer samples than those reported for lab experiments. The optimal parameters for PDS or DS depend on the slave device and number of transfer samples. For device 3, it is better to use CEM than the PDS as CT. For both slave devices, PDS at the maximum window size (13) and a high number of transfer samples gives the best results. In turn, when using PLSDA-CT (Figure 11) we can see that the best results are obtained with few components and medium number of transfer samples. However, good and more stable results with respect to the nLVs are obtained with many transfer samples.

Figures 12 and 13 show that there is a considerable decrease in the general performance of PDS, DS and PLSDA-CT methods when using non-standard transfer samples. CEM results, which are similar to the case of standard transfer samples above, are still the best for device 3 (as in Figure 10). However, for device 2 it becomes comparable to PDS with window size 1, while DS and PDS for window size 13 drop below the threshold. Window size equal to 1 for PDS does not give good results but it keeps the values mostly above the threshold level for both devices at the two cases of transfer samples used. Given the general behavior of DS in Figures 9 and 10, for a study with more data for which more transfer samples were available (thus bigger training set size) much better rates could be achieved. The general worsening of the results can also be seen for PLSDA-CT. However, AUC is still good for the smallest number of LVs.

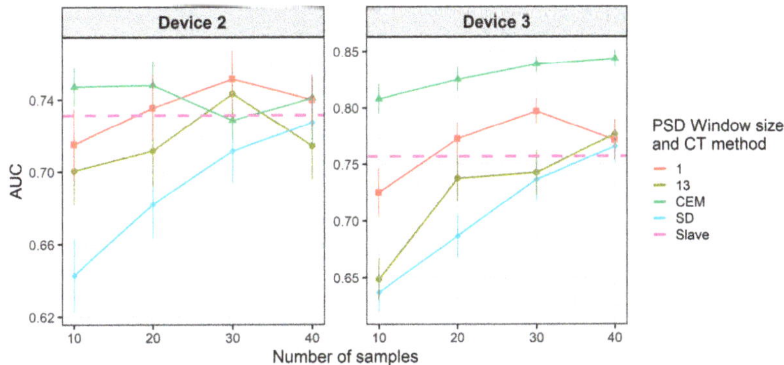

Figure 12. Classification results for CEM, DS and PDS using two-class non-standard transfer samples.

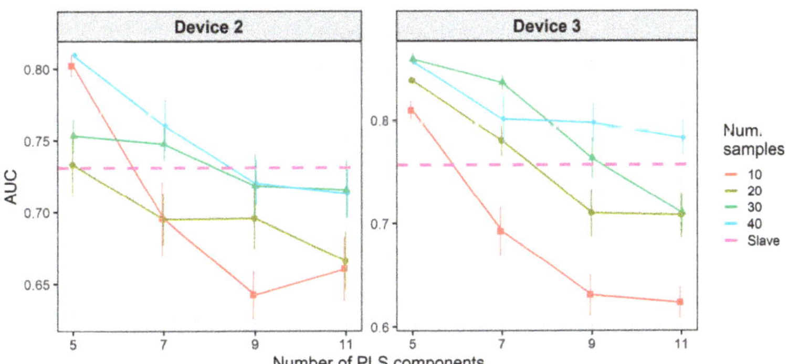

Figure 13. Classification results for PLSDA-CT using two-class non-standard transfer samples.

4.4. Calibration Transfer Using One-Class Transfer Samples

The following figures show results for the case of using transfer samples of only one class, "no-meal" in our case. Figure 14 shows results for CORAL, which behaves similarly although slightly worse than for the previous case (Figure 8). The high λ values give close results which are stable with respect to the number of transfer samples, while small λ values give increasingly better results with increasing number of transfer samples. In fact, best results for device 3 are given by the smallest λ but it is the contrary for device 2. Unfortunately, in this case none of the results given by PLSDA overcomes the threshold, therefore we do not show them here.

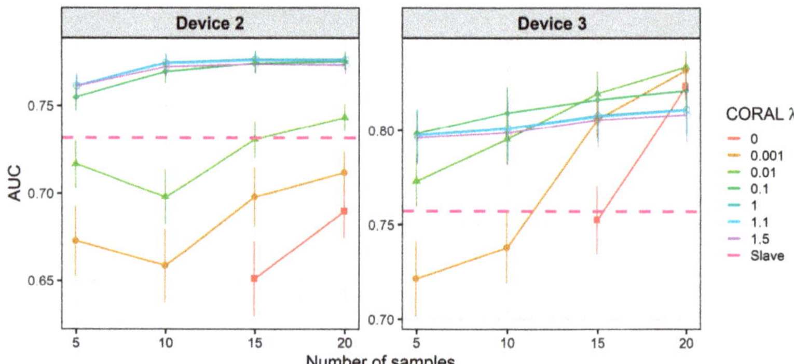

Figure 14. Classification results for CORAL using one-class transfer samples from class "no-meal".

Results for DS and PDS are not good when using one-class standard transfer samples, only PDS with window size equal to 1 gives AUC slightly over the threshold (Figure 15). However, PLSDA-CT still shows good results for small numbers of LVs but only for high number of transfer samples (Figure 16).

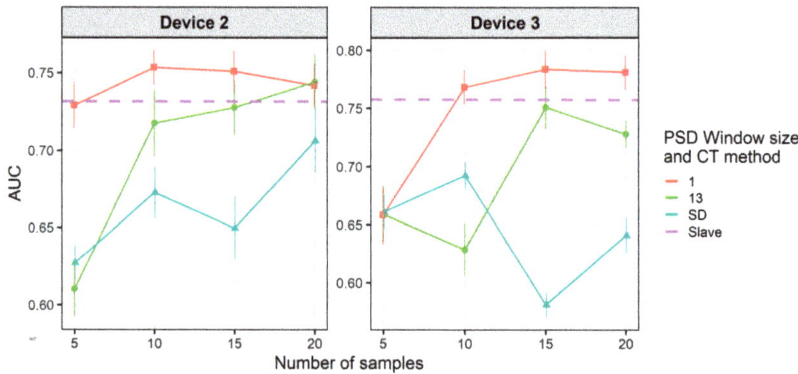

Figure 15. Classification results for DS and PDS using one-class standard transfer samples from class "no-meal".

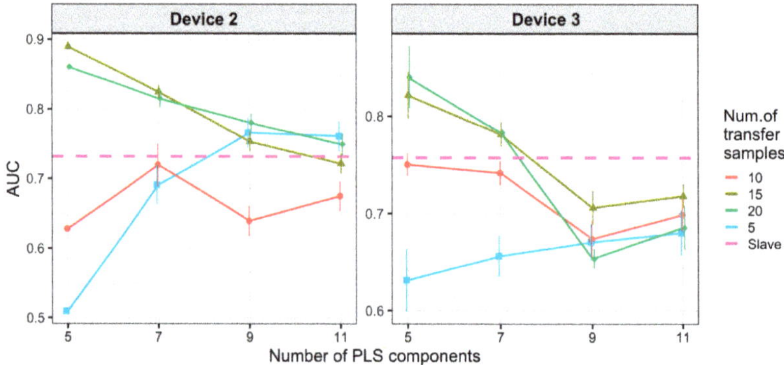

Figure 16. Classification results for PLSDA-CT using one-class standard transfer samples from class "no-meal".

Finally, the use of one-class non-standard transfer samples results in Figure 17, which shows a worsening in the performance of DS and PDS and a similar behavior of PLSDA-CT. In addition, PLSDA-CT shows a change in the trend for the smallest number of one-class transfer samples (Figures 16 and 18) of the performance, which increases with the LVs, with respect to two-class transfer samples (Figures 11 and 13) for which the performance decreases with the LVs.

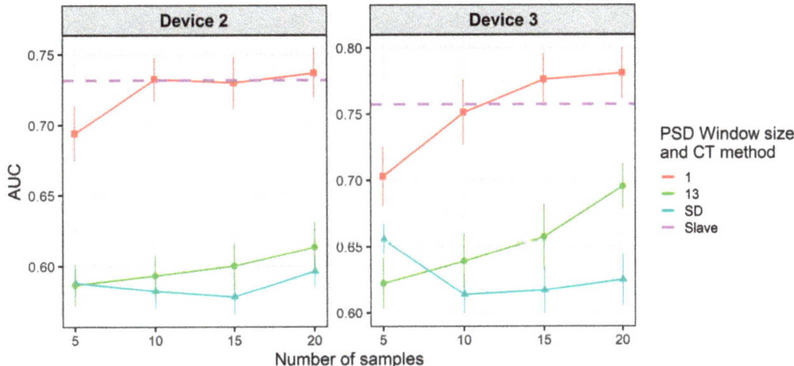

Figure 17. Classification results for DS and PDS using one-class non-standard transfer samples from class "no-meal".

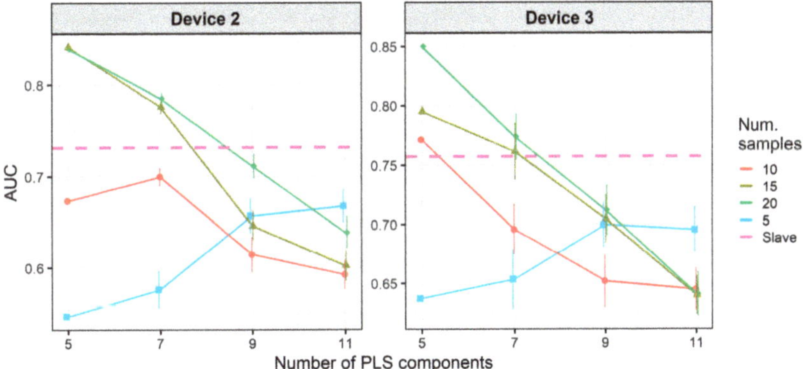

Figure 18. Classification results for PLSDA-CT using one-class non-standard transfer samples from class "no-meal".

In summary, we have shown the performance of several CT methods using labeled transfer samples (DS, PDS, PLSDA-CT) with samples from one or two classes, and using unlabeled transfer samples (CORAL, PLSDA) knowing that they contain one or two classes. Since CORAL is based on covariance and in our data set the covariance of both classes are not dramatically different (so we can use LDA as classifier), its performance using a transfer sample set with one class is not much worse than using two classes. On the contrary, PLSDA needs both classes in the transfer sample set to find a component on a suitable direction to be removed. In turn, due to their nature, DS and PDS are very sensitive not only to the classes present in the transfer set but also to the type of samples. Therefore, standard samples of both classes are necessary for correct performance. Finally, results for PLSDA-CT are more robust and stable for standard samples of both classes, but for a small number of components and a high number of transfer samples, PLSDA-CT is giving good results for all cases.

5. Conclusions

In this work, we have presented a novel, compact and easy-to-use breath analyzer platform with a modular sensing chamber and direct breath sampling unit. Furthermore, we have tested the performance of four calibration transfer methods in a breath analysis experiment using real human breath measurements to classify breath samples of subjects before and after eating a specific meal. In our study the breath measurement is taken about 4 h after the food intake; this leads to the conclusion that the sensors are affected by every food intake. This can be viewed as a potential disorder in studies with healthy and sick people and should be considered when designing an appropriate sampling protocol.

The measurements were made using three instruments. One of them (device 1) was selected as the master instrument, so that its measurements were used to build classification models along with transfer samples whenever their classes were known. The other two instruments were the slave instruments.

The four CT methods tested follow very different approaches, especially with regard to the transfer samples they use. The test of these CT methods is in fact focused on the transfer samples they need for an acceptable performance as a response for a practical problem that arrives in on-field experiments, in our case in experiments using real human breath measurements with gas sensor-based instruments. In such experiments, measurements of samples at different locations and with different instruments are usual. The problem arrives when transferring the calibration from the master to the remaining slave instruments, because labeled samples from the slaves are needed and sometimes, they are difficult to obtain. We wondered firstly if real sample measurements (instead of lab-samples) could be used as transfer samples, and if so, how many and whether or not they must contain all classes under study in the classification problem.

In the figures above, we have shown that real human breath measurements can be used as transfer samples, although in large numbers, much larger than in lab experiments, and with results that depend on the device. However, we could derive some general conclusions. First, in all cases that need labeled transfer samples, the best performance of all methods was obtained for two-class standard samples, and a decrease could be seen when the two-class samples were not standard. Methods like PLSDA, DS and PDS need transfer samples containing all the classes involved in the classification problem, although for PLSDA these samples do not need to be specifically known. However, PLSDA-CT gives good results for small LVs and large transfer samples which in our experiment could contain only one class. CORAL also shows good results for both one and two-class unlabeled transfer samples, although it depends on the device and a parameter. Therefore, calibration transfer methods such as CORAL and PLSDA-CT could be used in on-field experiments using transfer samples from the samples under study, without the need of laboratory samples specifically measured for calibration transfer tasks or for recalibration purposes.

Supplementary Materials: The following are available online, Figure S1: Internal view with description of the three main units of the Modular Breath Analyzer (MBA) platform for laboratory and clinical use, Table S1: List of integrated analog sensors, number of obtained signals from each sensor and used heater voltages (adapted with permission from Jaeschke et al., Copyright 2018 by the authors), Table S2: List of integrated digital sensors, number of obtained signals from each sensor and used heater voltages (adapted with permission from Jaeschke et al., Copyright 2018 by the authors).

Author Contributions: C.J., J.G., I.P., V.V. and M.L. (Marcis Leja) conceived and designed the experiments; C.J., J.G., I.P., M.L. (Martins Leja) and V.V. performed the experiments and collected the data; C.J. and M.P. analyzed the data; J.G., I.P., V.V., J.M. and M.L. (Marcis Leja) contributed infrastructure and materials; M.P. contributed analysis tools; J.G., I.P., V.V., M.L. (Marcis Leja) and B.M. proofread the article; C.J., M.P. and J.G. wrote the article. All authors have read and agreed to the published version of the manuscript.

Funding: The MBA platform is based on the VOLGACORE breath analyzer system which was developed within the VOLGACORE project which was funded by the German Federal Ministry of Education and Research (Grant number 13N13060) within the transnational EURONANOMED II research program. We gratefully acknowledge support from the European Union's Horizon 2020 Programme within the Project VOGAS (grant agreement No 824986) and post-doctorate grant co-funded by the European Regional Development Fund (No. 1.1.1.2/VIAA/2/18/270).

Institutional Review Board Statement: The study was conducted according to the guidelines of the Declaration of Helsinki and approved by the Ethics Committee of the Riga East University Support Fund, approval No. 4-A/18 from 8 February 2018.

Informed Consent Statement: Informed consent was obtained from all subjects involved in the study.

Data Availability Statement: The data presented in this study are available on request from the corresponding author. The original data are not publicly available due to privacy of participants.

Conflicts of Interest: The authors declare no conflict of interest.

References

1. Miekisch, W.; Schubert, J.K.; Noeldge-Schomburg, G.F.E. Diagnostic potential of breath analysis—Focus on volatile organic compounds. *Clin. Chim. Acta* **2004**, *317*, 25–39. [CrossRef]
2. Musteata, F.M. Recent progress in in-vivo sampling and analysis. *TrAC Trends Anal. Chem.* **2013**, *45*, 154–168. [CrossRef]
3. Buszewski, B.; Kesy, M.; Ligor, T.; Amann, A. Human exhaled air analytics: Biomarkers of diseases. *Biomed. Chromatogr.* **2007**, *21*, 553–566. [CrossRef] [PubMed]
4. Di Francesco, F.; Fuoco, R.; Trivella, M.G.; Ceccarini, A. Breath analysis: Trends in techniques and clinical applications. *Microchem. J.* **2005**, *79*, 405–410. [CrossRef]
5. Wikipedia. Breathing Webpage. Available online: http://en.wikipedia.org/wiki/Breathing (accessed on 4 May 2021).
6. Pauling, L.; Robinson, A.B.; Teranishi, R.; Cary, P. Quantitative analysis of urine vapor and breath by gas-liquid partition chromatography. *Proc. Natl. Acad. Sci. USA* **1971**, *68*, 2374–2376. [CrossRef] [PubMed]
7. Phillips, M. Breath tests in medicine. *Sci. Am.* **1992**, *267*, 74–79. [CrossRef] [PubMed]
8. Phillips, M.; Herrera, J.; Krishnan, S.; Zain, M.; Greenberg, J.; Cataneo, R.N. Variation in volatile organic compounds in the breath of normal humans. *J. Chromatogr. B Anal. Technol. Biomed. Life Sci.* **1999**, *729*, 75–88. [CrossRef]
9. Pleil, J.D.; Lindstrom, A.B. Exhaled human breath measurement method for assessing exposure to halogenated volatile organic compounds. *Clin. Chem.* **1997**, *43*, 723–730. [CrossRef] [PubMed]
10. Ma, W.; Liu, X.; Pawliszyn, J. Analysis of human breath with micro extraction techniques and continuous monitoring of carbon dioxide concentration. *Anal. Bioanal. Chem.* **2006**, *385*, 1398–1408. [CrossRef] [PubMed]
11. Kim, K.-H.; Jahan, S.A.; Kabir, E. A review of breath analysis for diagnosis of human health. *TrAC Trends Anal. Chem.* **2012**, *33*, 1–8. [CrossRef]
12. Singer, S.J.; Nicolson, G.L. The fluid mosaic model of the structure of cell membranes. *Science* **1972**, *175*, 720–731. [CrossRef] [PubMed]
13. Kneepkens, C.M.F.; Lepage, A.G.U.Y.; Roy, C.C. The potential of hydrocarbon breath test as a measure of lipid peroxidation. *Free Radic. Biol. Med.* **1994**, *17*, 127–160. [CrossRef]
14. Alberts, B.; Johnson, A.; Lewis, J.; Raff, M.; Roberts, K.; Walter, P. *Molecular Biology of the Cell*; Garland Science: New York, NY, USA, 2002.
15. Buszewski, B.; Rudnicka, J.; Ligor, T.; Walczak, M.; Jezierski, T.; Amann, A. Analytical and unconventional methods of cancer detection using odor. *TrAC Trends Anal. Chem.* **2012**, *38*, 1–12. [CrossRef]
16. Horváth, I.; Lázár, Z.; Gyulai, N.; Kollai, M.; Losonczy, G. Exhaled biomarkers in lung cancer. *Eur. Respir. J.* **2009**, *34*, 261–275. [CrossRef] [PubMed]
17. Bajtarevic, A.; Ager, C.; Pienz, M.; Klieber, M.; Schwarz, K.; Ligor, M.; Ligor, T.; Filipiak, W.; Denz, H.; Fiegl, M.; et al. Noninvasive detection of lung cancer by analysis of exhaled breath. *BMC Cancer* **2009**, *9*, 348. [CrossRef]
18. Tisch, U.; Haick, H. Nanomaterials for cross-reactive sensor arrays. *MRS Bull.* **2010**, *35*, 797–803. [CrossRef]
19. Ligor, M.; Ligor, T.; Bajtarevic, A.; Ager, C.; Pienz, M.; Klieber, M.; Denz, H.; Fiegl, M.; Hilbe, W.; Weiss, W.; et al. Determination of volatile organic compounds appearing in exhaled breath of lung cancer patients by solid phase microextraction and gas chromatography mass spectrometry. *Clin. Chem. Lab. Med.* **2009**, *47*, 550–560. [CrossRef] [PubMed]
20. Poli, D.; Carbognani, P.; Corradi, M.; Goldoni, M.; Acampa, O.; Balbi, B.; Bianchi, L.; Rusca, M.; Mutti, A. Exhaled volatile organic compounds in patients with non-small cell lung cancer: Cross sectional and nested short-term follow-up study. *Respir. Res.* **2005**, *6*, 71. [CrossRef]
21. Schubert, J.K.; Miekisch, W.; Birken, T.; Geiger, K.; Nöldge-Schomburg, G.F.E. Impact of inspired substance concentrations on the results of breath analysis in mechanically ventilated patients. *Biomarkers* **2005**, *10*, 138–152. [CrossRef] [PubMed]
22. Schubert, J.; Miekisch, W.; Nöldge-Schomburg, G. VOC breath markers in critically ill patients: Potentials and limitations. *Breath Anal. Clin. Diagn. Ther. Monit.* **2005**, 267–292.
23. Schubert, J.K.; Miekisch, W.; Geiger, K.; Nöldge-Schomburg, G.F.E.; Nöldge–Schomburg, G.F. Breath analysis in critically ill patients: Potential and limitations. *Expert Rev. Mol. Diagn.* **2004**, *4*, 619–629. [CrossRef] [PubMed]

24. Ligor, T.; Ligor, M.; Amann, A.; Ager, C.; Bachler, M.; Dzien, A.; Buszewski, B. The analysis of healthy volunteers' exhaled breath by the use of solid-phase microextraction and GC-MS. *J. Breath Res.* **2008**, *2*, 46006. [CrossRef] [PubMed]
25. Hakim, M.; Broza, Y.Y.; Barash, O.; Peled, N.; Phillips, M.; Amann, A.; Haick, H. Volatile organic compounds of lung cancer and possible biochemical pathways. *Chem. Rev.* **2012**, *112*, 5949–5966. [CrossRef] [PubMed]
26. Tisch, U.; Haick, H. Arrays of chemisensitive monolayer-capped metallic nanoparticles for diagnostic breath testing. *Rev. Chem. Eng.* **2010**, *26*, 171–179. [CrossRef]
27. Mazzone, P.J. Analysis of volatile organic compounds in the exhaled breath for the diagnosis of lung cancer. *J. Thorac. Oncol.* **2008**, *3*, 774–780. [CrossRef]
28. Montuschi, P.; Santonico, M.; Mondino, C.; Pennazza, G.; Mantini, G.; Martinelli, E.; Capuano, R.; Ciabattoni, G.; Paolesse, R.; Di Natale, C.; et al. Diagnostic performance of an electronic nose, fractional exhaled nitric oxide, and lung function testing in asthma. *CHEST J.* **2010**, *137*, 790–796. [CrossRef] [PubMed]
29. Miekisch, W.; Kischkel, S.; Sawacki, A.; Liebau, T.; Mieth, M.; Schubert, J.K. Impact of sampling procedures on the results of breath analysis. *J. Breath Res.* **2008**, *2*, 026007. [CrossRef]
30. Röck, F.; Barsan, N.; Weimar, U. Electronic nose: Current status and future trends. *Chem. Rev.* **2008**, *108*, 705–725. [CrossRef]
31. Hubers, A.J.; Brinkman, P.; Boksem, R.J.; Rhodius, R.J.; Witte, B.I.; Zwinderman, A.H.; Heideman, D.A.M.; Duin, S.; Koning, R.; Steenbergen, R.D.M.; et al. Combined sputum hypermethylation and eNose analysis for lung cancer diagnosis. *J. Clin. Pathol.* **2014**, *64*, 707–711. [CrossRef] [PubMed]
32. Dragonieri, S. An electronic nose distinguishes the exhaled breath of patients with pleural malignant mesothelioma from subjects with professional asbestos exposure. In Proceedings of the 30th International Congress on Occupational Health, Cancun, Mexico, 18–23 March 2012.
33. Dragonieri, S.; Annema, J.T.; Schot, R.; van der Schee, M.P.C.; Spanevello, A.; Carratú, P.; Resta, O.; Rabe, K.F.; Sterk, P.J. An electronic nose in the discrimination of patients with non-small cell lung cancer and COPD. *Lung Cancer* **2009**, *64*, 166–170. [CrossRef]
34. Chapman, E.A.; Thomas, P.S.; Stone, E.; Lewis, C.; Yates, D.H. A breath test for malignant mesothelioma using an electronic nose. *Eur. Respir. J.* **2012**, *40*, 448–454. [CrossRef]
35. Machado, R.F.; Laskowski, D.; Deffenderfer, O.; Burch, T.; Zheng, S.; Mazzone, P.J.; Mekhail, T.; Jennings, C.; Stoller, J.K.; Pyle, J.; et al. Detection of lung cancer by sensor array analyses of exhaled breath. *Am. J. Respir. Crit. Care Med.* **2005**, *171*, 1286–1291. [CrossRef]
36. McWilliams, A.; Beigi, P.; Srinidhi, A.; Lam, S.; MacAulay, C.E. Sex and smoking status effects on the early detection of early lung cancer in high-risk smokers using an electronic nose. *IEEE Trans. Biomed. Eng.* **2015**, *62*, 2044–2054. [CrossRef] [PubMed]
37. Di Natale, C.; Macagnano, A.; Martinelli, E.; Paolesse, R.; D'Arcangelo, G.; Roscioni, C.; Finazzi-Agrò, A.; D'Amico, A. Lung cancer identification by the analysis of breath by means of an array of non-selective gas sensors. *Biosens. Bioelectron.* **2003**, *18*, 1209–1218. [CrossRef]
38. Leunis, N.; Boumans, M.-L.; Kremer, B.; Din, S.; Stobberingh, E.; Kessels, A.G.H.; Kross, K.W. Application of an electronic nose in the diagnosis of head and neck cancer. *Laryngoscope* **2014**, *124*, 1377–1381. [CrossRef] [PubMed]
39. Chen, X.; Cao, M.; Li, Y.; Hu, W.; Wang, P.; Ying, K.; Pan, H. A study of an electronic nose for detection of lung cancer based on a virtual SAW gas sensors array and imaging recognition method. *Meas. Sci. Technol.* **2005**, *16*, 1535–1546. [CrossRef]
40. Yu, K.; Wang, Y.; Yu, J.; Wang, P. A portable electronic nose intended for home healthcare based on a mixed sensor array and multiple desorption methods. *Sens. Lett.* **2011**, *9*, 876–883. [CrossRef]
41. Mazzone, P.J.; Wang, X.-F.; Xu, Y.; Mekhail, T.; Beukemann, M.C.; Na, J.; Kemling, J.W.; Suslick, K.S.; Sasidhar, M. Exhaled breath analysis with a colorimetric sensor array for the identification and characterization of lung cancer. *J. Thorac. Oncol.* **2012**, *7*, 137–142. [CrossRef] [PubMed]
42. D'Amico, A.; Pennazza, G.; Santonico, M.; Martinelli, E.; Roscioni, C.; Galluccio, G.; Paolesse, R.; Di Natale, C. An investigation on electronic nose diagnosis of lung cancer. *Lung Cancer* **2010**, *68*, 170–176. [CrossRef] [PubMed]
43. Santonico, M.; Lucantoni, G.; Pennazza, G.; Capuano, R.; Galluccio, G.; Roscioni, C.; La Delfa, G.; Consoli, D.; Martinelli, E.; Paolesse, R.; et al. In situ detection of lung cancer volatile fingerprints using bronchoscopic air-sampling. *Lung Cancer* **2012**, *77*, 46–50. [CrossRef]
44. Wang, D.; Yu, K.; Wang, Y.; Hu, Y.; Zhao, C.; Wang, L.; Ying, K.; Wang, P. A hybrid electronic noses' system based on MOS-SAW detection units intended for lung cancer diagnosis. *J. Innov. Opt. Health Sci.* **2012**, *5*, 1150006. [CrossRef]
45. Shehada, N.; Brönstrup, G.; Funka, K.; Christiansen, S.; Leja, M.; Haick, H. ultrasensitive silicon nanowire for real-world gas sensing: Noninvasive diagnosis of cancer from breath volatolome. *Nano Lett.* **2015**, *15*, 1288–1295. [CrossRef]
46. Peled, N.; Hakim, M.; Bunn, P.A.; Miller, Y.E.; Kennedy, T.C.; Mattei, J.; Mitchell, J.D.; Hirsch, F.R.; Haick, H. Non-invasive breath analysis of pulmonary nodules. *J. Thorac. Oncol.* **2013**, *7*, 1528–1533. [CrossRef] [PubMed]
47. Hakim, M.; Billan, S.; Tisch, U.; Peng, G.; Dvrokind, I.; Marom, O.; Abdah-Bortnyak, R.; Kuten, A.; Haick, H. *Diagnosis of Head-and-Neck Cancer from Exhaled Breath*; Nature Publishing Group: Berlin, Germany, 2011; Volume 104.
48. Xu, Z.-Q.; Broza, Y.Y.; Ionsecu, R.; Tisch, U.; Ding, L.; Liu, H.; Song, Q.; Pan, Y.; Xiong, F.; Gu, K.; et al. A nanomaterial-based breath test for distinguishing gastric cancer from benign gastric conditions. *Br. J. Cancer* **2013**, *108*, 941–950. [CrossRef]
49. Amal, H.; Shi, D.-Y.; Ionescu, R.; Zhang, W.; Hua, Q.; Pan, Y.-Y.; Tao, L.; Liu, H.; Haick, H. Assessment of ovarian cancer conditions from exhaled breath. *Int. J. Cancer* **2015**, *136*, 614–622. [CrossRef]

50. Amal, H.; Leja, M.; Funka, K.; Skapars, R.; Sivins, A.; Ancans, G.; Liepniece-Karele, I.; Kikuste, I.; Lasina, I.; Haick, H. Detection of precancerous gastric lesions and gastric cancer through exhaled breath. *Gut* **2015**, *65*, 400–407. [CrossRef] [PubMed]
51. Gruber, M.; Tisch, U.; Jeries, R.; Amal, H.; Hakim, M.; Ronen, O.; Marshak, T.; Zimmerman, D.; Israel, O.; Amiga, E.; et al. Analysis of exhaled breath for diagnosing head and neck squamous cell carcinoma: A feasibility study. *Br. J. Cancer* **2014**, *111*, 790–798. [CrossRef] [PubMed]
52. Wang, X.R.; Lizier, J.T.; Berna, A.Z.; Bravo, F.G.; Trowell, S.C. Human breath-print identification by E-nose, using information-theoretic feature selection prior to classification. *Sens. Actuators B Chem.* **2015**, *217*, 165–174. [CrossRef]
53. De Vries, R.; Brinkman, P.; Van Der Schee, M.P.; Fens, N.; Dijkers, E.; Bootsma, S.K.; de Jongh, F.H.C.C.; Sterk, P.J. Integration of electronic nose technology with spirometry: Validation of a new approach for exhaled breath analysis. *J. Breath Res.* **2015**, *9*, 46001. [CrossRef] [PubMed]
54. Yan, K.; Zhang, D. Calibration transfer and drift compensation of e-noses via coupled task learning. *Sens. Actuators B Chem.* **2016**, *225*, 288–297. [CrossRef]
55. Fonollosa, J.; Fernández, L.; Gutiérrez-Gálvez, A.; Huerta, R.; Marco, S. Calibration transfer and drift counteraction in chemical sensor arrays using Direct Standardization. *Sens. Actuators B Chem.* **2016**, *236*, 1044–1053. [CrossRef]
56. Fernandez, L.; Guney, S.; Gutierrez-Galvez, A.; Marco, S. Calibration transfer in temperature modulated gas sensor arrays. *Sens. Actuators B Chem.* **2016**, *231*, 276–284. [CrossRef]
57. Yan, K.; Zhang, D. Improving the transfer ability of prediction models for electronic noses. *Sens. Actuators B Chem.* **2015**, *220*, 115–124. [CrossRef]
58. Workman, J.J. A Review of Calibration Transfer Practices and Instrument Differences in Spectroscopy. *Appl. Spectrosc.* **2018**, *72*, 340–365. [CrossRef]
59. Wold, S.; Antti, H.; Lindgren, F.; Öhman, J. Orthogonal signal correction of near-infrared spectra. *Chemom. Intell. Lab. Syst.* **1998**, *44*, 175–185. [CrossRef]
60. Bouveresse, E.; Hartmann, C.; Massart, D.L.; Last, I.R.; Prebble, K.A. Standardization of near-infrared spectrometric instruments. *Anal. Chem.* **1996**, *68*, 982–990. [CrossRef]
61. Wang, Y.; Veltkamp, D.J.; Kowalski, B.R. Multivariate instrument standardization. *Anal. Chem.* **1991**, *63*, 2750–2756. [CrossRef]
62. Wang, Y.; Lysaght, M.J.; Kowalski, B.R. Improvement of multivariate calibration through instrument standardization. *Anal. Chem.* **1992**, *64*, 562–564. [CrossRef]
63. Andrew, A.; Fearn, T. Transfer by orthogonal projection: Making near-infrared calibrations robust to between-instrument variation. *Chemom. Intell. Lab. Syst.* **2004**, *72*, 51–56. [CrossRef]
64. Feudale, R.N.; Woody, N.A.; Tan, H.; Myles, A.J.; Brown, S.D.; Ferré, J. Transfer of multivariate calibration models: A review. *Chemom. Intell. Lab. Syst.* **2002**, *64*, 181–192. [CrossRef]
65. Fearn, T. Standardisation and calibration transfer for near infrared instruments: A review. *J. Near Infrared Spectrosc.* **2001**, *9*, 229–244. [CrossRef]
66. Kramer, K.E.; Morris, R.E.; Rose-Pehrsson, S.L. Comparison of two multiplicative signal correction strategies for calibration transfer without standards. *Chemom. Intell. Lab. Syst.* **2008**, *92*, 33–43. [CrossRef]
67. Ni, W.; Brown, S.D.; Man, R. Stacked PLS for calibration transfer without standards. *J. Chemom.* **2011**, *25*, 130–137. [CrossRef]
68. Blank, T.B.; Sum, S.T.; Brown, S.D.; Monfre, S.L. Transfer of near-infrared multivariate calibrations without standards. *Anal. Chem.* **1996**, *68*, 2987–2995. [CrossRef]
69. Tan, H.; Sum, S.T.; Brown, S.D. Improvement of a standard-free method for near infrared calibration transfer. *Appl. Spectrosc.* **2002**, *56*, 1098–1106. [CrossRef]
70. Sjöblom, J.; Svensson, O.; Josefson, M.; Kullberg, H.; Wold, S. An evaluation of orthogonal signal correction applied to calibration transfer of near infrared spectra. *Chemom. Intell. Lab. Syst.* **1998**, *44*, 229–244. [CrossRef]
71. Bouveresse, E.; Massart, D.L.; Dardenne, P. Calibration transfer across near-infrared spectrometric instruments using Shenk's algorithm: Effects of different standardisation samples. *Anal. Chim. Acta* **1994**, *297*, 405–416. [CrossRef]
72. Du, W.; Chen, Z.-P.; Zhong, L.-J.; Wang, S.-X.; Yu, R.-Q.; Nordon, A.; Littlejohn, D.; Holden, M. Maintaining the predictive abilities of multivariate calibration models by spectral space transformation. *Anal. Chim. Acta* **2011**, *690*, 64–70. [CrossRef] [PubMed]
73. Fan, W.; Liang, Y.; Yuan, D.; Wang, J. Calibration model transfer for near-infrared spectra based on canonical correlation analysis. *Anal. Chim. Acta* **2008**, *623*, 22–29. [CrossRef] [PubMed]
74. Pan, S.J.; Tsang, I.W.; Kwok, J.T.; Yang, Q. Domain adaptation via transfer component analysis. *IEEE Trans. Neural Networks* **2011**, *22*, 199–210. [CrossRef]
75. Liang, Z.; Tian, F.; Zhang, C.; Sun, H.; Song, A.; Liu, T. Improving the robustness of prediction model by transfer learning for interference suppression of electronic nose. *IEEE Sens. J.* **2017**, *18*, 1111–1121. [CrossRef]
76. Malli, B.; Birlutiu, A.; Natschläger, T. Standard-free calibration transfer—An evaluation of different techniques. *Chemom. Intell. Lab. Syst.* **2017**, *161*, 49–60. [CrossRef]
77. Igne, B.; Roger, J.-M.; Roussel, S.; Bellon-Maurel, V.; Hurburgh, C.R. Improving the transfer of near infrared prediction models by orthogonal methods. *Chemom. Intell. Lab. Syst.* **2009**, *99*, 57–65. [CrossRef]
78. Ferré, J.; Brown, S.D. Reduction of model complexity by orthogonalization with respect to non-relevant spectral changes. *Appl. Spectrosc.* **2001**, *55*, 708–714. [CrossRef]
79. Wise, B.M.; Roginski, R.T. A calibration model maintenance roadmap. *IFAC-PapersOnLine* **2015**, *48*, 260–265. [CrossRef]

80. Jaeschke, C.; Glöckler, J.; El Azizi, O.; Gonzalez, O.; Padilla, M.; Mitrovics, J.; Mizaikoff, B. An innovative modular eNose system based on a unique combination of analog and digital metal oxide sensors. *ACS Sensors* **2019**, *4*, 2277–2281. [CrossRef] [PubMed]
81. Jaeschke, C.; Glöckler, J.; Padilla, M.; Mitrovics, J.; Mizaikoff, B. An eNose-based method performing drift correction for online VOC detection under dry and humid conditions. *Anal. Methods* **2020**, *12*, 4724–4733. [CrossRef] [PubMed]
82. Jaeschke, C.; Gonzalez, O.; Glöckler, J.J.; Hagemann, L.T.; Richardson, K.E.; Adrover, F.; Padilla, M.; Mitrovics, J.; Mizaikoff, B. A novel modular eNose system based on commercial MOX sensors to detect low concentrations of VOCs for breath gas analysis. *Proceedings* **2018**, *2*, 993. [CrossRef]
83. Herbig, J.; Titzmann, T.; Beauchamp, J.; Kohl, I.; Hansel, A. Buffered end-tidal (BET) sampling-a novel method for real-time breath-gas analysis. *J. Breath Res.* **2008**, *2*, 037008. [CrossRef] [PubMed]
84. Sun, B.; Feng, J.; Saenko, K. Correlation alignment for unsupervised domain adaptation. In *Guide to 3D Vision Computation*; Springer: Cham, Switzerland, 2017; pp. 153–171.
85. Zhao, Y.; Yu, J.; Shan, P.; Zhao, Z.; Jiang, X.; Gao, S. PLS Subspace-based calibration transfer for near-infrared spectroscopy quantitative analysis. *Molecules* **2019**, *24*, 1289. [CrossRef]
86. Bouveresse, E.; Massart, D.L. Improvement of the piecewise direct standardisation procedure for the transfer of NIR spectra for multivariate calibration. *Chemom. Intell. Lab. Syst.* **1996**, *32*, 201–213. [CrossRef]
87. Wise, B.M. *Introduction to Instrument Standardization and Calibration Transfer*; Eigenvector Research: Manson, WA, USA, 1996; pp. 1–28.
88. Kennard, R.W.; Stone, L.A. Computer aided design of experiments. *Technometrics* **1969**, *11*, 137. [CrossRef]

Article

Monitoring the Reaction of the Body State to Antibiotic Treatment against *Helicobacter pylori* via Infrared Spectroscopy: A Case Study

Kiran Sankar Maiti [1,2] and **Alexander Apolonski** [1,2,3,4,*]

1   Max Planck Institute for Quantum Optics, Hans-Kopfermann-Strasse 1, 85748 Garching, Germany; kiran.maiti@mpq.mpg.de
2   Department of Experimental Physics, Faculty of Physics, Ludwig-Maximilians-Universität München, Am Coulombwall 1, 85748 Garching, Germany
3   Institute of Automation and Electrometry SB RAS, 630090 Novosibirsk, Russia
4   Department of Physics, Novosibirsk State University, 630090 Novosibirsk, Russia
*   Correspondence: apolonskiy@lmu.de

Citation: Maiti, K.S.; Apolonski, A. Monitoring the Reaction of the Body State to Antibiotic Treatment against *Helicobacter pylori* via Infrared Spectroscopy: A Case Study. *Molecules* **2021**, *26*, 3474. https://doi.org/10.3390/molecules26113474

Academic Editors: Ben de Lacy Costello and Natalia Drabińska

Received: 5 April 2021
Accepted: 5 June 2021
Published: 7 June 2021

Publisher's Note: MDPI stays neutral with regard to jurisdictional claims in published maps and institutional affiliations.

Abstract: The current understanding of deviations of human microbiota caused by antibiotic treatment is poor. In an attempt to improve it, a proof-of-principle spectroscopic study of the breath of one volunteer affected by a course of antibiotics for *Helicobacter pylori* eradication was performed. Fourier transform spectroscopy enabled searching for the absorption spectral structures sensitive to the treatment in the entire mid-infrared region. Two spectral ranges were found where the corresponding structures strongly correlated with the beginning and end of the treatment. The structures were identified as methyl ester of butyric acid and ethyl ester of pyruvic acid. Both acids generated by bacteria in the gut are involved in fundamental processes of human metabolism. Being confirmed by other studies, measurement of the methyl butyrate deviation could be a promising way for monitoring acute gastritis and anti-*Helicobacter pylori* antibiotic treatment.

Keywords: breath; metabolites; volatile organic compound; acute gastritis; antibiotic treatment; treatment dynamics; microbiota; mid-infrared spectroscopy; short-chain fatty acid; alpha-keto acid; *Helicobacter pylori*

1. Introduction

A number of bacteria-related diseases increases as our understanding of the role of microbiota deepens (for detail, see Section 1 of the Supplementary Materials (SM)). One of the bacterium in stomach called *Helicobacter pylori* (*Hp*) attracts much attention [1] since its discovery in 1983 because, under some unknown circumstances it can lead to gastric problems including peptic ulcer disease [2]. It is agreed that should *Hp* be present in stomach and not eradicated in cases of related gastric problems, it can lead to gastric cancer. A common way of eradication includes a certain combination of antibiotics of broad spectra, called a quadruple course (QAC, detail in Methods). In 2011, QAC demonstrated 95% eradication success [3]. It has to be noted that the success rate of the course degrades with time because of *Hp* adaptation to the antibiotics [1,3]. The only reliable method to determine whether the bacterium is susceptible or resistant to a particular antimicrobial is to perform in vitro antimicrobial susceptibility testing [4].

Revealing the *Hp* presence via ^{13}C urease or gastroscopy tests represents two established techniques for practical monitoring. Measurements with the first technique can be done either by means of mass-spectrometry or optical spectroscopy, showing similar accuracy [5–7]. The second technique is invasive, with clinician-dependent outcome. In both cases of *Hp*-positive tests, a clinician usually prescribes QAC. The treatment does not imply an extra step to verify the remaining *Hp* in stomach. The duration of QAC should be optimal for a given case: being too short, it does not eradicate *Hp* but makes bacteria more resistant to the antibiotics, whereas being too long, it brings negative side effects due to general harm of antibiotics on microbiota. Therefore, monitoring the progress of QAC aimed at its optimal duration would be beneficial in each individual case. In regard to the antibiotic anti-*Hp* treatment, an important question should be posed: how does QAC affect other than *Hp* bacteria? The urease test cannot answer this question because of its specificity to *Hp*. It was reported that about 30% of bacterial species were influenced by the treatment with ciprofloxacin [8]. Although most bacterial groups and subgroups (called strains; diversity within the bacterial gene) recovered after treatment, several of them did not, even after six months. Controlling their recovering is an important practical task not solved so far, to the best of our knowledge [8,9]. For cases not related to *Hp*, the recovery process was recently modeled [10].

There are three more powerful techniques for detecting bacteria that are not (yet) of practical use for *Hp*: serum antibody test, quantitative polymerase chain reaction (qPCR) [11] and 16S rRNA sequencing [12]. The restricted practical application of the latter technique is caused by its high price and the fact that 16S rRNA sequencing has some limitations [12] including the accuracy [13] and the necessity to use one pipeline for an accurate comparison of the data. The sequencing applied to *Hp* in feces [14] and other bio-samples [15] already revealed detailed information about the *Hp* strains and migration in the stomach for a steady state in the body. We are not aware of any study with this technique focused on the dynamics of anti-*Hp* treatment. Variations of bacterial concentrations caused by an antibiotic treatment have already been analyzed [8] in feces, resulting in highly diverse effects for three volunteers. The authors found a substantial, but not full, return to the pretreatment feces composition within 4 weeks after the treatment.

Another, technically more practical, way of monitoring the state of microbiota could be to measure the products of its metabolism. Among them, volatile products extracted from headspace of urine, blood (including breath), or feces [16] have attracted much attention. The corresponding measuring techniques such as e-nose [17] and laser-induced breakdown spectroscopy [18] were recently tested. One study based on mass spectrometry gas chromatography was focused on analysis of breath variations caused by anti-*Hp* treatment [19]. The study revealed certain (but not significant) differences in the volatile organic compound (VOC) content prior to anti-*Hp* treatment and after. The authors concluded that their observation could be explained either by the action of antibiotics on the gut microbiome, or by the effect of the probiotics rather than the presence or absence of *Hp*.

Bearing the current situation in mind, we suggest that measurements prior, during, and after the antibiotic treatment should be beneficial. Being combined with the steady-state data of the individual [20], such a study could unambiguously reveal the effect of the treatment.

There is growing understanding that the response of a subject to antibiotic treatment is unique [8,10]. The response is the result of at least two major factors: the gut microbiome content in general and the history of previous antibiotic treatments. The first factor defines a list of bacteria present in the gut together with their strains. It is important to note that strains of the bacterium have different sensitivity to antibiotics [21]. So far, the second factor was demonstrated experimentally only for mice. The main conclusion was that antibiotics reduce or eliminate most products of bacterial metabolism including short-chain fatty acids (SCFA) [22].

In comparison to analysis of feces mentioned above, breath could have several advantages: (a) it allows to make more frequent monitoring of the metabolic state of bacteria and (b) it carries direct information about the gut microbiome state. To note, bacteria in feces become already modified in comparison to the gut state and, strictly speaking, should be considered as being measured in vitro [13].

The aim of this one-case study was to verify whether breath carries significant information about acute gastritis and the dynamic response of the body to anti-*Hp* treatment.

2. Working Hypotheses, Results, and Discussion

The results presented below are aimed at verifying four working hypotheses related to QAC. They can be formulated as follows: (1) a conventional anti-*Hp* course based on antibiotics of broad spectrum kills *Hp* as well as other bacteria in the entire gastro-intestinal system (inspired by [8]); (2) different types of bacteria have different resistivity and reaction to antibiotics, depending in the same time on the individual. The bacterial resistivity can be so high that the treatment cannot affect them; (3) bacterial groups affected by antibiotics of QAC but present in a parallel probiotic course, should show fast recovering; (4) in order to see the treatment dynamics, monitoring must be provided not only before and after the course but also during it. The expected corresponding outcomes of these hypotheses are the following (referring to the hypothesis numbering): (1) and (2) monotonic decay of the concentration of different types of bacteria on different time scales; (3) significantly different time scales of the microbiota recovering. For volatile products related to bacteria strains present in the Omni-biotic 10 probiotic course [23], one can expect their recovering on the time scale of the course. For volatile products of the bacteria out of the Omni-biotic 10 course, one can expect slower recovering on a long time scale of months [24,25] or even years [8,10]. The hypotheses define the time scale of the study we aimed for. Specifically, a period of several months before and after the treatment should be used to collect the data regarding the steady state of the body. The absence of the steady state would mean that the hypotheses we formulated cannot be verified. Deviations of VOCs originated from bacteria sensitive to QAC, must be synchronized with the beginning and the end of the treatment.

Figure 1 demonstrates variations of the absorption signals from breath samples before, during and after QAC. To note, the signals are proportional to concentrations of the corresponding VOCs. First, we see that in both plots corresponding to different spectral ranges, the steady state (i.e., the same absorbance level before and after QAC) does exist. A correlation of both signals with the beginning and the end of the treatment allowed us to surmise that they originate from bacteria. Figure 2 represents an extended illustration of the steady state found in [20], related here to the signal at 2972 cm^{-1} (the left plot in Figure 1). Analysis of the identified metabolites corresponding to the both plots also revealed their bacterial origin (Section 3.4).

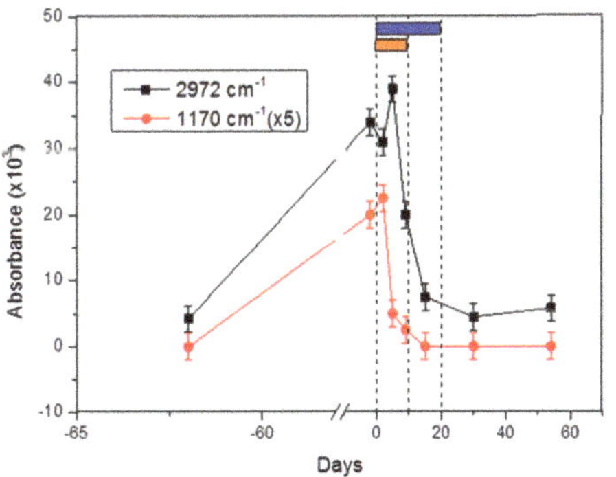

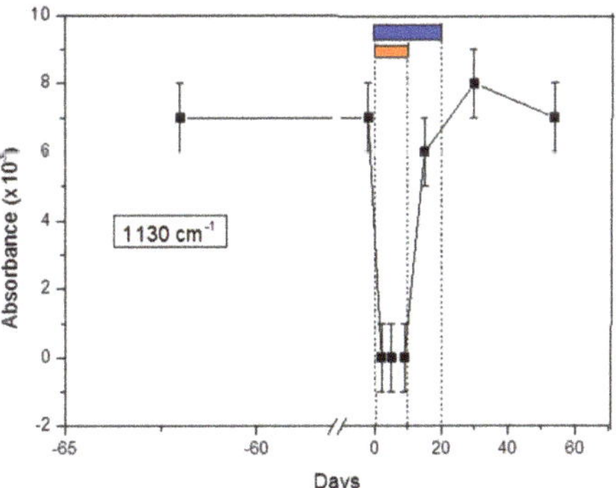

Figure 1. Absorbance variation (proportional to the concentration variation) of three spectral structures centered at 2972 cm^{-1}, 1170 cm^{-1} (**left**), and 1130 cm$^{\beta 1}$ (**right**) caused by QAC. The lines connecting the data points are used for better visibility. Vertical dash lines show the start of the QAC course (left line), the end of the antibiotic course (middle line) and the end of the probiotic course. The corresponding bars indicate the same: orange horizontal bar shows the antibiotic treatment in frame of QAC whereas blue bar—Omnibiotic 10 course taken in parallel to QAC. Absorbances at −62 and 58 days at the plots correspond to the steady state level of the corresponding VOC. Data points corresponding to dates earlier than −60 and longer than 60 days were collected and used only for analysis; they are not presented here in order to improve the visibility of the plots.

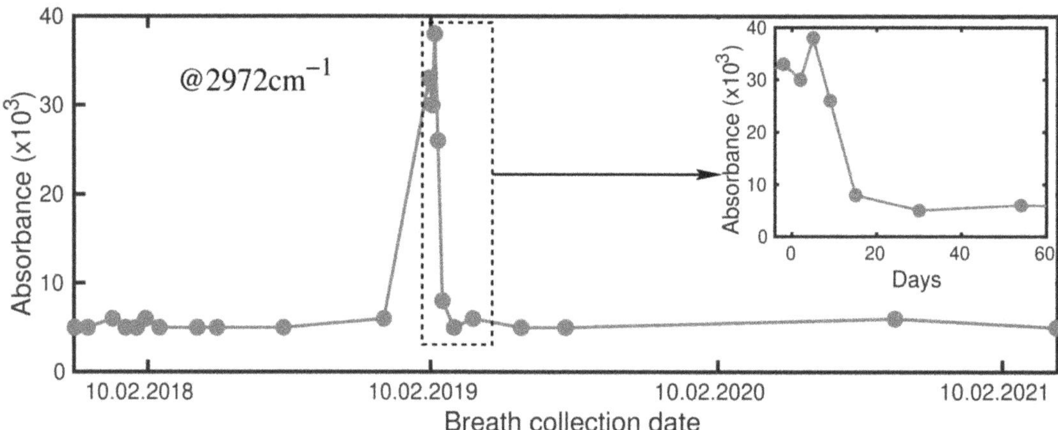

Figure 2. Variation of the absorbance centered at 2972 cm^{-1} during a period of 3.5 years, with the peak related to accute gastritis and QAC. Inset: the recovering dynamics via QAC.

Second, in both plots of Figure 1 we see a ten-fold variation of the absorbance caused by acute gastritis and QAC, significantly exceeding natural variations of the steady state. One can also see that the characteristic time scales of the detected signals vary from few days (the right plot) to approximately 20 days (the left plot). The result correlates with the available literature data [8,9]. Different time constants identified from the two plots, call for considering two classes of bacteria affected by QAC, namely semi-resistant (the left plot) and susceptible (the right plot).

2.1. Signals at 2972 cm^{-1} and 1170 cm^{-1}

Two signals on the left plot of Figure 1 demonstrate qualitatively similar variations from the steady state level, with well-synchronized onsets of their decays. The steady state level of the subject was defined by the data used in Figure 2. Time series data became possible in frame of another study [20] and here we used the extracted values corresponding to the period out of QAC. Significant elevation of the signal just before QAC started (day "−2") was attributed to acute gastritis. That day, because of extra pain in stomach, the subject visited a doctor who recommended immediate QAC. Because of their slow reaction to QAC, the signals were linked to the product (-s) of semi-resistant bacteria. Their decaying part was attributed to the main eradication effect of QAC. We do not attribute the elevated point on day "−2" to *Hp* because of two reasons: based on the anamnesis, the corresponding infection occurred many decades ago. The second reason is discussed in Section 2.6. The findings shown in Figure 1 support the first two working hypotheses.

2.2. Signal at 1130 cm^{-1}

The signal starts recovering to its steady-state level right at the end of the antibiotic part of QAC and finishes at the end of the probiotic part. It indicates that it is related to the bacteria present in the QAC probiotic part [23] described in the Methods. The abruptly decaying signal in the beginning of the treatment (day "0" to day "10") indicates that the effect of antibiotics on the corresponding bacteria is significantly stronger than the effect of probiotics. Potential candidates for such bacteria are discussed in Section 3 of Supplementary Materials.

2.3. Other Signals

We also analyzed the structures centered at 1189 and 1203 cm^{-1}. They found to be insensitive to QAC. The first one was attributed to the mixture of ethyl and propyl propionates in our previous study [20]. Propionic acid is the product of bacteria responsible for food fermentation (Section 3 of Supplementary Materials) and represents one of the main SCFA in the body. The second structure was attributed to ethyl vinyl ketone, an oxygenated hydrocarbon lipid molecule. The two types of molecules support the second working hypothesis.

2.4. Identification of the Molecules Responsible for the Spectral Signals

Figure 3 represents the structures of the absorption spectra that were used for the temporal analysis in Figure 1. It turned out that the three spectra correspond to two VOCs, namely methyl butyrate (1170 cm^{-1} and 2972 cm^{-1}) and ethyl pyruvate (1130 cm^{-1}). In plot (a), one can see that the fitting quality, in spite of weak absorbance and high noise, allowed to consider methyl butyrate and ethyl pyruvate responsible for the detected structures. Structures at 1130 cm^{-1} and 1170 cm^{-1} were revealed by digital removal of the above-lying structures that belong to carbon dioxide and aldehydes [26]. They were then identified using the two-step procedure described in Methods (without step 3). Fitting the structures with the spectra taken from the NIST database [27] led to the agreement within 3 cm^{-1} (0.3% inaccuracy). Plot (b) represents the difference between the inflammation (acute gastritis) and the healthy state. The subtraction procedure removed the structures with sharp spikes that belong to methane. We used the fact that methane concentration was constant during the entire period shown in Figure 2. Plot (c) presents the spectral structure centered around 3000 cm^{-1} that we revealed in the following way: the subtracting procedure was applied to the data taken in different days of QAC. In this step, we took the spectrum measured 30 days after the start of QAC as a reference, subtracting it from the spectra measured during the QAC period. A similar result was obtained when the spectra measured 60 and 192 days prior QAC were taken as the reference. Subtraction of one reference data from another gave a flat line supporting our suggestion that only the body reaction to QAC should be revealed via such a procedure. In general, identification of spectral structures around 3000 cm^{-1} is difficult: hundreds of VOCs have fundamental absorption bands there caused by C-H bonds. A synchronous variation of the entire structure between 2940 cm^{-1} and 3016 cm^{-1} for different days along QAC (Figure 3) supports a suggestion that in our case it mainly represents one VOC. By applying a three-step identification procedure (see the Methods section), we concluded with high probability that the structure belongs to methyl butyrate.

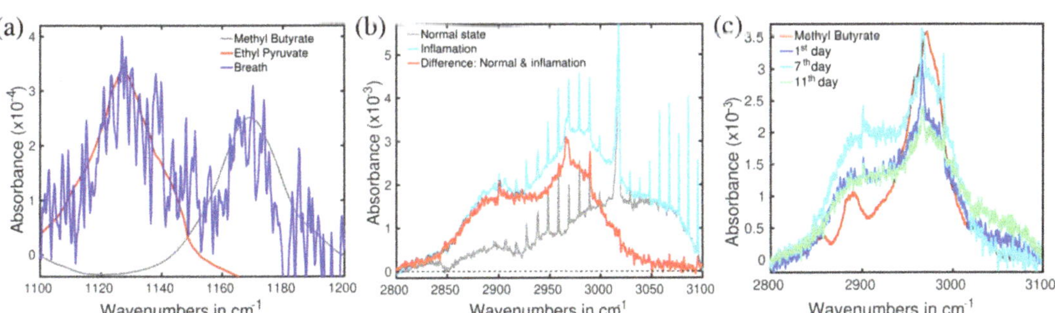

Figure 3. (a) Breath absorption spectra at 1130 cm^{-1} together with the ethyl pyruvate spectrum (red) taken from [27] as the best fitting candidate, and at 1170 cm^{-1} together with the methyl butyrate [27] (grey). The spectra correspond to day "−2" (the first elevated points on the left plot of Figure 1). Noisy signals are caused by the presence of residual water. (b) The difference (red) between the inflammation (day "0", turquoise blue) and normal state (day "10 Feburary 2018", gray). Sharp spikes in the turquoise blue and gray curves belong to methane. (c) Differential (see text) breath absorption spectrum taken during the antibiotic treatment together with the measured methyl butyrate absorption spectrum as the best fitting candidate (red).

Methyl butyrate—The detailed analysis of the spectral structures shown in Figure 3 provides four reasons to trust the identification: it (a) fits narrow, far-separated peaks at 1170 cm^{-1} and 2972 cm^{-1} with the accuracy of better than 6 cm^{-1} (i.e., relative inaccuracy 0.2%), (b) perfectly matches the main broad peak 2946–2992 cm^{-1}, (c) perfectly fits the total spectral structure of approximately 200 cm^{-1} width between 2840 and 3040 cm^{-1}, (d) demonstrates similar asymmetry of the absorption peak (the low-wavenumber tail). Nevertheless, we do not exclude contributions of other molecules to the experimental peak in the range 2850–2940 cm^{-1}. Methyl butyrate was reported in the compendium for breath and feces [16], whereas according to HMBD [28], it was previously found only in feces. It is a product of bacterial metabolism in the gut (detail in Section 3 of Supplementary Materials).

Ethyl pyruvate—The metabolite represents a derivative of another class of acids, namely alpha-keto acids. It has been identified in [28] but is absent in the compendium [16]. The latter could relate to the fact that pyruvic acid is not an end product but rather a source for other metabolites like SCFAs, carbohydrates, etc. (detail in Section 3 of Supplementary Materials).

2.5. Type of Vibrations Attributed to the Characteristic Spectral Structures of Ethyl Pyruvate Observed at 1130 cm^{-1} and Methyl Butyrate at 1170 cm^{-1}

Whereas the vibration around 3000 cm^{-1} is mainly defined by the C-H bond resulting in a difficulty of the molecular identification, the vibrations of ethyl pyruvate at 1130 cm^{-1} and methyl butyrate observed at 1170 cm^{-1} turned out to be very specific to these molecules. Numerical simulation (Section 3.4 of the Methods) allowed us to identify the corresponding vibration modes. In both cases, they were identified as a combination of C-O stretching and C-H bending modes involving the movement of the entire molecular skeleton. For methyl butyrate, the vibration was identified as a combination of a very strong backbone C-O stretching mode and the C-H bending mode in the methyl group attached to the C-O bond. These two modes affect the movement of the other part of the molecule, and all the C-H bonds undergo some kind of bending motion. The backbone of the molecule also shows bending motion restricted to the plane of the backbone.

For illustration, the retrieved absorption spectrum of methyl butyrate is shown in Figure 4, together with the measured spectrum. Their good qualitative agreement in the entire fingerprint region can be considered as another confirmation of validity of the identification. Shifts in positioning for the peaks centered at 1170 and 3000 cm^{-1}, observable for the calculated and measured spectra, represent a general problem for any quantum chemical calculation. We consider the agreement as another evidence of the power of gas phase spectroscopy in terms of accuracy of molecular identification. The identified complex vibration, unlike low-specific single C-H, C-O, or C=O modes used in analyses of biofluids in liquid phase, characterizes the molecule in a unique way because all the molecular skeleton is involved (see the arrows in the inset).

2.6. A Hypothesized Transportation Scheme and Parent Bacteria

A possible origin and transportation of ethyl pyruvate and methyl butyrate in the body until their extraction in the lung alveoli are discussed and illustrated in Section 4 of Supplementary Materials. Their parent bacteria are discussed in Section 3 of Supplementary Materials. Butyric and propionic acids and their derivatives, in our case methyl butyrate and ethyl/propyl propionate, are absent in the list of main metabolomic pathways of *Hp* (see figure 2 in [29]). As ethyl pyruvate demonstrates variations only during QAC but not after (the right plot of Figure 1), we also do not attribute this metabolite to the product of *Hp*. It allowed us to consider the identified metabolites as suitable for monitoring bacteria other than *Hp*, being affected by anti-*Hp* antibiotic treatment.

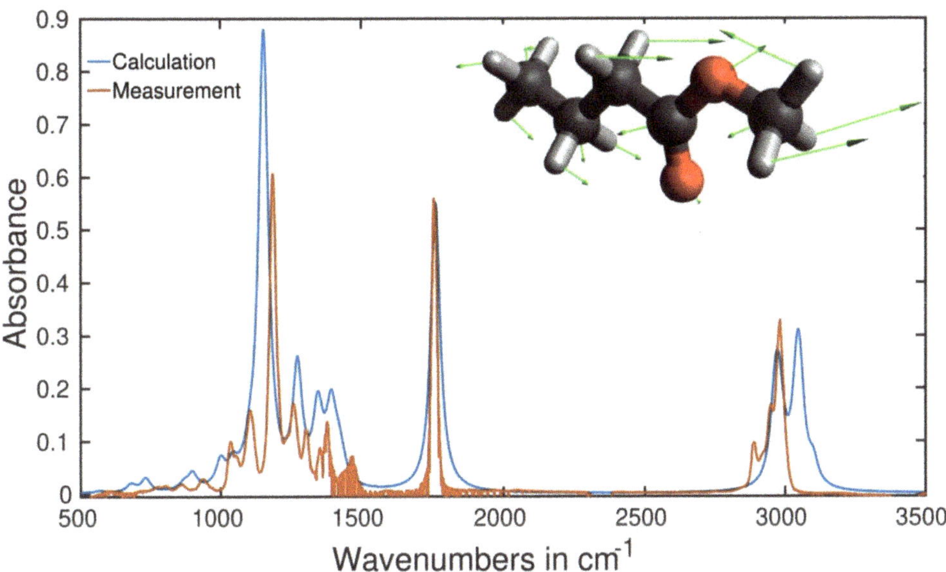

Figure 4. Orange curve: measured absorption spectrum of methyl butyrate; blue curve: result of numerical calculation (see Section 3.4). Inset: 3D structure of methyl butyrate. Green arrows show the movement of atoms in the complex vibration linked to the peak at 1170 cm^{-1}. Red balls: oxygen, large grey balls: carbon, small grey pins: hydrogen.

3. Materials and Methods

3.1. The Instrument

We used Bruker FTIR spectrometer Vertex 70 based on thermal source, operating in the mid-infrared spectral range of our interest 500–4000 cm^{-1} (2.5–20 μm). The spectral resolution 0.5 cm^{-1} was kept for all the measurements. Gas samples were collected in single-use Tedlar bags (Sigma Aldrich) and measured then by using a system that significantly suppresses the amount of water vapor [30]. The minimum detectable concentration (the detection noise) reached 50 ppb (part-per-billion, Section 2 in SM). This value corresponds to 1×10^{-4} of the absorbance at 1100 cm^{-1} in Figure 3a. The absorbance units were used throughout the text because they represent physical values measured in the experiment. Their transformation into concentrations can be done only if the corresponding spectral structure is identified as a certain VOC.

3.2. Subject of the Study and a Description of the Treatment Antibiotics Course

The breath content of a 64-year old men participating in this study was monitored for more than three and a half years, including several snapshots during and after the antibiotics course. The reason for the QAC course was acute gastritis. Gastritis started in his childhood, with different periods of pain intensity. His visit to gastroenterologist revealed a distinct epithelial area affected by inflammation corresponding to the *Hp* activity. The following QAC assisted by the probiotic course was monitored by five measurements shown in Figures 1 and 2. The QAC included: a 10-day course of pump inhibitor (PPI, esomeprazol), tetracycline, metronidazole, bismuth citrate accompanied by a 20 day of the probiotic yeast course (Omni-biotic 10). PPI decreases the acid production in stomach and possibly, adhesion of *Hp* to stomach epithelium. Omni-biotic 10 contains 10 human bacterial strains [23]: *Lactobacillus acidophilus W55, Lactobacillus salivarius W24, Lactobacillus acidophilus W37, Lactobacillus plantarum W62, Lactobacillus paracasei W72, Bididobacterium bifidium W23, Lactobacillus rhamnosus W71, Bifidobacterium lactic W18, Enterococcus faecium W54, Bifidobacterium longum W51.* No post-treatment healthy check was performed.

3.3. Identification Procedures of the Spectral Structures

Identification of molecules was done according to the three-step procedure developed in [26,31], with the help of theoretical hints developed in [32]. The hints make a link between a 3D molecular structure and the shape of the corresponding absorption spectral structure. It has to be noted that the identification is a high probability guess, bearing in mind a large number of molecules with the absorption bands in the range of our interest, especially around 3000 cm^{-1}. The range is defined by C-H vibrations common for all organic molecules. In short, the identification steps included comparison of the absorption spectra of VOCs from the compendium [16] with the experiment by using NIST database [27] (step 1). Candidates chosen in such a way were then measured with the spectrometer under identical technical parameters as breath samples (step 2). The final proof consisted in comparison of the remaining candidates with HMBD [28] (step 3). The accuracy of the identification of molecules achieved in our spectroscopic experiment approached 99.8% in terms of the peak position and characteristic width.

3.4. Numerical Simulation

Numerical simulation of the equilibrium structure of the identified gaseous metabolites has been performed based on the density functional theory with the Perdew–Burke–Ernzerho functional using Gaussian 09 computational chemistry software [33–36].

4. Conclusions and Future Work

To the best of our knowledge, we demonstrate the first signatures of breath-related dynamics of acute gastritis affected by the quadruple antibiotics course. The carriers of the dynamics were identified as volatile derivatives of SCFA and alpha-keto acid. They represent the products of bacterial metabolism in the gut. As SCFAs play an important role in the regulating energy metabolism and energy supply, maintaining the homeostasis of the intestinal environment [37–39], their variations found in this study reflect fundamental changes of the body state caused by acute gastritis and antibiotic treatment. A direct link between SCFA, alpha-keto acid, and acute gastritis has to be clarified in further studies.

In our opinion, at this promising stage it makes sense to continue work on (1) identifying new spectral ranges sensitive to QAC; and (2) verifying the observation reported here, with conventional techniques where possible. An example for the latter could be a biochemical analysis of feces, even it is not instant in comparison to breath. Collecting the statistical data of the reaction of patients treated with QAC, with its further analysis should be considered with a precaution: a reaction of a human body to the same medication is specific, especially for non-targeted bacteria (i.e., excluding *Hp* in our case). Diverse effects in reactions of individuals to the same antibiotic treatment observed in [8,10] may be considered as an illustration of this concern. It has to be noted that, for the targeted bacteria (*Hp* in our case), the observed diversity of human reactions is low. Instead of statistical analysis, we are in favor of implementing bio-passports introduced previously [20]. They should contain individual information about the state of microbiota together with all previous reactions (in the sense of this study) to any antibiotic treatment. In our opinion, such a strategy should replace the statistical approach in both cases of analysis and treatment by introducing a personalized approach, with higher expected outcome.

As the metabolism of each bacterium is complex depending on its strains, there is no direct link between the production of a certain SCFA or alpha-keto acid, and the bacterium. Nevertheless, if the ratio of concentrations of acetate, propionate and butyrate deviates significantly from the norm 60:20:20 in the gut [37,38], it may be considered as a sign of abnormality. In our case, the reconstructed propionate/butyrate ratio of concentrations before and after QAC was found to be equal to 1, supporting the evidence of norm [37,38].

Supplementary Materials: Figure S1: Left: measured 1/noise vs square root of number of scans. Right: measured 1/noise vs spectral resolution in cm−1, Figure S2: Working hypothesis of transportation of methyl butyrate and ethyl pyruvate in the body, with their release via exhaled air in lungs. References [40–60] are cited in the Supplementary Materials.

Author Contributions: Conceptualization, A.A. and K.S.M.; Methodology, A.A.; Software, K.S.M.; Formal analysis, A.A.; Investigation, A.A. and K.S.M.; Writing—original draft preparation, A.A.; Writing—review and editing, A.A. Both authors have read and agreed to the published version of the manuscript.

Funding: Munich Center for Advanced Photonics (MAP).

Institutional Review Board Statement: Ethical review and approval were waived for this study, because the necessary data were collected in frame of other studies [20,30,31]. They mostly included the data of a longitudinal study used in Figures 1 and 2.

Informed Consent Statement: Informed consent was obtained from the subject involved in the study.

Data Availability Statement: Data available on request from the authors.

Conflicts of Interest: The authors declare no conflict of interest.

Sample Availability: Samples of the compounds are not available from the authors.

References

1. Goderska, K.; Pena, S.A.; Alarcon, T. Helicobacter pylori treatment: Antibiotics or probiotics. *Appl. Microbiol. Biotechnol.* **2018**, *102*, 1–7. [CrossRef] [PubMed]
2. Kuipers, E.J. Helicobacter pylori and the risk and management of associated diseases: Gastritis, ulcer disease, atrophic gastritis and gastric cancer. *Aliment. Pharmacol. Ther.* **1997**, *11*, 71–88. [CrossRef] [PubMed]
3. Debraekeleer, A.; Remaut, H. Future perspectives for potential Helicobacter pylori eradication therapies. *Future Microbiol.* **2018**, *13*, 671–687. [CrossRef] [PubMed]
4. Váradi, L.; Luo, J.L.; Hibbs, D.E.; Perry, J.D.; Anderson, R.J.; Orenga, S.; Groundwater, P.W. Methods for the detection and identification of pathogenic bacteria: Past, present, and future. *Chem. Soc. Rev.* **2017**, *46*, 4818–4832. [CrossRef]
5. Saito, M.K.M.; Fukuda, S.; Kato, C.; Ohara, S.; Hamada, S.; Nagashima, R.; Obara, K.; Suzuki, M.; Honda, H.; Asaka, M.; et al. 13C-urea breath test, using a new compact nondispersive isotope-selective infrared spectrophotometer: Comparison with mass spectrometry. *J. Gastroenterol.* **2004**, *39*, 629–634.
6. Gisbert, J.P. Comparison between two 13C-urea breath tests for the diagnosis of Helicobacter pylori infection: Isotope ratio mass spectrometer versus infrared spectrometer. *Gastroenterol. Hepatol.* **2003**, *26*, 141–146. [CrossRef]
7. Ferwana, M.; Abdulmajeed, I.; Alhajiahmed, A.; Madani, W.; Firwana, B.; Hasan, R.; Altayar, O.; Limburg, P.J.; Murad, M.H.; Knawy, B. Accuracy of urea breath test in Helicobacter pylori infection: Meta-analysis. *World J. Gastroenterol.* **2015**, *21*, 1305–1314. [CrossRef]
8. Dethlefsen, L.; Huse, S.; Sogin, M.L.; Relman, D.A. The pervasive effects of an antibiotic on the human gut microbiota, as revealed by deep16S rRNA sequencing. *PLoS Biol.* **2008**, *6*, e280. [CrossRef]
9. Cully, M. Antibiotics alter the gut microbiome and host health. *Nat. Milest.* **2019**, *6*, S19.
10. Shaw, L.P.; Bassam, H.; Barnes, C.P.; Walker, A.S.; Klein, N.; Balloux, F. Modelling microbiome recovery after antibiotics using a stability landscape framework. *ISME J.* **2019**, *13*, 1845–1856. [CrossRef]
11. Logan, J.; Edwards, K.; Saunders, N. (Eds.) *Real-Time PCR: Current Technology and Applications*; Caister Academic Press: Norfolk, UK, 2009; ISBN 978-1-904455-39-4.
12. Poretsky, R.; Rodriguez-R, L.M.; Luo, C.; Tsementzi, D.; Konstantinidis, K.T. Strengths and limitations of 16S rRNA gene amplicon sequencing in revealing temporal microbial community dynamics. *PLoS ONE* **2014**, *9*, e93827. [CrossRef]
13. Hiergeist, A.; Reischl, U.; Gessner, A. Multicenter quality assessment of 16S ribosomal DNA-sequencing for microbiome analyses reveals high inter-center variability. *Int. J. Med Microbiol.* **2016**, *306*, 334–342. [CrossRef] [PubMed]
14. Shepherd, L.T.W.A.J.; Doherty, C.P.; McColl, K.E.L.; Williams, C.L. Helicobacter pylori in the faeces? *QJM Int. J. Med.* **1999**, *92*, 361–364.
15. Kahn, S.H.S.M.; Jiang, W.; Green, P.H.; Neu, H.C.; Chin, N.; Morotomi, M.; LoGerfo, P.; Weinstein, I.B. Direct detection and amplification of Helicobacter pylori ribosomal 16S gene segments from gastric endoscopic analysis. *Diagn. Microbiol. Infect. Dis.* **1990**, *13*, 473–479.
16. Costello, B.d.; Amann, A.; Al-Kateb, H.; Flynn, C.; Filipiak, W.; Khalid, T.; Osborne, D.; Ratcliffe, N.M. A review of the volatiles from the healthy human body. *J. Breath Res.* **2014**, *8*, 014001. [CrossRef]
17. Chan, D.K.; Leggett, C.L.; Wang, K.K. Diagnosing gastrointestinal illnesses using fecal headspace volatile organic compounds. *World J. Gastroenterol.* **2016**, *22*, 1639–1649. [CrossRef]

18. Wang, Q.; Teng, G.; Qiao, X.; Zhao, Y.; Kong, J.; Dong, L.; Cui, X. Importance evaluation of spectral lines in laser-induced breakdown spectroscopy for classification of pathogenic bacteria. *Biomed. Opt. Express* **2018**, *9*, 5837–5850. [CrossRef]
19. Leja, M.; Amal, H.; Lasina, I.; Skapars, R.; Sivins, A.; Ancans, G.; Tolmanis, I.; Vanags, A.; Kupcinskas, J.; Ramonaite, R.; et al. Analysis of the effects of microbiome-related confounding factors on the reproducibility of the volatolomic test. *J. Breath Res.* **2016**, *10*, 037101. [CrossRef]
20. Maiti, K.S.; Lewton, M.; Fill, E.; Apolonski, A. Human beings as islands of stability: Monitoring body states using breath profiles. *Sci. Reports* **2019**, *9*, 16167. [CrossRef]
21. Nahar, S.; Mukhopadhyay, A.K.; Khan, R.; Ahmad, M.M.; Datta, S.; Chattopadhyay, S.; Dhar, S.C.; Sarker, S.A.; Engstrand, L.; Berg, D.E.; et al. Antimicrobial Susceptibility of *Helicobacter pylori* Strains Isolated in Bangladesh. *J. Clin. Microbiol.* **2004**, *42*, 4856–4858. [CrossRef]
22. Theriot, C.M.; Koenigsknecht, M.J., Jr.; Hatton, E.G.; Nelson, A.M.; Li, B.; Huffnagle, G.B.; Li, J.Z.; Young, V.B. Antibiotic-induced shifts in the mouse gut microbiome and metabolome increase susceptibility to Clostridium difficile infection. *Nat. Commun.* **2014**, *5*, 3114. [CrossRef] [PubMed]
23. Available online: https://www.omnibioticlife.com/omnibiotic/ab-10/ (accessed on 6 June 2021).
24. Niv, Y. H pylori recurrence after successful eradication. *World J. Gastroenterol.* **2008**, *14*, 1477–1478. [CrossRef] [PubMed]
25. Gisbert, J.P.; Luna, M.; Gomez, B.; Herrerais, J.M.; Mones, J.; Castro-Fernandez, M.; Sanchez-Porbe, P.; Cosme, A.; Oliveras, D.; Pajares, J.M. Recurrence of Helicobacter pylori infection after several eradication therapies: Long-term follow-up of 1000 patients. *Aliment. Pharmacol. Ther.* **2006**, *23*, 713–719. [CrossRef] [PubMed]
26. Apolonski, A.; Maiti, K.S. Towards a standard operating procedure for revealing hidden volatile organic compounds in breath: The Fourier-transform IR spectroscopy case. *Appl. Opt.* **2021**, *60*, 4217–4224. [CrossRef]
27. Kramida, A.; Ralchenko, Y.; Reader, J.; NIST ASD Team. NIST Atomic Spectra Database (ver. 5.7.1), National Institute of Standards and Technology, Gaithersburg, MD, USA. 2019. Available online: https://physics.nist.gov/asd (accessed on 9 April 2017).
28. Wishart, D.S.; Feunang, Y.D.; Marcu, A.; Guo, A.C.; Liang, K.; Vázquez-Fresno, R.; Sajed, T.; Johnson, D.; Li, C.; Karu, N.; et al. HMDB 4.0: The Human metabolome database for 2018. *Nucleic Acids Res.* **2018**, *46*, D608–D617. [CrossRef]
29. Doing, P.; de Jonge, B.L.; Alm, R.A.; Brown, E.D.; Uria-Nickelsen, M.; Noonan, B.; Mills, S.D.; Tummino, P.; Carmel, G.S.; Guild, B.C.; et al. Helicobacter pylori physiology predicted from genomic comparison of two strains. *Microbiol. Mol. Biol. Rev.* **1999**, *63*, 675–707.
30. Maiti, K.S.; Lewton, M.; Fill, E.; Apolonski, A. Sensitive spectroscopic breath analysis by water condensation. *J. Breath Res.* **2018**, *12*, 046003. [CrossRef]
31. Maiti, K.S.; Roy, S.; Lampe, R.; Apolonski, A. Breath indeed carries significant information about a disease. Potential biomarkers of cerebral palsy. *J. Biophoton.* **2020**. [CrossRef]
32. Gelin, M.F.; Blokhin, A.P.; Ostrozhenkova, E.; Apolonski, A.; Maiti, K.S. Theory helps experiment to reveal VOCs in human breath. *Spectrochim. Acta Part A* **2021**, *258*, 119785. [CrossRef]
33. Frisch, M.J.; Trucks, G.W.; Schlegel, H.B.; Scuseria, G.E.; Robb, M.A.; Cheeseman, J.R.; Scalmani, G.; Barone, V.; Petersson, G.A.; Nakatsuji, H.; et al. *Gaussian 09 Citation, Revision, A.02*; Gaussian Inc.: Wallingford, CT, USA, 2016.
34. Kohn, W.; Sham, L.J. Self-Consistent Equations Including Exchange and Correlation Effects. *Phys. Rev.* **1965**, *140*, A1133. [CrossRef]
35. Perdew, J.P.; Burke, K.; Ernzerhof, M. Generalized Gradient Approximation Made Simple. *Phys. Rev. Lett.* **1996**, *77*, 3865. [CrossRef]
36. Perdew, J.P.; Burke, K.; Ernzerhof, M. Erratum to Generalized Gradient Approximation Made Simple. *Phys. Rev. Lett.* **1997**, *78*, E1396. [CrossRef]
37. Den Besten, G.; van Eunen, K.; Groen, A.K.; Venema, K.; Reijngoud, D.J.; Bakker, B.M. The role of short-chain fatty acids in the interplay between diet, gut microbiota, and host energy metabolism. *J. Lip. Res.* **2013**, *54*, 2325–2340. [CrossRef]
38. Cummings, J.H.; Pomare, E.W.; Branch, W.J.; Naylor, C.P.; Macfarlane, G.T. Short chain fatty acids in human large intestine, portal, hepatic and venous blood. *Gut* **1987**, *28*, 1221–1227. [CrossRef]
39. He, J.; Zhang, P.; Shen, L.; Niu, L.; Tan, Y.; Chen, L.; Zhao, Y.; Bai, L.; Hao, X.; Li, X.; et al. Short-chain fatty acids and their association with signalling pathways in inflammation, glucose and lipid metabolism. *Int. J. Mol. Sci.* **2020**, *21*, 6356. [CrossRef]
40. Berg, R.D. The Indigenous Gastrointestinal Microflora. *Trends Microbiol.* **1996**, *4*, 430–454. [CrossRef]
41. Bäumler, A.J.; Sperandio, V. Interactions Between the Microbiota and Pathogenic Bacteria in the Gut. *Nat. Cell Biol.* **2016**, *535*, 85–93. [CrossRef]
42. Yang, I.; Neil, S.; Suerbaum, S. Survival in Hostile Territory: The Microbiota of the Stomach. *FEMS Microbiol. Rev.* **2013**, *37*, 736–761. [CrossRef]
43. Westhoff, S.; van Wezel, G.P.E.; Rozen, D. Distance-Dependent Danger Responses in Bacteria. *Curr. Opin. Microbiol.* **2017**, *36*, 95–101. [CrossRef]
44. Garcia-Bayona, L.; Comstock, L.E. Bacterial Antagonism in Host-Associated Microbal Communities. *Science* **2018**, *361*, 1215. [CrossRef]
45. Lozupone, C.A.; Stombaugh, J.I.; Gordon, J.I.; Jansson, J.K.; Knight, R. Diversity, stability and resilience of the human gut microbiota. *Nature* **2012**, *489*, 220–230. [CrossRef] [PubMed]

46. Angers, A.; Kagkli, D.; Patak, A.; Petrillo, M.; Querci, M.; Rüdelsheim, P.; Smets, G.; Van den Eede, G. The Human Gut Microbiota (2018), JRC Technical Report. Available online: http://publications.jrc.ec.europa.eu/repository/Handle/JRC112042 (accessed on 6 June 2021).
47. Dickson, I. Stability and individuality of adult microbiota. *Nat. Milest.* **2019**, *7*, S11.
48. Quevrain, E.; Maubert, M.A.; Michon, C.; Chain, F.; Marquant, R.; Tailhades, J.; Miquel, S.; Carlier, L.; Bermúdez-Humarán, L.G.; Pigneur, B.; et al. Identification of an Anti-Inflammatory Protein from Faecalibacterium Prausnitzii, a Commensal Bacterium Deficient in Crohn's Disease. *Gut* **2016**, *65*, 415–425. [CrossRef] [PubMed]
49. Faith, J.J.; Guruge, J.L.; Charbonneau, M.; Subramanian, S.; Seedorf, H.; Goodman, A.L.; Clemente, J.C.; Knight, R.; Heath, A.C.; Leibel, R.L.; et al. The Long-Term Stability of the Human Gut Microbiota. *Science* **2013**, *341*, 1237439. [CrossRef]
50. Lange, K.; Buerger, M.; Stallmach, A.; Bruns, T. Effects of Antibiotics on Gut Microbiota. *Dig. Dis.* **2016**, *34*, 260–268. [CrossRef]
51. Delgado, S.; Flórez, A.B.; Mayo, B. Antibiotic Susceptibility of Lactobacillus and Bifidobacterium Species from the Human Gastrointestinal Tract. *Curr. Microbiol.* **2005**, *50*, 202–207. [CrossRef]
52. Ramsey, M.; Hartke, A.; Huycke, M. The Physiology and Metabolism of Enterococci. In *Enterococci: From Commensals to Leading Causes of Drug Resistant Infection*; Gilmore, M.S., Clewell, D.B., Ike, Y., Shankar, N., Eds.; Eye and Ear Infirmary: Boston, MA, USA, 2014.
53. Kristich, C.J.; Rice, L.B.; Arias, C.A. Enterococcal infection—treatment and Antibiotic Resistance. In *Enterococci: From Commensals to Leading Causes of Drug Resistant Infection*; Gilmore, M.S., Clewell, D.B., Ike, Y., Shankar, N., Eds.; Eye and Ear Infirmary: Boston, MA, USA, 2014.
54. Hickey, M.W.; Hillier, A.; Jago, G. Metabolism of Pyruvate and Citrate in Lactobacilli. *Aust. J. Biol. Sci.* **1983**, *36*, 487. [CrossRef]
55. Belenguer, A.; Duncan, S.H.; Calder, A.G.; Holtrop, G.; Louis, P.; Lobley, G.E.; Flint, H.J. Two Routes of Metabolic Cross-Feeding Between Bifidobacterium Adolescentis and Butyrate-Producing Anaerobes from the Human Gut. *Appl. Envir. Microbiol.* **2006**, *72*, 3593–3599. [CrossRef]
56. Ferrario, C.; Taverniti, V.; Milani, C.; Fiore, W.; Laureati, M.; De Noni, I.; Stuknyte, M.; Chouaia, B.; Riso, P.; Guglielmetti, S. Modulation of Fecal Clostridiales Bacteria and Butyrate by Probiotic Intervention with Lactobacillus Paracasei DG Varies Among Healthy Adults. *J. Nutr.* **2014**, *144*, 1787–1796. [CrossRef]
57. Venegas, D.P.; De la Fuente, M.K.; Landskron, G.; González, M.J.; Quera, R.; Dijkstra, G.; Harmsen, H.J.M.; Faber, K.N.; Hermoso, M.A. Short Chain Fatty Acids (SCFAs)-Mediated Gut Epithelial and Immune Regulation and Its Relevance for Inflammatory Bowel Diseases. *Front. Immunol.* **2019**, *10*, 277. [CrossRef]
58. Louis, P.; Flint, H.J. Diversity, Metabolism and Microbial Ecology of Butyrate-Producing Bacteria from the Human Large Intestine. *FEMS Microbiol. Lett.* **2009**, *294*, 1–8. [CrossRef]
59. Lofgren, J.L.; Whary, M.T.; Ge, Z.; Muthupalani, S.; Taylor, N.S.; Mobley, M.; Potter, A.; Varro, A.; Eibach, D.; Suerbaum, S.; et al. Lack of Commensal Flora in Helicobacter pylori–Infected INS-GAS Mice Reduces Gastritis and Delays Intraepithelial Neoplasia. *Gastroenterology* **2011**, *140*, 210–220.e4. [CrossRef]
60. Boyanova, L.; Kolarov, R.; Mitov, I. Antimicrobial Resistance and the Management of Anaerobic Infections. *Expert Rev. Anti Infect. Ther.* **2007**, *5*, 685–701. [CrossRef]

Article

Quantification of Volatile Aldehydes Deriving from In Vitro Lipid Peroxidation in the Breath of Ventilated Patients

Lukas Martin Müller-Wirtz [1,2,*], Daniel Kiefer [1], Sven Ruffing [1], Timo Brausch [1], Tobias Hüppe [1,2], Daniel I. Sessler [2,3], Thomas Volk [1,2], Tobias Fink [1,2], Sascha Kreuer [1,2] and Felix Maurer [1,2]

[1] CBR—Center of Breath Research, Department of Anaesthesiology, Intensive Care and Pain Therapy, Saarland University Medical Center, Homburg, 66421 Saarland, Germany; daniel.kiefer@uks.eu (D.K.); sven.ruffing@uks.eu (S.R.); timo.brausch@uks.eu (T.B.); tobias.hueppe@uks.eu (T.H.); thomas.volk@uks.eu (T.V.); tobias.fink@uks.eu (T.F.); sascha.kreuer@uks.eu (S.K.); felix.maurer@uks.eu (F.M.)
[2] Outcomes Research Consortium, Cleveland, OH 44195, USA; ds@or.org
[3] Department of Outcomes Research, Anesthesiology Institute, Cleveland Clinic, Cleveland, OH 44195, USA
* Correspondence: lukas.wirtz@uks.eu

Citation: Müller-Wirtz, L.M.; Kiefer, D.; Ruffing, S.; Brausch, T.; Hüppe, T.; Sessler, D.I.; Volk, T.; Fink, T.; Kreuer, S.; Maurer, F. Quantification of Volatile Aldehydes Deriving from In Vitro Lipid Peroxidation in the Breath of Ventilated Patients. Molecules 2021, 26, 3089. https://doi.org/10.3390/molecules26113089

Academic Editors: Ben de Lacy Costello and Natalia Drabińska

Received: 16 April 2021
Accepted: 19 May 2021
Published: 21 May 2021

Publisher's Note: MDPI stays neutral with regard to jurisdictional claims in published maps and institutional affiliations.

Abstract: Exhaled aliphatic aldehydes were proposed as non-invasive biomarkers to detect increased lipid peroxidation in various diseases. As a prelude to clinical application of the multicapillary column–ion mobility spectrometry for the evaluation of aldehyde exhalation, we, therefore: (1) identified the most abundant volatile aliphatic aldehydes originating from in vitro oxidation of various polyunsaturated fatty acids; (2) evaluated emittance of aldehydes from plastic parts of the breathing circuit; (3) conducted a pilot study for in vivo quantification of exhaled aldehydes in mechanically ventilated patients. Pentanal, hexanal, heptanal, and nonanal were quantifiable in the headspace of oxidizing polyunsaturated fatty acids, with pentanal and hexanal predominating. Plastic parts of the breathing circuit emitted hexanal, octanal, nonanal, and decanal, whereby nonanal and decanal were ubiquitous and pentanal or heptanal not being detected. Only pentanal was quantifiable in breath of mechanically ventilated surgical patients with a mean exhaled concentration of 13 ± 5 ppb. An explorative analysis suggested that pentanal exhalation is associated with mechanical power—a measure for the invasiveness of mechanical ventilation. In conclusion, exhaled pentanal is a promising non-invasive biomarker for lipid peroxidation inducing pathologies, and should be evaluated in future clinical studies, particularly for detection of lung injury.

Keywords: anesthesia; breath analysis; mechanical ventilation; lipid peroxidation; biomarker; volatile aldehydes; pentanal; MCC–IMS; ventilator-induced lung injury; volatile organic compounds

1. Introduction

Lipid peroxidation products are established markers of oxidative stress [1], and are potential non-invasive biomarkers for detection of various diseases. For example, increased aldehyde exhalation has been reported in patients suffering from pulmonary diseases such as lung cancer [2–5], chronic obstructive pulmonary disease [6], and COVID-19 [7,8]. Additionally, volatile aldehydes considerably increase in the blood of patients suffering from acute respiratory distress syndrome [9,10]. Recent animal experiments suggest that the volatile aldehyde pentanal may be a biomarker for ventilator-induced lung injury [11]. There is thus increasing evidence that monitoring of aldehyde exhalation may help detect diseases and acute injuries of the lung.

Analyses of liquid aliphatic aldehydes indicate that they originate from lipid peroxidation [12,13]. However, gaseous concentrations also depend on vapor pressure, which progressively decreases with longer aldehyde chain lengths. Most previous investigations that measured gaseous concentrations of volatile aliphatic aldehydes focused on single aldehydes to quantify lipid peroxidation in vitro [14,15]. To our knowledge, only a single study performed comparative measurements of various volatile aldehydes deriving from oxidizing synthetic lipid membranes [16]. Consequently, the relative contributions of various isolated polyunsaturated fatty acids to gaseous aldehyde generation remains unclear.

Monitoring aldehyde exhalation is particularly interesting in ventilated patients, as they often have baseline pulmonary diseases and are susceptible to ventilator-induced lung injury, which might be identified by aldehyde exhalation [11]. A potential complication, though, is that most materials in breathing circuits of modern anesthesia machines and airway devices are made from plastic which can emit volatile aldehydes [17], thus potentially interfering with breath analysis in ventilated patients.

Multicapillary column–ion mobility spectrometry (MCC–IMS) has been used to examine exhaled volatile organic compounds [18–20] and monitor exhaled propofol [21–23] in ventilated patients. However, volatile aldehydes are typically exhaled at concentrations of a few parts per billion [24,25], which raises the question of whether the MCC–IMS technique with our corresponding sampling setup is sensitive enough to quantify exhaled aldehydes. Moreover, cross-contaminations from ambient air or the ventilator may even exclude in vivo quantification. A pilot study is therefore needed to assess the clinical suitability of the MCC–IMS technique to monitor aldehyde exhalation.

Our primary aim was to assess clinical use of MCC–IMS for bedside online measurements of exhaled aliphatic aldehydes as a measure of lipid peroxidation. Secondarily, we aimed to identify the most promising aliphatic aldehydes for monitoring lipid peroxidation in mechanically ventilated patients under in vitro and in vivo conditions. We therefore: (1) identified the predominant volatile aliphatic aldehydes originating from in vitro peroxidation of various polyunsaturated fatty acids; (2) evaluated emittance of volatile aldehydes from parts of the breathing circuit; (3) conducted a pilot study using MCC–IMS for in vivo quantification of exhaled aldehydes in mechanically ventilated patients.

2. Results

2.1. Calibration

All calibrations showed a good linear fit (R^2: 0.97 to 0.99; Supplementary Materials, File 1, Figure S2). Limits of detection and quantification were 0.008 and 0.011 Volt.

2.2. Volatile Aldehydes Originating from In Vitro Lipid Peroxidation

Pentanal, hexanal, heptanal, and nonanal were detected in the headspace of an animal-sourced mixture of oxidizing polyunsaturated fatty acids (PUFA-Mix, Figure 1). Pentanal and hexanal emerged from all polyunsaturated fatty acids, with pentanal and hexanal predominating. Nonanal emerged from the mixture of polyunsaturated fatty acids and heptanal from arachidonic acid (Figure 1). Unquantifiable traces of octanal were identified in all probes. Decanal was not detected.

2.3. Volatile Aldehydes Emitted By Plastic Parts of the Breathing Circuit

Parts of the breathing circuit emitted hexanal, octanal, nonanal, and decanal with nonanal and decanal originating from all assessed parts (Table 1). Pentanal and heptanal were not detected.

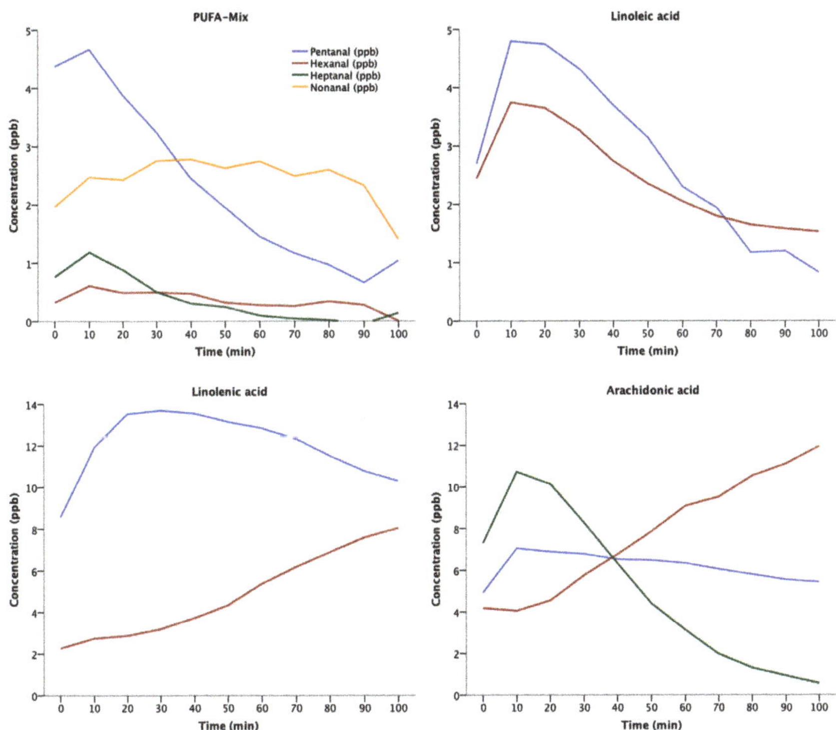

Figure 1. Volatile aldehydes produced by oxidation of polyunsaturated fatty acids. A total of 30 μL of isolated or mixed polyunsaturated fatty acids were injected into a perfluoroalkoxy alkane flask and oxidized under a constant flow of 100 mL/min synthetic air (21% O2). Headspace samples were analyzed by means of MCC–IMS. Measurement series were performed once; therefore, raw data are presented. PUFA-Mix is the animal-sourced mixture of polyunsaturated fatty acids.

Table 1. Evaporation of volatile aldehydes by parts of the breathing circuit.

Material	Detected Aldehydes	Concentration (ppb)
Endotracheal tube	Octanal	7.0 ± 1.4
	Nonanal	12.5 ± 0.7
	Decanal	2.5 ± 0.4
Humidity and moisture exchanging filter	Nonanal	0.1 ± 0.4
	Decanal	1.7 ± 0.1
Breathing bag	Hexanal	0.5 ± 0.1
	Nonanal	2.0 ± 0.3
	Decanal	0.7 ± 0.1
Breathing tubes	Hexanal	0.2 ± 0.1
	Nonanal	1.4 ± 0.2
	Decanal	0.6 ± 0.2
Test lung	Nonanal	unquantifiable traces
	Decanal	0.8 ± 0.2

Data are presented as means $\pm$ SD.

2.4. Volatile Aldehydes in the Breath of Ventilated Patients

All patients screened for eligibility underwent pancreaticoduodenectomies, as this operation is scheduled to last at least 4 h at our medical center. Twelve surgical adult patients undergoing elective pancreaticoduodenectomy were assessed. No patient was excluded.

Two third of the patients had a malignant tumor and/or arterial hypertension, and half of the patients suffered from diabetes mellitus. Patients were ventilated on average for about six hours (Table 2).

Table 2. Patient characteristics and ventilation parameters.

Patient Characteristics	
Patients included/screened for eligibility	12/12
Age (years)	67 ± 11
Sex (male/female)	8 (67)/4 (33)
Height (cm)	170 ± 8
Weight (kg)	69 ± 13
ASA physical status (I/II/III)	0/6 (50)/6 (50)
Malignant tumor	8 (67)
Arterial hypertension	8 (67)
Diabetes mellitus	6 (50)
Mechanical ventilation time (min)	344 ± 102
Ventilation Parameters	
Tidal volume (mL)	452 ± 82
Respiratory rate (breaths·min^{-1})	12 ± 1
Minute volume (L·min^{-1})	5.4 ± 1.1
Inspiratory pressure (mbar)	15.3 ± 2.1
Positive end expiratory pressure (mbar)	5.1 ± 0.5
Mechanical power (J·min^{-1})	8.3 ± 2.6

Data are presented either as means ± SD, or as frequencies (%). Ventilation parameters repeatedly measured over time were summarized with single means.

Only exhaled pentanal was quantifiable in breath. Unquantifiable traces of nonanal were detected and other aldehydes were not detected at all. Ventilators were contaminated with a mean pentanal concentration of 1.2 ± 1.1 ppb. Exhaled pentanal concentrations over time are presented in Figure 2. In three patients, the exhaled pentanal concentration did not substantially exceed ventilator contaminations who were therefore excluded from further statistical analyses. None of the patients with limited pentanal exhalation had malignant tumors. In contrast, all but one of the remaining nine patients had malignancy. Excluding the three patients with limited pentanal exhalation, the overall mean exhaled pentanal concentration was 13 ± 5 ppb.

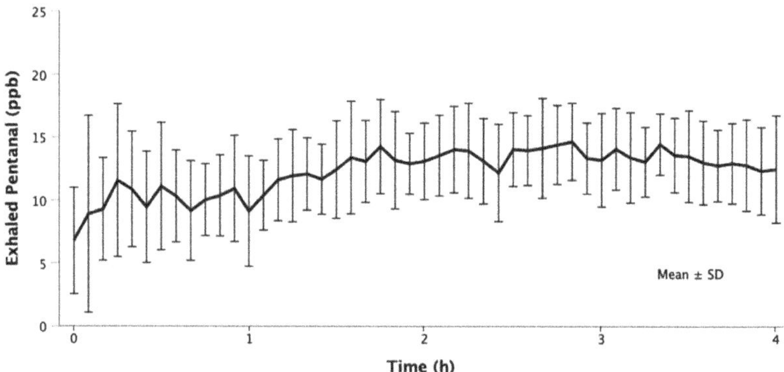

Figure 2. Pentanal in the breath of surgical patients during mechanical ventilation. Twelve surgical patients undergoing prolonged mechanical ventilation (≥4 h) were assessed. In three patients, the exhaled pentanal concentration did not substantially exceed ventilator contaminations who were therefore excluded from graphical presentation. The presented concentrations are corrected for baseline ventilator contaminations.

An exploratory analysis revealed a significant association of exhaled pentanal with tidal volume, minute volume, and mechanical power, but not with inspiratory pressure (Figure 3, Table 3).

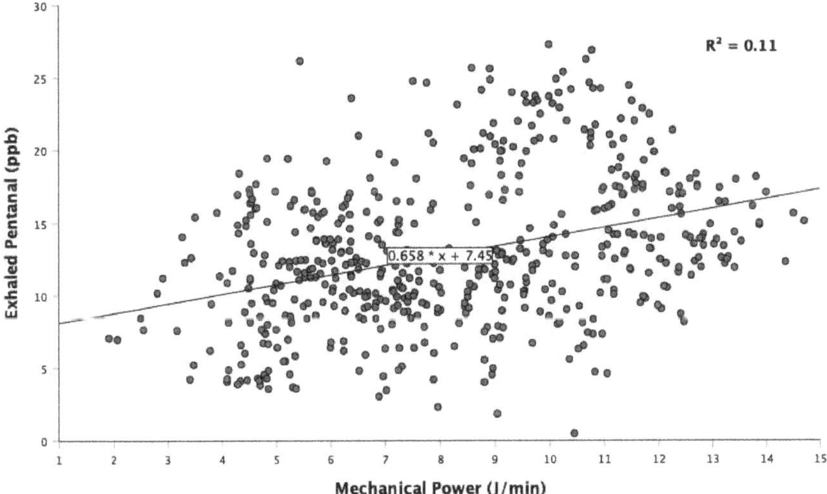

Figure 3. Exhaled pentanal versus mechanical power. A linear generalized estimating equations regression model and the marginal R^2 were calculated to assess the relationship of exhaled pentanal with mechanical power. Subjects included/total: $n = 9/12$, data pairs: $n = 547$. Data from three patients with limited pentanal exhalation were excluded.

Table 3. Association of exhaled pentanal with ventilation parameters.

Parameter	Regression Coefficient	95% Confidence Interval	R^2	p
Tidal volume (mL)	0.01	0.003–0.018	0.02	0.004
Minute volume (L·min^{-1})	2.0	0.6–3.3	0.05	0.004
Inspiratory pressure (mbar)	0.2	-0.3–0.6	0.04	0.463
Mechanical power (J·min^{-1})	0.7	0.3–1.1	0.11	0.001

Univariable linear generalized estimating equations regression models were calculated to assess the association of exhaled pentanal (dependent variable) with ventilation parameters and mechanical power (independent variable). Subjects: $n = 9$, data pairs: $n = 547$. Data from 3 patients with limited pentanal exhalation were excluded.

3. Discussion

This study presents the preparatory analytical work and a clinical pilot study for online monitoring of aldehyde exhalation to quantify lipid peroxidation in ventilated surgical patients by means of MCC–IMS. As an overall finding, pentanal represents the most promising exhaled volatile aliphatic aldehyde. Pentanal is predominantly a product of lipid peroxidation rather than evaporating from parts of the breathing circuit and was the only quantifiable volatile aldehyde in the breath of ventilated surgical patients with our measurement setup.

Oxidation of mixed polyunsaturated fatty acids confirmed that the aliphatic aldehydes pentanal, hexanal, heptanal, and nonanal are gaseous products of lipid peroxidation. Only traces of octanal were detected and no decanal was detected, probably because they were generated in limited quantities and their vapor pressures are low [26]. In contrast, pentanal and hexanal were ubiquitous. Pentanal dominated early phases of in vitro lipid peroxidation, whereas hexanal increased over time and approached or even exceeded pentanal concentrations. Our findings are consistent with a previous study that assessed volatile aldehydes emerging from oxidizing phospholipid membranes and similarly showed that pentanal dominated early in the process of lipid peroxidation, whereas hexanal dominated later [16]. Results from an analysis of fluid aliphatic aldehydes, originating from several polyunsaturated fatty acids oxidized by air, confirm that pentanal and hexanal are the predominant products of lipid peroxidation, with hexanal concentrations being twice those of pentanal after 48 h of peroxidation [12].

Pentanal thus dominates early and hexanal dominates later phases of lipid peroxidation, which can be explained by their chemical properties. Specifically, pentanal has twice the vapor pressure than hexanal (pentanal: 26 mmHg at 20 °C and hexanal: 11.3 mmHg at 25 °C), and therefore evaporates more quickly [27,28]. However, pentanal is more reactive then hexanal. Consequently, autoxidation of pentanal may, over time, exceed its generation rate. Hexanal may therefore be the primary and more stable product of lipid peroxidation, but pentanal may be a better biomarker by virtue of responding quickly to oxidative stress.

Plastic components of the breathing circuit emitted hexanal, octanal, nonanal, and decanal, which is consistent with previous analyses showing that all are emitted by polypropylene and polyethylene [17,29]—the two most commonly used materials for plastic components. The largest source of volatile aldehydes was the endotracheal tube, which is made from polyvinylchloride. In addition to octanal and decanal, the endotracheal tube emitted considerable amounts of nonanal, which is consistent with a previous analysis of volatile organic compound profiles emitted by polyvinylchloride materials [30]. Under the influence of heat and moisture from the body and breathing gases, release of these volatile organic compounds from plastic components may be unpredictable. Thus, even when corrected for baseline contamination, measurements of hexanal, octanal, nonanal, and decanal in the breath of ventilated patients might be compromised by non-organic sources. In contrast, pentanal and heptanal were not emitted by plastic breathing circuit components and are thus presumably better biomarkers for lipid peroxidation in ventilated patients.

We finally evaluated aldehyde exhalation in twelve surgical patients during prolonged mechanical ventilation. Aside from traces of nonanal, pentanal was the only volatile aldehyde we detected. Ventilators were contaminated with low amounts of pentanal, possibly representing residuals from previously ventilated patients since the breathing circuit was apparently not the source. The overall exhaled pentanal concentrations in ventilated patients were in the low parts-per-billion range, consistent with previous reports from spontaneously breathing healthy volunteers [24,25]. However, we measured slightly higher exhaled pentanal concentrations in ventilated patients, possibly consequent to intubation, which increases the sampled proportion of alveolar air. Another reason could be that mechanical ventilation induces pulmonary lipid peroxidation and therefore increase pentanal exhalation, as previously shown in animals [11,31].

In three patients, the exhaled pentanal concentrations were roughly at the concentration of ventilator contamination, whereas exhaled concentrations in the others averaged 13 ppb. Interestingly, none of the patients with low concentrations had cancer, whereas eight of the nine others did. While possibly spurious, the results are consistent with previous reports that cancer promotes exhalation of pentanal [2–4,32,33].

Although we did not actively vary ventilation parameters, our explorative analysis revealed a significant linear relationship between exhaled pentanal and mechanical power—a clinical measure for the invasiveness of mechanical ventilation [34,35]. The higher the mechanical power dissipated to the lungs, the higher was the exhaled pentanal concentration, which is consistent with our previous findings in 150 ventilated rats [11]. We therefore previously proposed that exhaled pentanal results from stretched lung tissue, which exposes lipids in cell membranes to oxidation. Intuitively, higher minute volumes may dilute exhaled pentanal. Instead, higher minute volumes were associated with increased pentanal exhalation, further supporting the hypothesis that mechanical ventilation induces pulmonary lipid peroxidation measurable by exhaled pentanal—a potential biomarker for ventilator-induced lung injury. More experimental and clinical studies are needed to identify various causes of pentanal exhalation, which is critical to its potential use as a biomarker.

A limitation of our study is that a mixture of exhaled and inspired gases can lead to cross-contaminations or diluted concentrations. An integration of carbon dioxide or flow triggered sampling could help sample isolated exhaled gas and thus increase the proportion of alveolar gas in breath samples [36]. Furthermore, activated charcoal filters, originally designed to eliminate residual volatile anesthetics emitted from anesthesia workstations, are now available [37]. Using an activated charcoal filter between the anesthesia machine and the inspiratory limb of the rebreathing circuit would presumably eliminate contamination from within the machine. We also note that our study population is small. While sufficient to confirm applicability of our measurement setup to patients, larger studies will be necessary to confirm the association of pentanal exhalation and mechanical power.

In summary, pentanal and hexanal are the predominant volatile aldehydes deriving from lipid peroxidation under in vitro conditions, and therefore represent promising breath biomarkers for oxidative stress. Emission of volatile aldehydes from plastic parts of the breathing circuit may bias breath analysis for hexanal, octanal, nonanal, and decanal but not for pentanal. Future studies should quantify exhaled pentanal in mechanically ventilated patients with various pathologies and assess its potential as a biomarker for ventilator-induced lung injury.

4. Materials and Methods

4.1. Calibration

The detailed experimental setup and procedure of the calibration is presented in the supplement (Supplementary Materials, File 1). In short, hexane-diluted aldehyde standards were pipetted into a closed flask made from inert perfluoroalkoxy alkane. Evaporation was accelerated by an electrically driven fan inside the flask and the resulting gaseous mixture was sampled by the MCC–IMS (B&S Analytik, Dortmund, Germany). The composition of the liquid hexane-diluted aldehyde standards needed to generate specific gaseous concentrations inside the flask were calculated according to the ideal gas law (Supplementary Materials, File 2).

4.2. Volatile Aldehydes Originating from In Vitro Lipid Peroxidation

We used the same technical setup as for calibration (Supplementary Materials, File 1). A total of 30 μL of an animal-sourced mixture of polyunsaturated fatty acids and three isolated polyunsaturated fatty acids—linoleic, linolenic, and arachidonic acid (analytical standard, Merck, Darmstadt, Germany)—were oxidized in a cleaned flask under 100 mL/min flow of highly purified synthetic air (oxygen content: 21%; Alphagaz 1, Air Liquide, Paris, France) and constant fanning. Headspace gas was sampled at 10-min intervals by the MCC–IMS. Signal intensities between the limit of detection and quantification were considered as unquantifiable traces.

4.3. Volatile Aldehydes Emitted by Plastic Parts of the Breathing Circuit

An endotracheal tube (Ruesch®, Teleflex, Kernen, Germany) and a humidity and moisture exchanging filter (Gibeck Humid-Vent®; Teleflex, Kernen, Germany) were placed in the cleaned flask used for calibration and lipid peroxidation. The flask was sealed and flushed with purified nitrogen for 5 min and subsequently with highly purified synthetic air for 1 min. Breathing tubes and bag (Anesthesia set VentStar®, disposable, basic, 2 L, 1.8 m/1.5 m, latex-free, Draeger, Lübeck, Germany) and a test lung (Draeger SelfTestLung™) were flushed from the inside using a similar procedure. Headspace gas was sampled from materials placed in the flask and from the inside of the breathing tube, breathing bag, and test lung by the MCC–IMS at 5-min intervals for at least 20 min. All measurements were performed in a room maintained at 20 °C with an air purification system (CamCleaner City M, Camfil, Reinfeld, Germany).

8. Ruszkiewicz, D.M.; Sanders, D.; O'Brien, R.; Hempel, F.; Reed, M.J.; Riepe, A.C.; Bailie, K.; Brodrick, E.; Darnley, K.; Ellerkmann, R.; et al. Diagnosis of COVID-19 by analysis of breath with gas chromatography-ion mobility spectrometry—A feasibility study. *EClinicalMedicine* **2020**, *29–30*, 100609. [CrossRef]
9. Lichtenstern, C.; Hofer, S.; Möllers, A.; Snyder-Ramos, S.; Spies-Martin, D.; Martin, E.; Schmidt, J.; Motsch, J.; Bardenheuer, H.J.; Weigand, M.A. Lipid peroxidation in acute respiratory distress syndrome and liver failure. *J. Surg. Res.* **2011**, *168*, 243–252. [CrossRef]
10. Weigand, M.A.; Snyder-Ramos, S.A.; Möllers, A.G.; Bauer, J.; Hansen, D.; Kochen, W.; Martin, E.; Motsch, J. Inhaled nitric oxide does not enhance lipid peroxidation in patients with acute respiratory distress syndrome. *Crit. Care Med.* **2000**, *28*, 3429–3435. [CrossRef]
11. Müller-Wirtz, L.M.; Kiefer, D.; Maurer, F.; Floss, M.A.; Doneit, J.; Hüppe, T.; Shopova, T.; Wolf, B.; Sessler, D.I.; Volk, T.; et al. Volutrauma Increases Exhaled Pentanal in Rats: A Potential Breath Biomarker for Ventilator-Induced Lung Injury. *Anesth. Analg.* **2021**. Published Ahead-of-Print. [CrossRef] [PubMed]
12. Yoshino, K.; Sano, M.; Fujita, M.; Tomita, I. Production of Aliphatic Aldehydes on Peroxidation of Various Types of Lipids. *Chem. Pharm. Bull.* **1991**, *39*, 1788–1791. [CrossRef]
13. Reinheckel, T.; Noack, H.; Lorenz, S.; Wiswedel, I.; Augustin, W. Comparison of protein oxidation and aldehyde formation during oxidative stress in isolated mitochondria. *Free Radic. Res.* **1998**, *29*, 297–305. [CrossRef] [PubMed]
14. Frankel, E.N.; Hu, M.-L.; Tappel, A.L. Rapid headspace gas chromatography of hexanal as a measure of lipid peroxidation in biological samples. *Lipids* **1989**, *24*, 976–981. [CrossRef]
15. Frankel, E.N.; Tappel, A.L. Headspace gas chromatography of volatile lipid peroxidation products from human red blood cell membranes. *Lipids* **1991**, *26*, 479–484. [CrossRef] [PubMed]
16. Shestivska, V.; Olšinová, M.; Sovová, K.; Kubišta, J.; Smith, D.; Cebecauer, M.; Španěl, P. Evaluation of lipid peroxidation by the analysis of volatile aldehydes in the headspace of synthetic membranes using selected ion flow tube mass spectrometry. *Rapid Commun. Mass Spectrom.* **2018**, *32*, 1617–1628. [CrossRef]
17. Rebeyrolle-Bernard, P.; Etiévant, P. Volatile compounds extracted from polypropylene sheets by hot water: Influence of the temperature of sheets injection. *J. Appl. Polym. Sci.* **1993**, *49*, 1159–1164. [CrossRef]
18. Hüppe, T.; Lorenz, D.; Wachowiak, M.; Maurer, F.; Meiser, A.; Groesdonk, H.; Fink, T.; Sessler, D.I.; Kreuer, S. Volatile organic compounds in ventilated critical care patients: A systematic evaluation of cofactors. *BMC Pulm. Med.* **2017**, *17*, 116. [CrossRef]
19. Hüppe, T.; Klasen, R.; Maurer, F.; Meiser, A.; Groesdonk, H.-V.; Sessler, D.I.; Fink, T.; Kreuer, S. Volatile Organic Compounds in Patients With Acute Kidney Injury and Changes During Dialysis. *Crit. Care Med.* **2019**, *47*, 239–246. [CrossRef]
20. Wirtz, L.M.; Kreuer, S.; Volk, T.; Hüppe, T. Moderne Atemgasanalysen. *Med. Klin.-Intensivmed. Notf.* **2019**, *114*, 655–660. [CrossRef]
21. Perl, T.; Carstens, E.; Hirn, A.; Quintel, M.; Vautz, W.; Nolte, J.; Jünger, M. Determination of serum propofol concentrations by breath analysis using ion mobility spectrometry. *Br. J. Anaesth.* **2009**, *103*, 822–827. [CrossRef] [PubMed]
22. Kreuder, A.-E.; Buchinger, H.; Kreuer, S.; Volk, T.; Maddula, S.; Baumbach, J.I. Characterization of propofol in human breath of patients undergoing anesthesia. *Int. J. Ion Mobil. Spectrom.* **2011**, *14*, 167–175. [CrossRef]
23. Müller-Wirtz, L.M.; Maurer, F.; Brausch, T.; Kiefer, D.; Floss, M.; Doneit, J.; Volk, T.; Sessler, D.I.; Fink, T.; Lehr, T.; et al. Exhaled Propofol Concentrations Correlate With Plasma and Brain Tissue Concentrations in Rats. *Anesth. Analg.* **2021**, *132*, 110–118. [CrossRef] [PubMed]
24. Huang, J.; Kumar, S.; Hanna, G.B. Investigation of C3-C10 aldehydes in the exhaled breath of healthy subjects using selected ion flow tube-mass spectrometry (SIFT-MS). *J. Breath Res.* **2014**, *8*, 037104. [CrossRef]
25. McCartney, M.M.; Thompson, C.J.; Klein, L.R.; Ngo, J.H.; Seibel, J.D.; Fabia, F.; Simms, L.A.; Borras, E.; Young, B.S.; Lara, J.; et al. Breath carbonyl levels in a human population of seven hundred participants. *J. Breath Res.* **2020**, *14*, 046005. [CrossRef]
26. National Center for Biotechnology Information. PubChem Database. C5 to C10 Aliphatic Aldehydes. Available online: https://pubchem.ncbi.nlm.nih.gov (accessed on 28 February 2021).
27. National Center for Biotechnology Information. PubChem Database. Hexanal, CID=6184. Available online: https://pubchem.ncbi.nlm.nih.gov/compound/Hexanal (accessed on 20 January 2020).
28. National Center for Biotechnology Information. PubChem Database. Pentanal, CID=8063. Available online: https://pubchem.ncbi.nlm.nih.gov/compound (accessed on 20 January 2020).
29. Bravo, A.; Hotchkiss, J.H. Identification of volatile compounds resulting from the thermal oxidation of polyethylene. *J. Appl. Polym. Sci.* **1993**, *47*, 1741–1748. [CrossRef]
30. Panseri, S.; Chiesa, L.M.; Zecconi, A.; Soncini, G.; De Noni, I. Determination of Volatile Organic Compounds (VOCs) from wrapping films and wrapped PDO Italian cheeses by using HS-SPME and GC/MS. *Molecules* **2014**, *19*, 8707–8724. [CrossRef] [PubMed]
31. Sun, Z.T.; Yang, C.Y.; Miao, L.J.; Zhang, S.F.; Han, X.P.; Ren, S.E.; Sun, X.Q.; Cao, Y.N. Effects of mechanical ventilation with different tidal volume on oxidative stress and antioxidant in lung. *J. Anesth.* **2015**, *29*, 346–351. [CrossRef]
32. Li, J.; Peng, Y.; Liu, Y.; Li, W.; Jin, Y.; Tang, Z.; Duan, Y. Investigation of potential breath biomarkers for the early diagnosis of breast cancer using gas chromatography–mass spectrometry. *Clin. Chim. Acta* **2014**, *436*, 59–67. [CrossRef]
33. Kumar, S.; Huang, J.; Abbassi-Ghadi, N.; Mackenzie, H.A.; Veselkov, K.A.; Hoare, J.M.; Lovat, L.B.; Španěl, P.; Smith, D.; Hanna, G.B. Mass Spectrometric Analysis of Exhaled Breath for the Identification of Volatile Organic Compound Biomarkers in Esophageal and Gastric Adenocarcinoma. *Ann. Surg.* **2015**, *262*, 981–990. [CrossRef]

34. Zhang, Z.; Zheng, B.; Liu, N.; Ge, H.; Hong, Y. Mechanical power normalized to predicted body weight as a predictor of mortality in patients with acute respiratory distress syndrome. *Intensive Care Med.* **2019**, *45*, 856–864. [CrossRef]
35. Cressoni, M.; Gotti, M.; Chiurazzi, C.; Massari, D.; Algieri, I.; Amini, M.; Cammaroto, A.; Brioni, M.; Montaruli, C.; Nikolla, K.; et al. Mechanical power and development of ventilator-induced lung injury. *Anesthesiology* **2016**, *124*, 1100–1108. [CrossRef] [PubMed]
36. Schubert, J.K.; Spittler, K.-H.; Braun, G.; Geiger, K.; Guttmann, J. CO_2-controlled sampling of alveolar gas in mechanically ventilated patients. *J. Appl. Physiol.* **2001**, *90*, 486–492. [CrossRef] [PubMed]
37. Müller-Wirtz, L.M.; Godsch, C.; Sessler, D.I.; Volk, T.; Kreuer, S.; Hüppe, T. Residual volatile anesthetics after workstation preparation and activated charcoal filtration. *Acta Anaesthesiol. Scand.* **2020**, *64*, 759–765. [CrossRef]
38. Chiumello, D.; Gotti, M.; Guanziroli, M.; Formenti, P.; Umbrello, M.; Pasticci, I.; Mistraletti, G.; Busana, M. Bedside calculation of mechanical power during volume- and pressure-controlled mechanical ventilation. *Crit. Care* **2020**, *24*, 417. [CrossRef]
39. Dixon, W.J. Analysis of Extreme Values. *Ann. Math. Stat.* **1950**, *21*, 488–506. [CrossRef]
40. Zheng, B. Summarizing the goodness of fit of generalized linear models for longitudinal data. *Stat. Med.* **2000**, *19*, 1265–1275. [CrossRef]

molecules

Article

Differential Response of Pentanal and Hexanal Exhalation to Supplemental Oxygen and Mechanical Ventilation in Rats

Lukas M. Müller-Wirtz [1,2,*], Daniel Kiefer [1], Joschua Knauf [1], Maximilian A. Floss [1], Jonas Doneit [1], Beate Wolf [1], Felix Maurer [1,2], Daniel I. Sessler [2,3], Thomas Volk [1,2], Sascha Kreuer [1,2] and Tobias Fink [1,2]

[1] CBR—Center of Breath Research, Department of Anaesthesiology, Intensive Care and Pain Therapy, Saarland University Medical Center, Homburg, 66421 Saarland, Germany; daniel.kiefer@uks.eu (D.K.); joschua.knauf@gmx.de (J.K.); max.floss@outlook.de (M.A.F.); jonas.doneit@gmx.de (J.D.); beate.wolf177@yahoo.de (B.W.); felix.maurer@uks.eu (F.M.); thomas.volk@uks.eu (T.V.); sascha.kreuer@uks.eu (S.K.); tobias.fink@uks.eu (T.F.)
[2] Outcomes Research Consortium, Cleveland, OH 44195, USA; ds@or.org
[3] Department of Outcomes Research, Anesthesiology Institute, Cleveland Clinic, Cleveland, OH 44195, USA
*   Correspondence: lukas.wirtz@uks.eu

Abstract: High inspired oxygen during mechanical ventilation may influence the exhalation of the previously proposed breath biomarkers pentanal and hexanal, and additionally induce systemic inflammation. We therefore investigated the effect of various concentrations of inspired oxygen on pentanal and hexanal exhalation and serum interleukin concentrations in 30 Sprague Dawley rats mechanically ventilated with 30, 60, or 93% inspired oxygen for 12 h. Pentanal exhalation did not differ as a function of inspired oxygen but increased by an average of 0.4 (95%CI: 0.3; 0.5) ppb per hour, with concentrations doubling from 3.8 (IQR: 2.8; 5.1) ppb at baseline to 7.3 (IQR: 5.0; 10.8) ppb after 12 h. Hexanal exhalation was slightly higher at 93% of inspired oxygen with an average difference of 0.09 (95%CI: 0.002; 0.172) ppb compared to 30%. Serum IL-6 did not differ by inspired oxygen, whereas IL-10 at 60% and 93% of inspired oxygen was greater than with 30%. Both interleukins increased over 12 h of mechanical ventilation at all oxygen concentrations. Mechanical ventilation at high inspired oxygen promotes pulmonary lipid peroxidation and systemic inflammation. However, the response of pentanal and hexanal exhalation varies, with pentanal increasing by mechanical ventilation, whereas hexanal increases by high inspired oxygen concentrations.

Keywords: mechanical ventilation; anesthesia; supplemental oxygen; oxygen toxicity; lipid peroxidation; volatile aldehydes; pentanal; hexanal; volatile organic compounds

Citation: Müller-Wirtz, L.M.; Kiefer, D.; Knauf, J.; Floss, M.A.; Doneit, J.; Wolf, B.; Maurer, F.; Sessler, D.I.; Volk, T.; Kreuer, S.; et al. Differential Response of Pentanal and Hexanal Exhalation to Supplemental Oxygen and Mechanical Ventilation in Rats. *Molecules* **2021**, *26*, 2752. https://doi.org/10.3390/molecules26092752

Academic Editors: Natalia Drabińska and Ben de Lacy Costello

Received: 19 April 2021
Accepted: 5 May 2021
Published: 7 May 2021

Publisher's Note: MDPI stays neutral with regard to jurisdictional claims in published maps and institutional affiliations.

1. Introduction

High inspired oxygen concentrations may cause toxicities, including oxidative stress, hyperoxic vasoconstriction, and resorption atelectasis [1,2]. Furthermore, reactive oxygen species promoted by high oxygen concentrations attack cell components, including lipids, proteins, and DNA—all of which provoke local and systemic inflammation [3]. While prolonged hyperoxia undoubtedly causes lung damage, the extent of hyperoxia-induced injury during short-term mechanical ventilation, such as might occur during surgery, is controversial [2].

The lowest inspired oxygen concentration used for intraoperative mechanical ventilation is about 30%, usually resulting in only slight hyperoxemia because mechanical ventilation causes a degree of shunt and dead-space ventilation. Nevertheless, higher inspired oxygen concentrations are frequently used, either out of necessity to maintain a suitable arterial oxygen saturation, or simply to provide pulmonary oxygen reserve in the case of an airway problem. To assess the clinical tradeoff between additional safety and potential hyperoxia-induced lung injury, the effects of different inspired oxygen concentrations on pulmonary oxidative stress and systemic inflammation are thus of considerable interest.

Lipid peroxidation is a major mechanism by which oxygen causes toxicity [4]. The process releases volatile products, including the two aldehydes, pentanal and hexanal [5], both of which have been proposed as possible breath biomarkers for pulmonary pathologies, such as lung cancer [6,7] and ventilator-induced lung injury [8]. Hyperoxia produces reactive oxygen species in lung tissue with a consequent increase in lipid peroxidation [9,10]. Other volatile lipid peroxidation products, especially ethane and pentane, consistently increase in the breath of hyperoxic animals and humans [11–14]. However, the influence of inspired oxygen on pentanal and hexanal exhalation in mechanically ventilated subjects remains to be determined for a valid interpretation of these newly proposed biomarkers.

We, therefore, evaluated pentanal and hexanal exhalation in rats mechanically ventilated for 12 h with various inspired oxygen concentrations. We simultaneously determined interleukin serum concentrations as a measure of systemic inflammation. Specifically, we tested the primary hypothesis that high- and medium-inspired oxygen concentrations provoke more pentanal and hexanal exhalation than lower concentrations in rats. Secondarily, we tested the hypothesis that high inspired oxygen concentrations increase serum cytokine concentrations.

2. Results

2.1. Experimental Conditions

All animals survived the study period and were included. The median weight of the rats was 343 (IQR: 336; 351) g, and all survived the 12 h observation period. Heart rate and mean arterial pressure decreased over the observation period, but similarly at each inspired oxygen concentration (Supplementary File 1; Figure S1). Blood gas values, hemoglobin, electrolytes, glucose, and lactate remained within physiological ranges (Supplementary File 1, Table S1). Median minute ventilation was 180 (IQR: 174; 184) ml/min, median peak pressure was 10.9 (IQR: 10.6; 11.1) cmH_2O, and median tidal volume was 2.8 (IQR: 2.7; 2.8) ml over all groups. Median arterial partial pressures of oxygen differed markedly among the groups, as expected (Figure 1).

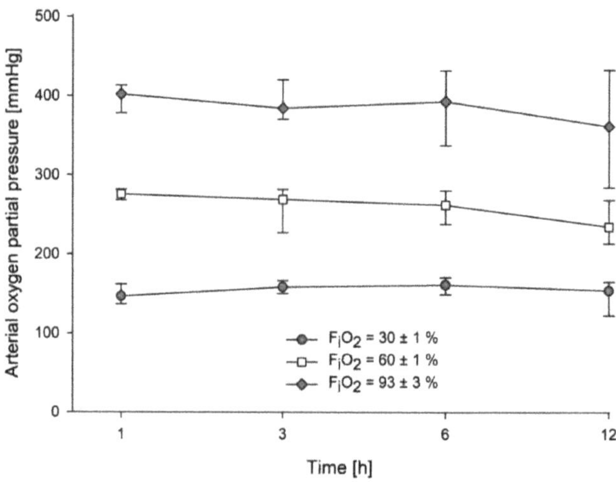

Figure 1. Arterial oxygen partial pressure. Data presented as medians and interquartile ranges. F_iO_2 = fraction of inspired oxygen.

2.2. Breath Analysis

Exhaled pentanal did not differ as a function of inspired oxygen but increased over all groups by an average of 0.4 (95%CI: 0.3; 0.5) ppb per hour of mechanical ventilation. The median exhaled pentanal concentration therefore almost doubled from 3.8 (IQR: 2.8;

5.1) ppb at baseline to 7.3 (IQR: 5.0; 10.8) ppb after 12 h of mechanical ventilation (Table 1, Figure 2).

Table 1. Influence of inspired oxygen and mechanical ventilation on pentanal and hexanal exhalation.

Pentanal			
Parameter	**Regression Coefficient**	**95% Confidence Interval**	*p*
F_iO_2 = 93%	0.03	−1.4–1.4	0.967
F_iO_2 = 60%	0.67	−1.1–2.4	0.454
F_iO_2 = 30%	0	-	-
Ventilation time [h]	0.4	0.3–0.5	<0.001
Hexanal			
Parameter	**Regression Coefficient**	**95% Confidence Interval**	*p*
F_iO_2 = 93%	0.09	0.002–0.172	0.046
F_iO_2 = 60%	0.03	−0.06–0.116	0.506
F_iO_2 = 30%	0	-	-
Ventilation time [h]	−0.01	−0.016–(−0.007)	<0.001

Linear generalized estimating equations (GEE)—regression was performed. The regression coefficient of ventilation time refers to one hour of mechanical ventilation. F_iO_2 = fraction of inspired oxygen.

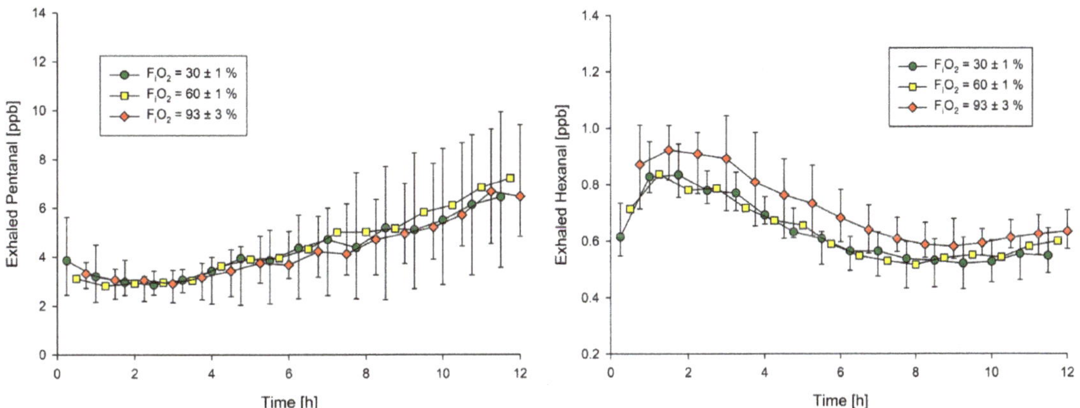

Figure 2. Pentanal and hexanal exhalation over 12 h of mechanical ventilation. Data presented as medians and interquartile ranges. F_iO_2 = fraction of inspired oxygen. Exhaled concentrations were measured at 15-minute intervals, with every third value displayed and error bars omitted for 60% to enhance clarity.

Exhaled hexanal was slightly higher in rats exposed to 93% inspired oxygen, with average concentrations being 0.09 (95%CI: 0.002; 0.172) ppb higher than in rats ventilated with 30% inspired oxygen. Exhaled hexanal initially increased in all groups, reached a maximum after around 2 h, and stabilized at a lower plateau after approximately 6 to 12 h; concentrations did not increase over time (Table 1, Figure 2).

2.3. Systemic Inflammation

IL-6 concentrations did not significantly differ as a function of inspired oxygen fraction (*p* = 0.888), whereas IL-10 concentrations averaged 2.5 (95%CI: 0.2; 4.8) pg/mL higher at 60% and 7.2 (95%CI: 2.8; 11.5) pg/mL higher at 93% than with 30% inspired oxygen (*p* = 0.035, *p* = 0.001; Figure 3). Interleukin serum concentrations across all groups increased significantly between 1 and 12 h of mechanical ventilation (IL-6: *p* = 0.002, IL-10: *p* = 0.035, Figure 3).

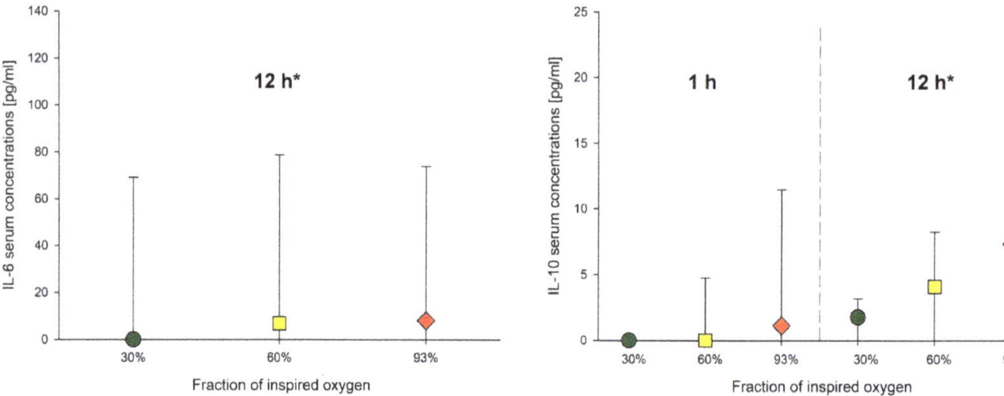

Figure 3. Cytokine serum concentrations. Data presented as medians and interquartile ranges. IL-6 at 1 h was 0 for all animals. IL-6 concentrations did not significantly differ by inspired oxygen ($p = 0.888$), whereas IL-10 concentrations were significantly greater at 60% and 93% compared to 30% inspired oxygen (30% vs. 60%: $p = 0.035$; 30 vs. 93%: $p = 0.001$). Interleukin serum concentrations across all groups increased significantly between 1 and 12 h of mechanical ventilation (* IL-6: $p = 0.002$, * IL-10: $p = 0.035$).

3. Discussion

We expected the exhalation of both pentanal and hexanal to increase at high inspired oxygen concentrations. Instead, responses differed, with hexanal exhalation increasing by high inspired oxygen concentrations, whereas pentanal exhalation gradually increased over 12 h of mechanical ventilation, but unrelated to inspired oxygen.

Hyperoxia produces reactive oxygen species in lung tissue with a consequent increase in lipid peroxidation [9,10]. Hexanal exhalation was consistently greater at high inspired oxygen concentrations. Likewise, previous studies reported increased exhalation of ethane and pentane in hyperoxic individuals [11–14]. Thus, available data suggest that high-inspired oxygen concentrations induce pulmonary lipid peroxidation. Breath analysis may, therefore, facilitate the early detection of hyperoxic lung injury, and may be especially helpful for investigating specific treatments that potentially reduce hyperoxic lung injury.

Mechanical ventilation increased pentanal exhalation over time, with concentrations almost doubling over just 12 h. As lung distension promotes injury [15,16], we used tidal volumes of 8 mL/kg, which falls within the broadly accepted range of 6–9 mL/kg commonly used in humans [17] and rodents [18,19]. However, even with moderate tidal volumes, mechanical ventilation damages the cell membrane and activates cellular repair mechanisms [20], which include transferring lipids to the cell membrane to enlarge the cell surface and help maintain its integrity [21]. Sufficiently high tidal volumes can overcome cellular repair mechanisms, leading to membrane defects, followed by apoptosis or necrosis [20]. Cell death seems unlikely with the ventilator settings we used, but sublethal cellular membrane damage may expose polyunsaturated fatty acids to oxidative processes, and thus increase pentanal exhalation. This theory is consistent with in vitro studies, showing that pentanal is the predominant volatile aldehyde generated by oxidizing lipid membranes exposed to mechanical stress through sonication [22] and by oxidizing isolated polyunsaturated fatty acids [23].

The observed gradual increase in pentanal exhalation over 12 h of mechanical ventilation is consistent with our previous finding that exhaled pentanal is highly sensitive to volutrauma [8]. Based on the biological background presented above, we previously postulated that stretch-induced cell membrane damage exposes membrane lipids to oxidative processes, thereby increasing the exhalation of pentanal. In contrast, hexanal concentrations did not increase over 12 h, although hexanal and pentanal are likely generated by similar mechanisms [5]. The relatively low vapor pressure of hexanal may have contributed [24,25], but it is non-obvious why hexanal did not increase. Future studies may clarify reasons for the differential response of pentanal and hexanal to mechanical ventilation.

Inspired oxygen concentrations greater than 30% did not increase exhaled pentanal, probably because even that amount of oxygen is sufficient to oxidize all available membrane lipids from which pentanal arises. Our lowest inspired oxygen concentration was 30% since lower concentrations are rarely used for mechanical ventilation. Our results, therefore, well characterize the effects of supplemental oxygen during mechanical ventilation. The current study consistently showed that pentanal increases due to mechanical ventilation and adds the important finding that exhaled pentanal is a potential breath biomarker that can be interpreted independently from inspired oxygen.

Pentanal and hexanal are potential biomarkers of lung, breast, and gastrointestinal cancers in humans [6,7,26,27]. Cancer causes cell death, for example, by tumor growth, exceeding supply with nutrients, attacks by immune cells, or by the inflammation and destruction of surrounding tissues. Cell death is accompanied by cell membrane breakdown, exposing lipids to oxidation, and may thus prompt the exhalation of aldehydes. As might, therefore, be expected, lung cancer increases pentanal and hexanal exhalation [6,7]. Furthermore, acute respiratory distress syndrome increases both pentanal and hexanal concentrations in blood [28]. Taken together, pentanal and hexanal seem to be general markers of cell membrane damage, and our findings suggest that exhaled pentanal also increases when lung tissue is mechanically stressed.

Interleukin serum concentrations increased over 12 h of mechanical ventilation, with only IL-10 being greater at higher inspired oxygen concentrations. Helmerhorst et al. reported that IL-6 serum concentrations remain similar at various inspired oxygen concentrations over 12 h of mechanical ventilation in mice, but IL-10 concentrations in bronchoalveolar fluid substantially increase [16]. However, others reported no influence of oxygen on the mRNA expression of IL-10 in mouse lungs [29,30]. Consistent with inflammation accruing over time, lung tissue IL-6 mRNA increases after 48 h of hyperoxia [31]. High inspired oxygen concentrations induce cytokine gene expression in alveolar macrophages of surgical patients, reflecting local pulmonary inflammation [32]. However, IL-6 was less affected by hyperoxia than other cytokines, and IL-10 was not measured [32]. Taken together, the available data suggest that high-inspired oxygen induces a slight systemic inflammatory response during up to 12 h of mechanical ventilation.

The most obvious limitation of our study is that lipid peroxidation, cell membrane components, and antioxidative capacities presumably differ among species. However, lipid peroxidation is fundamental to oxidative stress. It is, therefore, likely that results in humans are qualitatively similar, although presumably quantitatively different. We did not conduct a formal sample-size estimate because the treatment effect was non-obvious. We also note that anesthetic drugs could have influenced our results, as propofol has antioxidative properties and may inhibit lipid peroxidation processes [33–35]. We did not include a control group ventilated with 21% oxygen because, in previous studies, this concentration resulted in hypoxemia and even death. Finally, our results are specific to mechanical ventilation; results likely differ with spontaneous ventilation.

In summary, mechanical ventilation and high inspired oxygen promote pulmonary lipid peroxidation and systemic inflammation. However, the response of pentanal and hexanal exhalation varies, with pentanal increasing in response to mechanical ventilation and hexanal increasing in response to high concentrations of inspired oxygen. Our results suggest that exhaled pentanal, a potential biomarker for lung injury, can be interpreted independent of the inspired oxygen concentration during mechanical ventilation.

4. Materials and Methods

4.1. Animals

Experiments were conducted in accordance with the German Animal Welfare Act and with approval from the responsible Institutional Animal Care and Use Committee

(Landesamt für Soziales, Saarland, Saarbrücken, Germany, No. 28/2018, date of approval: 01.08.2018).

Thirty male Sprague Dawley rats (280–380 g body weight, age 8–10 weeks) were obtained from Charles River Laboratory International (Sulzfeld, Germany) and kept in our institutional animal facility under controlled conditions (temperature $20 \pm 2\,°C$ and $50 \pm 5\%$ relative humidity). The rats had free access to water and standard pellet food. Monitoring and preparation were performed, as previously described [36]. No specific inclusion or exclusion criteria were applied. Animals were included as long as the experimental protocol was adequately followed and in the absence of poor welfare signs (e.g., wounds, secretion of harderian gland, signs of dehydration, diarrhea, isolation from others, itching).

4.2. Anesthesia

Anesthesia was induced with sevoflurane (Baxter, Unterschleißheim, Germany) and maintained with intravenous propofol (Fresenius Kabi, Bad Homburg, Germany) starting at 25 mg/kg/h with hourly reductions of 0.5 mg/h until a minimal rate of 15 mg/kg/h was reached. Ketamine (Rotexmedica, Trittau, Germany) was added with 25 mg/kg/h throughout the experiment for analgesia. Neuromuscular blockade was induced by a bolus of 10 mg/kg rocuronium (Grünenthal, Stolberg, Germany) and maintained by a continuous infusion of 25 mg/kg/h rocuronium. Animals were observed for 12 h and then killed by exsanguination.

4.3. Ventilation

Oxygen was produced by a concentrator (Compact 525, Devilbiss, NY, USA) with a maximum output of $93 \pm 3\%$, mixed with generated nitrogen (Genius, Peak Scientific, Inchinnan, Scotland, UK) and purified by activated charcoal filtration. Inspired oxygen was constantly monitored and adapted to maintain concentrations at 30% or 60% within a range of $\pm 1\%$ (sensor: GGA 370, device: GMH 3695, Greisinger, Regenstauf, Germany). For the highest oxygen group, the maximum output of the oxygen concentrator was used, resulting in a concentration of $93 \pm 3\%$. Ten rats each were randomly assigned to three different fractions of inspired oxygen: 30%, 60%, and 93%.

Animals were randomized 1:1:1 based on a computer-generated list. Investigators were not blinded during the experiments, but allocation was concealed during post-experimental analysis. Animals were ventilated with a tidal volume of 8 mL/kg, a respiratory rate of 63 breaths/min, and a PEEP of 2 cmH_2O (VentStar small animal ventilator, RWD Life Sciences, Shenzhen, China). The respiratory rate was reduced by 10% when the partial pressure of carbon dioxide was less than 28 mmHg. Similarly, the respiratory rate was increased by 10% when partial pressure exceeded 45 mmHg.

4.4. Breath and Blood Samples

Blood for gas analyses was sampled after 1, 3, 6, and 12 h to monitor ventilation (Radiometer ABL 800 Basic, Willich, Germany). Ten milliliters of exhaled air were sampled and analyzed with two multi capillary columns—ion mobility spectrometers (MCC-IMS by B&S Analytik, Dortmund, Germany) in 15-minute intervals, as previously described [36]. The MCC-IMS was calibrated by pentanal and hexanal standards ranging from 0.1 to 50 ppb (analytical standard, Merck, Darmstadt, Germany), as previously described [23]. Arterial blood samples of 600 µL were collected after 1 and 12 h. Plasma was stored at $-75\,°C$. Interleukin 6 and 10 serum concentrations were measured by enzyme-linked immunosorbent assay (ELISA). Positive controls of each cytokine were measured routinely with each assay (ELISA Antibodies BD OptEIA; BD Biosciences Pharmingen, San Diego, CA, USA).

4.5. Statistics

Statistical analyses were carried out with R 4.0.2 (R Core Team, 2020) using the packages *geepack* (Højsgaard, Halekoh, and Yan, 2006) and *broom* (*v0.7.5*; Robinson, Hayes and

Couch, 2021). Figures were created with SigmaPlot 12.5 (Systat Software GmbH, Erkrath, Germany). Normality was assessed by visual inspection of histograms and quantile-quantile plots. Most data were not normally distributed. Therefore, all results are presented as medians and interquartile ranges. Influences of inspired oxygen and mechanical ventilation time on aldehyde exhalation were assessed by linear generalized estimating equations regression to account for within-subject correlations. The influence of inspired oxygen on interleukin concentrations was assessed by linear generalized estimating equations regression combined with a Wald statistic. Interleukin concentrations after 1 and 12 h were compared over all groups by a Wilcoxon signed-rank test. A two-sided $p < 0.05$ was considered statistically significant. There was no a priori sample size estimate since the expected effect sizes and the clinical significance of increases in aldehyde exhalation through supplemental inspired oxygen are essentially unknown.

Supplementary Materials: The following are available online, Supplementary File 1 (including Figure S1: Vital parameters and Table S1: Results of blood gas analysis); Supplementary File 2: Pentanal and hexanal exhalation; Supplementary File 3: Interleukin serum concentrations.

Author Contributions: Conceptualization, L.M.M.-W., S.K., and T.F.; methodology, L.M.M.-W., D.K., F.M., B.W., and T.F.; software, L.M.M.-W., and J.K.; validation, L.M.M.-W., J.K., and F.M.; formal analysis, L.M.M.-W., and J.K.; investigation, L.M.M.-W., J.K., M.A.F., J.D., F.M., and B.W.; resources, S.K., and T.V.; data curation, L.M.M.-W., J.K., M.A.F., and J.D.; writing—original draft preparation, L.M.M.-W.; writing—review and editing, D.I.S., S.K., and T.F.; visualization, L.M.M.-W., D.I.S. and T.F; supervision, D.I.S., S.K., T.V. and T.F.; project administration, L.M.M.-W., S.K., and T.F.; funding acquisition, L.M.M.-W., S.K., and T.V. All authors have read and agreed to the published version of the manuscript.

Funding: Support was mainly provided from institutional and departmental sources. This work was awarded and financially supported by the Professor Hans Köhler Price 2018 (Society of the Friends of the Saarland University Medical Center). The APC was funded by the Deutsche Forschungsgemeinschaft (DFG, German Research Foundation) and Saarland University.

Institutional Review Board Statement: Experiments were conducted with approval from the responsible Institutional Animal Care and Use Committee (Landesamt für Soziales, Saarland, Saarbrücken, Germany, No. 28/2018, date of approval: 01.08.2018) and in accordance with the German Animal Welfare Act.

Informed Consent Statement: Not applicable.

Data Availability Statement: Data are contained within the article and Supplementary Material. The data on pentanal and hexanal exhalation are available in Supplementary File 2 and on interleukin serum concentrations in Supplementary File 3.

Acknowledgments: We acknowledge Hans Köhler and the Society of the Friends of the Saarland University Medical Center for supporting this work. We acknowledge support by the Deutsche Forschungsgemeinschaft (DFG, German Research Foundation) and Saarland University within the funding programme Open Access Publishing. This study contains data taken from the thesis presented by Joschua Knauf as part of the requirements for the obtention of the degree "Doctor of Medicine" at Saarland University Medical Center and Saarland University Faculty of Medicine.

Conflicts of Interest: The authors declare no conflict of interest.

Sample Availability: Samples of the compounds are not available from the authors.

References

1. Winter, P.M.; Smith, G. The toxicity of oxygen. *Anesthesiology* **1972**, *37*, 210–241. [CrossRef]
2. Weenink, R.P.; de Jonge, S.W.; van Hulst, R.A.; Wingelaar, T.T.; van Ooij, P.-J.A.M.; Immink, R.V.; Preckel, B.; Hollmann, M.W. Perioperative Hyperoxyphobia: Justified or Not? Benefits and Harms of Hyperoxia during Surgery. *J. Clin. Med.* **2020**, *9*, 642. [CrossRef]
3. Kallet, R.H.; Matthay, M.A. Hyperoxic acute lung injury. *Respir. Care* **2013**, *58*, 123–141. [CrossRef]
4. Zielinski, Z.A.M.; Pratt, D.A. Lipid Peroxidation: Kinetics, Mechanisms, and Products. *J. Org. Chem.* **2017**, *82*, 2817–2825. [CrossRef] [PubMed]

5. Yoshino, K.; Sano, M.; Fujita, M.; Tomita, I. Production of Aliphatic Aldehydes on Peroxidation of Various Types of Lipids. *Chem. Pharm. Bull.* **1991**, *39*, 1788–1791. [CrossRef] [PubMed]

6. Fuchs, P.; Loeseken, C.; Schubert, J.K.; Miekisch, W. Breath gas aldehydes as biomarkers of lung cancer. *Int. J. Cancer* **2010**, *126*, 2663–2670. [CrossRef]

7. Ulanowska, A.; Kowalkowski, T.; Trawińska, E.; Buszewski, B. The application of statistical methods using VOCs to identify patients with lung cancer. *J. Breath Res.* **2011**, *5*, 046008. [CrossRef]

8. Müller-Wirtz, L.M.; Kiefer, D.; Maurer, F.; Floss, M.A.; Doneit, J.; Hüppe, T.; Shopova, T.; Wolf, B.; Sessler, D.I.; Volk, T.; et al. Volutrauma Increases Exhaled Pentanal in Rats: A Potential Breath Biomarker for Ventilator-Induced Lung Injury. *Anesth. Analg.* **2021**. [CrossRef]

9. Freeman, B.A.; Crapo, J.D. Hyperoxia increases oxygen radical production in rat lungs and lung mitochondria. *J. Biol. Chem.* **1981**, *256*, 10986–10992. [CrossRef]

10. Freeman, B.A.; Topolosky, M.K.; Crapo, J.D. Increases Oxygen Radical Rat Lung Homogenates. *Arch. Biochem. Biophys.* **1982**, *216*, 477–484. [CrossRef]

11. Loiseaux-Meunier, M.N.; Bedu, M.; Gentou, C.; Pepin, D.; Coudert, J.; Caillaud, D. Oxygen toxicity: Simultaneous measure of pentane and malondialdehyde in humans exposed to hyperoxia. *Biomed. Pharmacother.* **2001**, *55*, 163–169. [CrossRef]

12. Morita, S.; Snider, M.T.; Inada, Y. Increased N-pentane Excretion in Humans: A Consequence of Pulmonary Oxygen Exposure. *Anesthesiology* **1986**, *64*, 730–733. [CrossRef] [PubMed]

13. Habib, M.P.; Katz, M.A. Source of ethane in expirate of rats ventilated with 100% oxygen. *J. Appl. Physiol.* **1989**, *66*, 1268–1272. [CrossRef] [PubMed]

14. Habib, M.P.; Eskelson, C.; Katz, M.A. Ethane Production Rate in Rats Exposed to High Oxygen Concentration. *Am. Rev. Respir. Dis.* **1988**, *137*, 341–344. [CrossRef] [PubMed]

15. Serpa Neto, A.; Cardoso, S.O.; Manetta, J.A.; Pereira, V.G.M.; Espósito, D.C.; Pasqualucci, M.D.O.P.; Damasceno, M.C.T.; Schultz, M.J. Association between use of lung-protective ventilation with lower tidal volumes and clinical outcomes among patients without acute respiratory distress syndrome: A meta-analysis. *JAMA* **2012**, *308*, 1651–1659. [CrossRef] [PubMed]

16. Helmerhorst, H.J.F.; Schouten, L.R.A.; Wagenaar, G.T.M.; Juffermans, N.P.; Roelofs, J.J.T.H.; Schultz, M.J.; de Jonge, E.; van Westerloo, D.J. Hyperoxia provokes a time- and dose-dependent inflammatory response in mechanically ventilated mice, irrespective of tidal volumes. *Intensive Care Med. Exp.* **2017**, *5*, 27. [CrossRef]

17. Güldner, A.; Kiss, T.; Serpa Neto, A.; Hemmes, S.N.T.; Canet, J.; Spieth, P.M.; Rocco, P.R.M.; Schultz, M.J.; Pelosi, P.; Gama de Abreu, M. Intraoperative Protective Mechanical Ventilation for Prevention of Postoperative Pulmonary Complications. *Anesthesiology* **2015**, *123*, 692–713. [CrossRef] [PubMed]

18. Sun, Z.T.; Yang, C.Y.; Miao, L.J.; Zhang, S.F.; Han, X.P.; Ren, S.E.; Sun, X.Q.; Cao, Y.N. Effects of mechanical ventilation with different tidal volume on oxidative stress and antioxidant in lung. *J. Anesth.* **2015**, *29*, 346–351. [CrossRef]

19. Setzer, F.; Oschatz, K.; Hueter, L.; Schmidt, B.; Schwarzkopf, K.; Schreiber, T. Susceptibility to ventilator induced lung injury is increased in senescent rats. *Crit. Care* **2013**, *17*, R99. [CrossRef]

20. Vlahakis, N.E.; Hubmayr, R.D. Cellular stress failure in ventilator-injured lungs. *Am. J. Respir. Crit. Care Med.* **2005**, *171*, 1328–1342. [CrossRef]

21. Vlahakis, N.E.; Schroeder, M.A.; Pagano, R.E.; Hubmayr, R.D. Deformation-induced lipid trafficking in alveolar epithelial cells. *Am. J. Physiol. Cell Mol. Physiol.* **2001**, *280*, L938–L946. [CrossRef]

22. Shestivska, V.; Olšinová, M.; Sovová, K.; Kubišta, J.; Smith, D.; Cebecauer, M.; Španěl, P. Evaluation of lipid peroxidation by the analysis of volatile aldehydes in the headspace of synthetic membranes using selected ion flow tube mass spectrometry. *Rapid Commun. Mass Spectrom.* **2018**, *32*, 1617–1628. [CrossRef] [PubMed]

23. Müller-Wirtz, L.M.; Kiefer, D.; Ruffing, S.; Brausch, T.; Hüppe, T.; Sessler, D.I.; Volk, T.V.; Fink, T.; Kreuer, S.; Maurer, F. Quantification of volatile aldehydes from in vitro lipid peroxidation and in breath of ventilated patients. *Molecules* **2021**. (under review).

24. National Center for Biotechnology Information. PubChem Database. Pentanal, CID=8063. Available online: https://pubchem.ncbi.nlm.nih.gov/compound (accessed on 20 January 2020).

25. National Center for Biotechnology Information. PubChem Database. Hexanal, CID=6184. Available online: https://pubchem.ncbi.nlm.nih.gov/compound/Hexanal (accessed on 20 January 2020).

26. Kumar, S.; Huang, J.; Abbassi-Ghadi, N.; Mackenzie, H.A.; Veselkov, K.A.; Hoare, J.M.; Lovat, L.B.; Španěl, P.; Smith, D.; Hanna, G.B. Mass Spectrometric Analysis of Exhaled Breath for the Identification of Volatile Organic Compound Biomarkers in Esophageal and Gastric Adenocarcinoma. *Ann. Surg.* **2015**, *262*, 981–990. [CrossRef] [PubMed]

27. Phillips, M.; Cataneo, R.N.; Ditkoff, B.A.; Fisher, P.; Greenberg, J.; Gunawardena, R.; Kwon, C.S.; Tietje, O.; Wong, C. Prediction of breast cancer using volatile biomarkers in the breath. *Breast Cancer Res. Treat.* **2006**, *99*, 19–21. [CrossRef]

28. Weigand, M.A.; Snyder-Ramos, S.A.; Möllers, A.G.; Bauer, J.; Hansen, D.; Kochen, W.; Martin, E.; Motsch, J. Inhaled nitric oxide does not enhance lipid peroxidation in patients with acute respiratory distress syndrome. *Crit. Care Med.* **2000**, *28*, 3429–3435. [CrossRef]

29. Johnston, C.J.; Wright, T.W.; Reed, C.K.; Finkelstein, J.N. Comparison of adult and newborn pulmonary cytokine mRNA expression after hyperoxia. *Exp. Lung Res.* **1997**, *23*, 537–552. [CrossRef] [PubMed]

30. Shea, L.M.; Beehler, C.; Schwartz, M.; Shenkar, R.; Tuder, R.; Abraham, E. Hyperoxia activates NF-kappaB and increases TNF-alpha and IFN-gamma gene expression in mouse pulmonary lymphocytes. *J. Immunol.* **1996**, *157*, 3902–3908.
31. Bhandari, V.; Elias, J.A. Cytokines in tolerance to hyperoxia-induced injury in the developing and adult lung. *Free Radic. Biol. Med.* **2006**, *41*, 4–18. [CrossRef] [PubMed]
32. Kotani, N.; Hashimoto, H.; Sessler, D.I.; Muraoka, M.; Hashiba, E.; Kubota, T.; Matsuki, A. Supplemental intraoperative oxygen augments antimicrobial and proinflammatory responses of alveolar macrophages. *Anesthesiology* **2000**, *93*, 15–25. [CrossRef]
33. Eriksson, O.; Pollesello, P.; Saris, N.-E.L. Inhibition of lipid peroxidation in isolated rat liver mitochondria by the general anaesthetic propofol. *Biochem. Pharmacol.* **1992**, *44*, 391–393. [CrossRef]
34. Kahraman, S.; Kilinç, K.; Dal, D.; Erdem, K. Propofol attenuates formation of lipid peroxides in tourniquet-induced ischaemia-reperfusion injury. *Br. J. Anaesth.* **1997**, *78*, 279–281. [CrossRef] [PubMed]
35. Murphy, P.G.; Myers, D.S.; Davies, M.J.; Webster, N.R.; Jones, J.G. The antioxidant potential of propofol (2,6-diisopropylphenol). *Br. J. Anaesth.* **1992**, *68*, 613–618. [CrossRef] [PubMed]
36. Müller-Wirtz, L.M.; Maurer, F.; Brausch, T.; Kiefer, D.; Floss, M.; Doneit, J.; Volk, T.; Sessler, D.I.; Fink, T.; Lehr, T.; et al. Exhaled Propofol Concentrations Correlate with Plasma and Brain Tissue Concentrations in Rats. *Anesth. Analg.* **2021**, *132*, 110–118. [CrossRef] [PubMed]

 *molecules*

Communication

Breathing Rhythm Variations during Wash-In Do Not Influence Exhaled Volatile Organic Compound Profile Analyzed by an Electronic Nose

Silvano Dragonieri [1,*], Vitaliano Nicola Quaranta [2], Pierluigi Carratù [3], Teresa Ranieri [1], Enrico Buonamico [1] and Giovanna Elisiana Carpagnano [1]

[1] Respiratory Diseases, University of Bari, 70121 Bari, Italy; teresa.ranieri@uniba.it (T.R.); enricobuonamico@gmail.com (E.B.); elisiana.carpagnano@uniba.it (G.E.C.)
[2] Pulmonology, Di Venere Hospital, 70131 Bari, Italy; vitalianonicola.40@gmail.com
[3] Internal Medicine "A. Murri", University of Bari, 70121 Bari, Italy; pierluigi.carratu@uniba.it
* Correspondence: silvano.dragonieri@uniba.it

Abstract: E-noses are innovative tools used for exhaled volatile organic compound (VOC) analysis, which have shown their potential in several diseases. Before obtaining a full validation of these instruments in clinical settings, a number of methodological issues still have to be established. We aimed to assess whether variations in breathing rhythm during wash-in with VOC-filtered air before exhaled air collection reflect changes in the exhaled VOC profile when analyzed by an e-nose (Cyranose 320). We enrolled 20 normal subjects and randomly collected their exhaled breath at three different breathing rhythms during wash-in: (a) normal rhythm (respiratory rate (RR) between 12 and 18/min), (b) fast rhythm (RR > 25/min) and (c) slow rhythm (RR < 10/min). Exhaled breath was collected by a previously validated method (Dragonieri et al., J. Bras. Pneumol. 2016) and analyzed by the e-nose. Using principal component analysis (PCA), no significant variations in the exhaled VOC profile were shown among the three breathing rhythms. Subsequent linear discriminant analysis (LDA) confirmed the above findings, with a cross-validated accuracy of 45% (p = ns). We concluded that the exhaled VOC profile, analyzed by an e-nose, is not influenced by variations in breathing rhythm during wash-in.

Keywords: volatile organic compounds; e-nose; electronic nose; breath analysis; breathing rhythm

Citation: Dragonieri, S.; Quaranta, V.N.; Carratù, P.; Ranieri, T.; Buonamico, E.; Carpagnano, G.E. Breathing Rhythm Variations during Wash-In Do Not Influence Exhaled Volatile Organic Compound Profile Analyzed by an Electronic Nose. *Molecules* **2021**, *26*, 2695. https://doi.org/10.3390/molecules26092695

Academic Editors: Giuseppina Paola Parpinello and Natalia Drabińska

Received: 26 March 2021
Accepted: 3 May 2021
Published: 4 May 2021

Publisher's Note: MDPI stays neutral with regard to jurisdictional claims in published maps and institutional affiliations.

1. Introduction

The recent evolutions in sensor manufacturing and software advances have generated new promising devices for detecting and quantifying the numerous volatile organic compounds (VOCs) which originate from our metabolism [1]. Among these instruments, electronic noses (e-noses) imitate mammalian olfaction in order to obtain reproducible measurements of VOC profiles in human mediums such as urine, blood, or breath [2]. In addition, exhaled breath analysis by e-nose can be used as a noninvasive biomarker of various metabolic pathways occurring in health and illness. Interestingly, an increasing number of studies have revealed the potential for the application of VOC profiling in numerous respiratory and systemic diseases [3].

Recently, a European Respiratory Society (ERS) task force document established guidelines in order to standardize all the methodological concerns for breath sampling and analysis by e-noses [4]. In these guidelines, it is unmistakably indicated that, when investigating exhaled VOCs, non-disease, patient-related factors, such as breathing manoeuvers, airway caliber, food and beverages intake, physical exercise and pregnancy, should always be considered [4].

Among these, intra-/inter-individual subjects' own respiratory physiology-associated variations may represent important confounders in exhaled VOC profiling [5,6]. In particular, the conditioning of inspiratory air and the expiratory breathing maneuvers may both influence the VOC pattern [7]. Indeed, the control of breathing is mainly automatic, and its regulation is driven by the autonomic nervous regulation of the respiratory center in the human brain [8]. Therefore, modifications of ventilatory patterns may result in exhaled alveolar concentrations of VOCs, the exhalations of which are dependent on minute ventilation and/or on CO_2 exhalation. Moreover, ventilatory variations are known to modify arterial CO_2 pressure levels, cardiac output and pulse pressure in humans [8,9]. Although a 5 min steady-state washing-in with VOC-filtered air is suggested, based on recommendations for helium washing during lung volume measurements, it is not clear whether it needs to be modified in certain types of patients.

For the above reason, the aim of the current study was to assess whether variations in breathing rhythms during the wash-in phase reflect changes in the exhaled VOC profile when analyzed by an e-nose.

2. Results

The characteristics of the study population are described in Table 1.

Table 1. Clinical characteristics of the study population.

Parameter	Value
Subjects (n.)	20
M\F (n.)	11\9
Age (y.)	37.1 ± 10.2
FEV1%pred.	103.6 ± 10.7
BMI	25.52 ± 2.4
(ex)-smokers (n.)	0
comorbidities(n.)	0

Values are intended as mean ± SD.

The two-dimensional principal component analysis plot showed that the exhaled VOC profiles obtained for the three breathing rhythms could not be discriminated from each other (Figure 1). The CDA of the data showed a CVA of 45.1%, indicating that the difference was not significant (p = ns). Similarly, ANOVA of the main four principal components showed no significant differences among the three groups (p = ns for all, see Table 2). Therefore, the Cyranose 320's 32-sensor outputs from the sensor array were not significantly different among the three breathing rhythms.

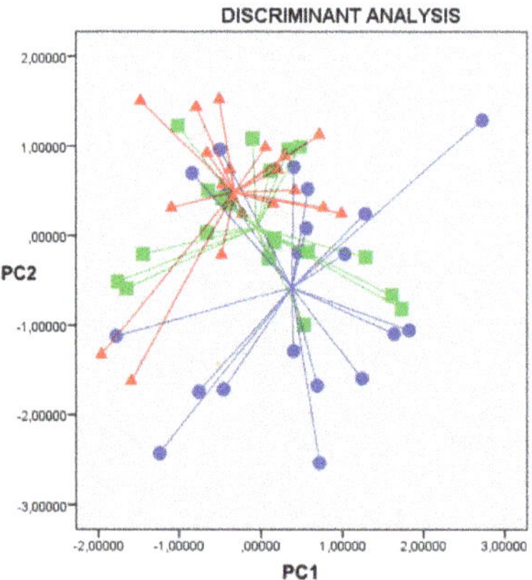

Figure 1. Two-dimensional principal component analysis plot, showing that exhaled VOC profiles among normal ventilation (blue circles), hyperventilation (green squares) and hypoventilation (red triangles) during wash-in are indistinguishable from each other. Cross validated accuracy was 45.1% (p = ns). X axis = Principal component 1; Y axis = Principal component 2.

Table 2. ANOVA of the main four principal components among the three breathing rhythms.

	Normal Rhythm	Fast Rhythm	Slow Rhythm	p
PC1	-0.131 ± 1.045	0.050 ± 0.955	0.081 ± 1.035	0.773
PC2	0.374 ± 1.113	-0.054 ± 0.983	-0.320 ± 0.801	0.084
PC3	0.577 ± 1.178	0.091 ± 0.667	0.485 ± 0.814	0.182
PC4	-0.003 ± 1.162	0.008 ± 1.031	-0.005 ± 0.831	0.999

3. Discussion

According to our results, it appears that the exhaled VOC profile measured by our e-nose is stable during variations in wash-in breathing rhythm.

To the best of our knowledge, this is the first study which specifically investigates e-nose analyzed exhaled breath VOC composition in relation to variation in breathing rhythm in a population of well-characterized, healthy subjects.

Research into the effects of ventilatory variations on exhaled breath composition is essential for a better comprehension of the physiological and metabolic phenotype of healthy subjects, and for implementing exhaled VOC profiling in routine pulmonary medicine.

It is known that a number of VOCs are exhalation flow-dependent, such as acetone, ethanol, pentane and isoprene [10–12]. In addition, alterations in exhaled flow, breath hold and dead space significantly modify e-nose assessed exhaled breath patterns with e-nose, thus influencing their ability to discriminate breathprints [13].

Very recently, Sukul et al. analyzed 25 healthy subjects and detected changes in a selection of the most abundant, endogenous and bloodborne VOCs when respiratory rhythms were switched between spontaneous and/or paced breathing [14]. Such changes were closely related to minute ventilation and end-tidal CO2 exhalation [14].

A number of limitations must be taken into account. Firstly, there were a relatively small number of enrolled subjects. However, our sample size with 20 individuals in our proof-of-concept study appeared to be suitable to merit further investigations including larger cohorts and a validation group.

Secondly, although we carefully monitored respiratory rates during sampling, we arbitrarily chose breathing rhythm intervals for each group, and therefore we may have missed some important information.

Thirdly, e-nose analysis does not quantify levels of single VOCs. Incontrovertibly, future studies should incorporate chemical analytical techniques, such as gas chromatography coupled to mass spectrometry (GC-MS) to identify specific discriminant compounds.

How can we explain our results? Human-exhaled breath contains over 3000 VOCs deriving from physiologic and pathophysiological mechanisms, operating via metabolic pathways [15]. In accordance with the findings of previous studies, our results suggest that, although breathing rhythm modifies the individual components of exhaled breath, the overall VOC profile, as measured by an e-nose, does not differ among groups with different breathing rhythms.

What are the implications of our findings? It appears that Cyranose 320 signature patterns output from the 32-sensor array were similar among the three breathing rhythms. Our data indicate that careful breathing rate monitoring during breath collection might not be necessary in future studies using a Cyranose 320. Hence, future research (possibly including patients with functional airways obstruction and restriction) should apply these models into larger clinical trials in order to confirm our findings and to investigate other possible confounding factors. Moreover, these studies must include several types of e-noses, using different technology, in order to assess the interchangeability of devices.

4. Materials and Methods

4.1. Patients

We enrolled 20 healthy, non-smoking subjects (11 males, 9 females), with a negative anamnesis of chest symptoms and systemic diseases and who were not taking any medications. The age range was 28–55. Lung function was normal for all participants. None of the subjects experienced upper or lower respiratory tract infections in the four weeks before testing, nor during the day of sampling. Subject characteristics are shown in Table 1.

A series of 3 exhaled breath measurements were performed on all subjects, and their exhaled breath was collected at three different breathing rhythms during wash-in phase: (a) normal rhythm (Respiratory Rate (RR) between 12 and 18/min), (b) fast rhythm (RR > 25/min) and (c) slow rhythm (RR < 10/min).

All participants were volunteers and were enrolled from hospital members.

The current study was previously approved by the local ethics committee (protocol number 46403/15) and all participants signed an informed consent before taking part in the study.

4.2. Study Design

We performed a longitudinal study. All measurements were obtained during two visits. During the first visit, subjects were carefully checked for inclusion/exclusion criteria and a flow-volume spirometry was performed (MasterscreenPneumo, Jaeger, Wurzburg, Germany).

During the second visit, exhaled breath was collected as described above and immediately analyzed by the e-nose. All participants were randomized to perform a different order of breathing rhythms during the wash-in phase: a-b-c, a-c-b, c-b-a, c-a-b, b-a-c, b-c-a. Intervals between each measurement were at least 2 h. Subjects were asked to refrain from eating and drinking, as well as from engaging in vigorous physical exercise, for at least 3 h before visit two. Breath was collected as follows: first a wash-in phase of 5 min through a 3-way non-rebreathing valve connected to an inspiratory VOC filter (A2; North Safety, Middelburg, The Netherlands) to reduce the effect of environmental VOCs, then subjects exhaled a single vital capacity volume into a Tedlar bag connected to the e-nose.

4.3. Electronic Nose

A commercially available e-nose was used (Cyranose 320, Sensigent, Irwindale, CA, USA). It consists of a nano-composite array of 32 organic polymer sensors. The polymers swell when exposed to VOC combinations, which changes their electrical resistance. Raw data are captured as changes in resistance of each of the 32 sensors in an onboard database, producing a distribution (breathprint) that describes the VOC mixture and that can be used for pattern-recognition algorithms. The operating parameters were as follows: Baseline purge: 30 s (pump speed: low); sampling time: 60 s (pump speed: medium), purging time: 200 s (pump speed: high), total run time: 300 s, temperature 42 °C. Post-run purges between samples: 5 min. In addition, pre-conditioning for the sensor array prior to running samples consisted of a 5 min exposure to the room air to assure stability of sensor outputs, followed by a "blank measurement", as indicated in the operating instructions manual.

4.4. Statistical Analysis

The sample size was estimated based on data deriving from previous studies [16]. The raw data of breath samples were analyzed by SPSS software, version 18.0. The same program was used for the random assignment of breathing sequences. Principal component analysis (PCA) and successive linear canonical discriminant analysis (CDA) were calculated, thus providing the cross-validated accuracy percentage (CVA%), which estimates how accurately a predictive model will perform in practice. Furthermore, ANOVA of the main four principal components (which captured 96.3% of the total variance) was performed among the three breathing rhythms. A p-value of <0.05 was considered to be statistically significant.

Author Contributions: Conceptualization, S.D. and P.C.; methodology, S.D.; software, V.N.Q.; validation, E.B., P.C. and G.E.C.; formal analysis, V.N.Q.; investigation, T.R., E.B.; resources, S.D.; data curation, T.R., S.D.; writing—original draft preparation, S.D.; writing—review and editing, P.C.; visualization, S.D.; supervision, G.E.C.; project administration, S.D.; funding acquisition, G.E.C. All authors have read and agreed to the published version of the manuscript.

Funding: This research received no external funding.

Institutional Review Board Statement: The study was conducted according to the guidelines of the Declaration of Helsinki, and approved by the Ethics Committee of Policlinico di Bari (protocol code 43406/2015).

Informed Consent Statement: Informed consent was obtained from all subjects involved in the study.

Data Availability Statement: Dataset can be available upon reasonable request.

Conflicts of Interest: The authors declare no conflict of interest.

References

1. Haick, H.; Broza, Y.Y.; Mochalski, P.; Ruzsanyi, V.; Amann, A. Assessment, origin, and implementation of breath volatile cancer markers. *Chem. Soc. Rev.* **2014**, *43*, 1423–1449. [CrossRef] [PubMed]
2. Wilson, A.D.; Baietto, M. Advances in Electronic-Nose Technologies Developed for Biomedical Applications. *Sensors* **2011**, *11*, 1105–1176. [CrossRef] [PubMed]
3. Dragonieri, S.; Pennazza, G.; Carratu, P.; Resta, O. Electronic Nose Technology in Respiratory Diseases. *Lung* **2017**, *195*, 157–165. [CrossRef] [PubMed]
4. Horváth, I.; Barnes, P.J.; Loukides, S.; Sterk, P.J.; Högman, M.; Olin, A.-C.; Amann, A.; Antus, B.; Baraldi, E.; Bikov, A.; et al. A European Respiratory Society technical standard: Exhaled biomarkers in lung disease. *Eur. Respir. J.* **2017**, *49*, 1600965. [CrossRef] [PubMed]
5. Kindig, N.B.; Hazlett, D.R. The effects of breathing pattern in the estimation of pulmonary diffusing capacity. *Q. J. Exp. Physiol. Cogn. Med. Sci.* **1974**, *59*, 311–329. [CrossRef] [PubMed]
6. Duffin, J. The fast exercise drive to breathe. *J. Physiol.* **2013**, *592*, 445–451. [CrossRef] [PubMed]
7. Dragonieri, S.; Schot, R.; Mertens, B.J.; Le Cessie, S.; Gauw, S.A.; Spanevello, A.; Resta, O.; Willard, N.P.; Vink, T.J.; Rabe, K.F.; et al. An electronic nose in the discrimination of patients with asthma and controls. *J. Allergy Clin. Immunol.* **2007**, *120*, 856–862. [CrossRef] [PubMed]
8. Modarreszadeh, M.; Bruce, E.N. Ventilatory variability induced by spontaneous variations of $PaCO_2$ in humans. *J. Appl. Physiol.* **1994**, *1985*, 2765–2775. [CrossRef] [PubMed]
9. Dornhorst, A.C.; Howard, P.; Leathart, G.L. Respiratory Variations in Blood Pressure. *Circulation* **1952**, *6*, 553–558. [CrossRef] [PubMed]
10. Boshier, P.R.; Priest, O.H.; Hanna, G.B.; Marczin, N. Influence of respiratory variables on the on-line detection of exhaled trace gases by PTR-MS. *Thorax* **2011**, *66*, 919–920. [CrossRef] [PubMed]
11. Bikov, A.; Paschalaki, K.; Logan-Sinclair, R.; Horváth, I.; Kharitonov, S.A.; Barnes, P.J.; Usmani, O.S.; Paredi, P. Standardised exhaled breath collection for the measurement of exhaled volatile organic compounds by proton transfer reaction mass spectrometry. *BMC Pulm. Med.* **2013**, *13*, 43. [CrossRef] [PubMed]
12. Lärstad, M.A.E.; Torén, K.; Bake, B.; Olin, A.-C. Determination of ethane, pentane and isoprene in exhaled air-effects of breath-holding, flow rate and purified air. *Acta Physiol.* **2007**, *189*, 87–98. [CrossRef]
13. Bikov, A.; Hernadi, M.; Korosi, B.Z.; Kunos, L.; Zsamboki, G.; Sutto, Z.; Tarnoki, A.D.; Tarnoki, D.L.; Losonczy, G.; Horvath, I. Expiratory flow rate, breath hold and anatomic dead space influence electronic nose ability to detect lung cancer. *BMC Pulm. Med.* **2014**, *14*, 1–9. [CrossRef] [PubMed]
14. Sukul, P.; Schubert, J.K.; Zanaty, K.; Trefz, P.; Sinha, A.; Kamysek, S.; Miekisch, W. Exhaled breath compositions under varying respiratory rhythms reflects ventilatory variations: Translating breathomics towards respiratory medicine. *Sci. Rep.* **2020**, *10*, 1–16. [CrossRef] [PubMed]
15. Ibrahim, W.; Carr, L.; Cordell, R.; Wilde, M.J.; Salman, D.; Monks, P.S.; Thomas, P.; Brightling, C.E.; Siddiqui, S.; Greening, N.J. Breathomics for the clinician: The use of volatile organic compounds in respiratory diseases. *Thorax* **2021**, *76*, 514–521. [CrossRef] [PubMed]
16. Dragonieri, S.; Quaranta, V.N.; Carratu, P.; Ranieri, T.; Resta, O. Influence of age and gender on the profile of exhaled volatile organic compounds analyzed by an electronic nose. *J. Bras. Pneumol.* **2016**, *42*, 143–145. [CrossRef] [PubMed]

molecules

Article

Comprehensive Two-Dimensional Gas Chromatography–Mass Spectrometry Analysis of Exhaled Breath Compounds after Whole Grain Diets

Kaisa Raninen [1,2,*], Ringa Nenonen [1], Elina Järvelä-Reijonen [1], Kaisa Poutanen [3], Hannu Mykkänen [1] and Olavi Raatikainen [1]

[1] Institute of Public Health and Clinical Nutrition, University of Eastern Finland, P.O. Box 1627, FI-70211 Kuopio, Finland; ringa.nenonen@martat.fi (R.N.); elina.jarvela-reijonen@uef.fi (E.J.-R.); hannu.mykkanen@uef.fi (H.M.); olavi.raatikainen@uef.fi (O.R.)
[2] SIB Labs, University of Eastern Finland, P.O. Box 1627, FI-70211 Kuopio, Finland
[3] VTT Technical Research Centre of Finland, P.O. Box 1000, FI-02044 VTT Espoo, Finland; kaisa.poutanen@vtt.fi
* Correspondence: kaisa.raninen@uef.fi

Citation: Raninen, K.; Nenonen, R.; Järvelä-Reijonen, E.; Poutanen, K.; Mykkänen, H.; Raatikainen, O. Comprehensive Two-Dimensional Gas Chromatography–Mass Spectrometry Analysis of Exhaled Breath Compounds after Whole Grain Diets. *Molecules* **2021**, *26*, 2667. https://doi.org/10.3390/molecules26092667

Academic Editors: Natalia Drabińska and Ben de Lacy Costello

Received: 30 March 2021
Accepted: 30 April 2021
Published: 2 May 2021

Publisher's Note: MDPI stays neutral with regard to jurisdictional claims in published maps and institutional affiliations.

Abstract: Exhaled breath is a potential noninvasive matrix to give new information about metabolic effects of diets. In this pilot study, non-targeted analysis of exhaled breath volatile organic compounds (VOCs) was made by comprehensive two-dimensional gas chromatography–mass spectrometry (GCxGC-MS) to explore compounds relating to whole grain (WG) diets. Nine healthy subjects participated in the dietary intervention with parallel crossover design, consisting of two high-fiber diets containing whole grain rye bread (WGR) or whole grain wheat bread (WGW) and 1-week control diets with refined wheat bread (WW) before both diet periods. Large interindividual differences were detected in the VOC composition. About 260 VOCs were detected from exhaled breath samples, in which 40 of the compounds were present in more than half of the samples. Various derivatives of benzoic acid and phenolic compounds, as well as some furanones existed in exhaled breath samples only after the WG diets, making them interesting compounds to study further.

Keywords: exhaled breath; whole grain; rye; comprehensive two-dimensional gas chromatography–mass spectrometry; dietary fiber

1. Introduction

Whole grain (WG) cereals are an important source of dietary fiber (DF) and micronutrients and are therefore acknowledged as part of the healthy diet in dietary recommendations [1,2]. Epidemiological studies and their meta-analyses have consistently shown high intake of WG to lower risk of chronic diseases and mortality [3–5], and associate negatively with obesity [6,7], type 2 diabetes [8–10], cardiovascular disease [11–13], and certain cancers [14,15]. However, the underlying physiological mechanisms are complex and unclear. Phenolic compounds in the fiber matrix of bran [16] are one proposed element for the protective effects of WG. Alkylresorcinols are present in the outer layers of wheat and rye grains and are known to be absorbed by humans. They have been detected in plasma and urine, and hence have been studied as a promising biomarker for WG wheat and rye in the diet [17].

People have variable metabolic responses to diets because of individual physiology and gut microflora [18]. Therefore, there has been an increasing interest in nutrigenomics, proteomics, and metabolomics to monitor metabolism from a wider perspective. Volatomic analysis of exhaled breath, used mainly for searching noninvasive biomarkers for diseases [19–21], could be used to characterize volatile organic compounds (VOCs) relating to various diets or specific foods, such as WG cereals. This research could lead to new information on the metabolic effects of WG foods and their association with health effects.

Exhaled breath is a potential noninvasive matrix to monitor metabolic changes induced by dietary modifications [22,23]. Foods contain numerous molecules which after digestion or metabolism by gut microflora are absorbed to the circulation. If they have a suitable boiling point, vapor pressure, and solubility, they can be excreted to exhaled breath. Currently, more than 3400 compounds have been identified in exhaled breath [24,25], with an average exhaled breath sample containing about 200 detected compounds [26]. However, there is wide interindividual variation since typically only a few dozen of the compounds are detected in every exhaled breath sample.

Although diet is known to cause variation in exhaled compounds [27], so far only a few pilot studies have been conducted to monitor the effects of diets on exhaled breath VOCs. Pioneering research in developing methods for exhaled breath analysis was done by Smith and Španěl who explored the effects of a meal [28] and glucose ingestion [29] to exhaled breath VOCs, as well as the effects of ketogenic diet on breath acetone levels [30]. Galassetti, Blake and colleagues studied exhaled breath compounds relating to diabetes [31,32] and monitored the effects of high-fat meals on exhaled breath VOCs [33]. They also studied exhaled breath VOC profiles relating to blood glucose [34–36] and lipid levels [37]. van Schooten et al. applied the breath analysis to monitor gastrointestinal diseases [38–40] and demonstrated a distinctive exhaled breath VOC profile after a gluten-free diet [41].

We have earlier demonstrated changes in exhaled breath VOC profiles by aspiration ion mobility spectrometry (AIMS) in diets differing in DF content (low-fiber diet vs. high-fiber) and type of bread (white wheat bread vs. sourdough fermented whole grain rye bread vs. white wheat bread enriched with modified rye fiber) [42,43]. However, the AIMS technology, regarded as a type of electronic nose, cannot identify the compounds responsible for the changes. Gas chromatography–mass spectrometry (GC-MS) is a standard technology for identifying VOCs [24,44] and multidimensional chromatography techniques, such as comprehensive two-dimensional GC-MS (GCxGC-MS), are utilized especially for characterization of compounds in metabolomic research, due to their increased separation capability [45]. However, they are not yet utilized for monitoring exhaled breath VOCs regarding to diets.

In this study, our aim was to pilot exhaled breath analysis with GCxGC-MS to explore VOCs relating to WG diets.

2. Results

About 260 VOCs were detected in 32 exhaled breath samples from 9 persons; of these VOCs, 40 were common, being present in more than half of the breath samples (Table 1). Carbon dioxide, isoprene, acetone, ethanol, 1-butanol, 2-propanol, benzene, benzaldehyde, methyl vinyl ketone, 2-butanone, phenol, hexanoic acid, and acetonitrile were found in all samples, but large individual differences existed in the other compounds. Additionally, 86 VOCs were tentatively identified by their MS spectra (Table S1), whereas about 170 detected compounds remained unidentified.

Some derivatives of benzoic acid and phenolic compounds were detected in exhaled breath samples only after the WG diets (Table 2). Phthalic acid or phthalic anhydride (similarity index, SI 93 for both compounds) was found in 57% of the exhaled breath samples during the whole grain rye bread diet (WGR), in 11% of breath samples during the whole grain wheat bread diet (WGW), and in 6% of the background room air (BG) samples, but in none of the samples collected after the control diets containing refined wheat bread (WW). Benzoic acid was detected in 29% of breath samples during the WGR diet and in 11% of breath samples during the WGW diet, but in none of the exhaled breath samples during the WW diets and in 6% of BG samples. Furthermore, diphenyl ethanedione and benzamide were detected only after the WG diets, diphenyl ethanedione in 29% of breath samples during the WGR and benzamide in one participant during the WGR and the WGW.

Table 1. Common [1] volatile compounds in the exhaled breath samples collected from the study participants after the diet periods and their presence in the background room air samples.

Compounds	Detected in % of Samples in			
	WGR	**WGW**	**WW**	**BG**
Carbon dioxide	100	100	100	100
Ethanol	100	100	100	100
Hexanoic acid	100	100	100	100
Acetophenone	100	100	100	100
1-Butanol	100	100	100	97
Benzene	100	100	100	97
Benzaldehyde	100	100	100	97
Methyl vinyl ketone	100	100	100	97
2-Butanone	100	100	100	97
Acetone	100	100	100	94
Phenol	100	100	100	94
2-Propanol	100	100	100	90
Acetonitrile	100	100	100	87
Isoprene	100	100	100	42
2,3-Butanedione	100	89	100	68
Toluene	100	33	81	74
3-Pentanol/2-Propanol, 2-methyl	86	100	100	97
Butanal	86	100	100	87
Hexanal	86	100	88	87
Heptanal	86	100	81	74
Octanal	86	89	94	23
Benzaldehyde, 2/4-methyl	86	89	75	90
Pentanal	86	89	69	68
n-Hexane	86	67	88	58
Nonanal	71	78	100	94
Acetaldehyde	71	56	81	77
3,4-Dimethyl heptane	71	33	75	35
D-Limonene	71	33	69	10
1,3-Pentadiene	71	33	69	0
Benzene, 1,4-dimethyl-	71	22	56	26
Dimethyl sulfide	57	78	81	0
Decanal	57	67	56	84
Methyl cyclopentane	57	56	63	42
6-Methyl-5-hepten-2-one	57	56	88	68
1 Propanol	57	44	81	39
Octane	57	44	56	17
Ethyl acetate	57	33	75	71
Styrene	57	33	69	29
p-Cymene	57	33	69	3
Heptane	43	56	75	71

[1] present in >50% of the analyzed exhaled breath samples, WGR = whole grain rye bread diet (n = 7), WGW = whole grain wheat bread diet (n = 9), WW = refined wheat bread diet (n = 16), BG = background room air (n = 31).

Some furanones (γ-lactones) were also identified in the exhaled breath only during the WG diets: 5-dodecyldihydro-(3H)-furanone (in two participants after WGR), dihydro-4-hydroxy-2(3H)-furanone (in one participant after WGR and WGW) and dihydro-5-tetradecyl-2(3H)-furanone (in one participant after WGR and WGW).

We also detected several unidentified compounds having mass spectrum fragments 105, 77 and 51, which are typical for benzoic acid derivatives, and 107, 121, 135, and 149, typical for alkylphenols (potential degradation products of alkylresorcinols). However, none of the unidentified compounds were detected only in a particular diet period.

Table 2. Volatile organic compounds detected only in the exhaled breath samples after WG diets.

Compounds	RT	SI	Detected in % of Samples in			
			WGR	WGW	WW	BG
Phthalic acid/Phthalic anhydride	40–69	93	57	11	0	6
Benzoic acid	30–69	94	29	11	0	6
Diphenyl ethanedione	61.7	92	29	0	0	0
5-Dodecyldihydro-(3H)-furanone	63.7	92	29	0	0	0
Benzamide	65.5	94	14	11	0	0
Dihydro-4-hydroxy-2(3H)-furanone	57.5	85	14	11	0	0
Dihydro-5-tetradecyl-2(3H)-furanone	67.7	89	14	11	0	0

RT = retention time (min), SI = similarity index, WGR = whole grain rye bread diet (n = 7), WGW = whole grain wheat bread diet (n = 9), WW = white wheat bread diet (n = 16), BG = background room air (n = 31).

3. Discussion

We piloted exhaled breath analysis with GCxGC-MS to detect VOCs relating to WG diets. With this technology and the chosen method, about 260 compounds were detected from exhaled breath samples, and of these, 40 VOCs were present in more than half of the exhaled breath samples. Some benzoic acid and phenol derivatives, as well as furanone compounds, were detected more frequently after the whole grain diets.

GCxGC-MS-technology has better sensitivity and separation of compounds as compared to the traditional GC-MS, which make it suitable for non-targeted analysis of exhaled breath compounds. We have earlier analyzed exhaled breath VOCs by traditional GC-MS having the same column and same kind of sampling protocol [46], and about 40 VOCs were detected in the exhaled breath samples collected from healthy men. In the current GCxGC-MS protocol, the total number of detected compounds was approximately seven times more (about 260 compounds). The comprehensive GCxGC-MS technology is based on cryogenic modulator: effluent from the first column is trapped in the modulator for a given period (for 8 s in our method) before being released into the second column. This increases the sensitivity of the method remarkably as compared to traditional GC-MS.

GCxGC-MS technology also improved the separation of compounds as compared to traditional GC-MS. For example, we found two compounds giving almost identical mass spectra with isoprene, having only slightly different retention times. These are probably cis-1,3-pentadiene and trans-1,3-pentadiene which have been detected earlier in exhaled breath samples [47,48], or 1,4-pentadiene, which has been associated with smoking [49,50]. However, our participants were non-smokers. It is noteworthy that these compounds can be erroneously identified as isoprene, and therefore interfere the quantification of isoprene if they are not separated in the analysis. Exhaled breath isoprene has been studied extensively as a potential biomarker compound for cholesterol synthesis, though with controversial results [51,52]. It is possible that these compounds have interfered the quantification of isoprene in some studies.

Although the GCxGC-MS has advantages in sensitivity and selectivity, it also has drawbacks. Because of its sensitivity, the signal is easily overloaded when both the quantification and identification are challenged. This technology is suitable mainly for detecting compounds from challenging matrixes (having a multitude of compounds to be separated), but it is not very convenient for their quantification. The quantitative method should be optimized for each compound of interest separately, including calibration with breath mimicking conditions. Therefore, in this study we did not quantify the detected compounds. Exhaled breath VOCs might have multiple sources, and therefore it would be more relevant to monitor the changes in their levels rather than searching for specific biomarker compounds. However, nontargeted volatomic analysis can be used to select the relevant target compounds to monitor.

In total, 86 VOCs were tentatively identified from exhaled breath samples while about 150 VOCs remained unidentified, as their MS spectra were not found in the MS libraries. This indicates that there might still exist numerous unidentified molecules in the exhaled breath because GC-MS is a standard technology for identifying volatile compounds, and identification is mainly based on the MS libraries.

There is no analytical method available to monitor all the compounds in exhaled breath. For example, breath sampling method and thermal desorption (TD) adsorbents select the compounds, and GC column determine which compounds are chromatographically separated and can be detected. In this study, we chose the polar Nukol column for the first separative column because we have found it suitable for detecting endogenous gut-related exhaled breath VOCs [46], and non-polar Zebron ZB-35HT Inferno column for the second column due to its chemically different stationary phase compared to Nukol. By choosing other columns or TD adsorbent, different compounds could have been detected. It is noteworthy that most GC-MS analyses for exhaled breath VOCs are made by using general purpose nonpolar methylpolysiloxane columns containing 5% phenyl. With our protocol, we detected some common exhaled breath compounds such as isoprene, acetone, and ethanol, but we were unable to detect, for example, ammonia and methane (too small to detect with MS SCAN 35–300 m/z), or short-chain fatty acids (not enough sensitivity with MS SCAN mode [46]), although these compounds would be interesting in the perspective of nutrition and gut health [53–55]. We detected some compounds (Table 1; 3,4-dimethyl heptane, 3-pentanol, and methyl cyclopentane), which have not been reported in exhaled breath before. However, it should be pointed out that in our study, the identification was done only by the spectral library match, and was not confirmed with standard molecules (i.e., tentative identification) or retention indices. Mass spectra can be almost identical for some compounds, for example, for structural isomers (e.g., 2-methylbutane and n-pentane) or compounds with same structure with different length of alkyl chain (e.g., undecanal and tetradecanal). Therefore, the identification of VOCs in this study must be considered with caution.

Some benzoic acid and phenolic derivatives, as well as furanones, were detected from exhaled breath samples only after whole grain diets. It is possible that these VOCs are degradation products of phenolic compounds such as phenolic acids, alkylresorcinols and lignans from the DF complex in the bran. Phenolic compounds can be metabolized to various compounds by colonic fermentation and metabolism [56]; for example, benzoic acid can be formed from rye phenolics [57]. The compounds were detected mostly in the same exhaled breath samples, which indicate the same origin for these compounds.

Benzoic acid is known to be related to various foods but considered to have relatively low levels in the alveolar exhaled breath in the fasting state, since it is metabolized by liver and kidneys to hippurate within a few hours after oral dosing [58]. However, benzoic acid is formed also from whole grains in gut fermentation, which may explain the elevated levels in the fasting state in some individuals during WG diets. The exhaled breath samples were taken in the fasting state, but the fermentation rate may have been varied based on individual orocecal transit time and timing of eating WG. Estimation of the fermentation rate by breath hydrogen measurements [59] would be relevant when studying fermentation-related exhaled breath VOCs.

It´s noteworthy that the GC parameters used were not optimal for benzoic acid and phthalic acid/anhydride. Both compounds had wide tailing chromatograph peaks. This did not interfere with the identification of compounds, but may have affected sensitivity in their detection, and partly explains why these compounds were seen only in a minority of exhaled breath samples. Other GC parameters or technology should be used for analyzing these compounds more accurately.

Furanones are known to be formed in chemical reactions during charbroiling and seed oil cooking, and in Maillard reaction between sugars and amino acids [60,61]. They can also be metabolized from grain lignans such as matairesinol or 7-hydroxymatairesinol, or from enterolactone, a mammalian lignan, which is formed in the large intestine from plant lignans [62]. All these lignans have dihydro-2(3H)-furanone in their molecule structure. Enterolactone is considered a biomarker for high lignan intake in the diet [63], but high interindividual variation has been found in its absorption and metabolism [64]. To our knowledge, 5-dodecyldihydro-(3H)-furanone (CAS 730-46-1, also known as γ-palmitolactone), dihydro-4-hydroxy-2(3H)-furanone (CAS 5469-16-9, 3-hydroxy-γ-butyrolactone) or dihydro-5-tetradecyl-2(3H)-furanone (CAS 502-26-1, γ-stearolactone) have not been detected from exhaled breath before, unlike some other furanones [25,41]. Dihydro-5-tetradecyl-2(3H)-furanone has been detected from skin [25]. It would be interesting to monitor exhaled breath phenolic and furanone compounds and their levels in relation to different dietary sources, for example rye, using optimized analysis methodology for those compounds.

In this study, the randomized crossover protocol was used because the inter-individual variation of breath VOCs is known to be high [26]. In a crossover protocol, it is more likely that detected differences in breath VOCs are due to dietary changes because the other lifestyle factors are rather constant. Furthermore, most of the study participants were students or staff members of the Faculty of Health Sciences in the University of Eastern Finland and therefore likely to pay more attention to their eating than the average population in Finland. The participants consumed plenty of fruits and vegetables and received plenty of DF, and probably also phenolic compounds, also from sources other than the study breads. Therefore, the supply of DF remained higher than expected during the WW diets, being in the level of dietary recommendations. However, the total amount of consumed fruits and vegetables remained stable during the study, and there was a significant difference in the DF levels between WG and WW diets, as intended.

In conclusion, the GCxGC-MS technology, being sensitive and selective, offered some advantage for detecting exhaled breath VOCs. Benzoic acid derivatives, phenolic compounds, and furanones are potential compounds in monitoring metabolic effects of whole grains in exhaled breath. However, based on earlier reports by us [42,43] and others [22], it seems that it would be more relevant to monitor changes in the levels of multiple compounds or in VOC profiles rather than individual compounds when monitoring diet-related changes in exhaled breath VOCs.

4. Materials and Methods

4.1. Protocol

A randomized crossover manner dietary intervention was performed with 9 participants. They followed high-fiber diets containing either whole grain rye bread (WGR) or whole grain wheat bread (WGW) for 1 week in randomized order, and there were 1-week periods with refined (white) wheat bread (WW) before both test periods (Figure 1). At the end of the diet periods, exhaled breath samples in fasting state and parallel background air samples (room air samples, BG) were analyzed with GCxGC-MS technology. The RYEBREATH study was approved by the Ethics Committee of the Hospital District of Northern Savo (University of Eastern Finland, Hannu Mykkänen, 40/2015).

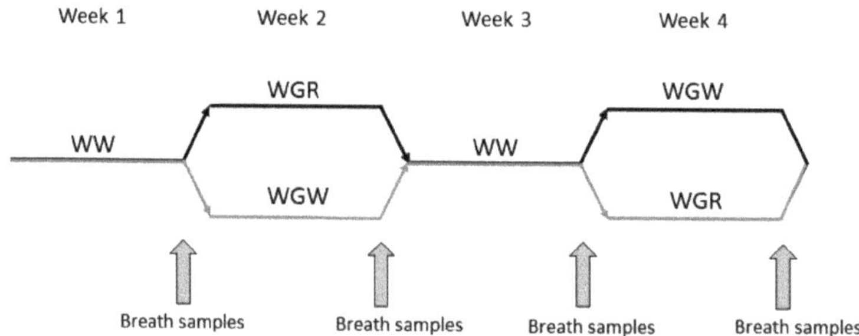

Figure 1. Study design of a randomized crossover manner dietary intervention with 9 participants. WW = control diet containing refined wheat bread; WGR = whole grain rye bread diet; WGW = whole grain wheat bread diet. Exhaled breath and parallel background room air samples were taken at the end of the diet periods.

4.2. Study Participants

The participants were recruited into the RYEBREATH study with the campus advertisements in the University of Eastern Finland. They were healthy non-smoking Finnish men (2) and women (7) aged 21 to 59 years (average 31 years) and with BMI (body mass index) between 18.7 and 29 kg/m^2 (average 23 kg/m^2). The participants were advised to maintain their body weight and habitual lifestyle throughout the study, except the devised dietary modification for cereal content. All the participants provided written informed consent prior to participating in the study.

4.3. Diets

Participants followed three diets differing in consumed grain products. They were advised to consume 5–7 slices of white wheat bread per day during WW periods, 5–7 slices of whole grain rye bread during WGR period and 7–8 slices of whole grain wheat bread during WGW period. The commercial breads used in each period were: white toasts Vaasan Iso Paahto (DF 0.9 g/slice) and Oululainen Reilu Vehnä (DF 1.2 g/slice) during the WW periods, whole grain rye breads Fazer Real Ruis (DF 1.2 g/slice) and Porokylän leipomo PikkuKartano (DF 1.6 g/slice) during the WGR period, and wholegrain wheat breads Fazer Täysjyvä Paahto (DF 1.5 g/slice) and Vaasan Täysjyvä Isopaahto (DF 2.5 g/slice) during the WGW period. The study subjects were advised to avoid whole grain products during the WW periods and not to consume any rye except during the WGR period. Food items which typically increase gut fermentation and fermentative gases in the intestines, such as beans, cabbages, and xylitol products, were avoided throughout the intervention. A master´s student in clinical nutrition advised the participants weekly on the practical management of the diets. The participants filled in 4-day food records during each diet period and recorded the eaten amount of the test breads in a daily questionnaire. The food records were analyzed for nutrient intakes using the Diet32 software (version 1.4.6.3, Aivo Finland Oy, Turku, Finland).

The intakes of energy, protein, fat, and carbohydrates during the test diet periods were maintained at the same level during the intervention (Table 3). Only intake of DF was significantly different between the WW and WG periods. The participants consumed breads on average 159 g/day during the WW periods, 208 g/day during WGR, and 200 g/day during WGW, which covered 18% of energy intake in WW1, 19% in WW2, 24% in WGR, and 23% in WGW.

Table 3. Mean daily intakes [1] of energy and nutrients during the 1-week diet periods (n = 9).

	WW1	WW2	WGR	WGW	*p*-Value [2]
Energy, MJ	9.0 ± 1.7	9.3 ± 1.6	9.0 ± 1.8	9.0 ± 1.5	0.865
Carbohydrates, E%	42 ± 2	41 ± 4	42 ± 3	41 ± 4	0.706
Protein, E%	20 ± 3	20 ± 3	19 ± 2	20 ± 3	0.254
Fat, E%	35 ± 4	36 ± 5	35 ± 4	35 ± 7	0.954
Dietary fiber, g	24 ± 8	25 ± 8	36 ± 6 *	34 ± 1 *	<0.001

[1] Values are means ± SD; [2] Statistical significance of the difference among the diet periods analyzed with Friedman's test; * Different from WW periods, Wilcoxon´s test, *p* = 0.008; WW = diet with white wheat bread; WGR = whole grain rye bread diet; WGW = whole grain wheat bread diet; E% = percentage of total energy intake.

4.4. Exhaled Breath Analysis

End-tidal exhaled breath samples were taken with Bio-VOC® samplers (Markes International Ltd., UK), which are made for capturing the last part of exhaled breath from the alveoli concentrated with VOCs excreted from the circulation (Figure 2a). Participants were trained to give an adequate sample. Before the sampling, participants brushed their teeth with toothpaste and rinsed the mouth effectively with water to stabilize the microbial fermentation in the mouth. They were sitting still without talking and breathing normally for a few minutes before sampling to standardize the ventilation. Then they gave a constant deep blow through the sampler. Exhaled breath samples were injected immediately after sampling into TD liners (fritted glass liner packed with Tenax GR, mesh 80–100, GL Sciences, Eindhoven, The Netherlands) using the Bio-VOC® sampler as a gas syringe and adapter (self-made by sculpting from PTFE rod) to connect the sampler and the liner tightly (Figure 2b). The TD liner was closed with a storage cap (Brass Liner Blanking Cap, GL Sciences, Eindhoven, the Netherlands) and analyzed within 2–5 h (Figure 2c). The internal standard (1 µL 0.22 µg/µL acetone-d6 (Euriso-top, Saint-Aubin, France) in Milli-Q ultrapure water (Millipore, Bedford, MA, USA)) was injected to the TD liners 30–60 min before sampling by using gas tight syringe and the Bio-VOC® sampler. The background room air samples were taken before breath samples by injecting the room air with the Bio-VOC® sampler to the TD liners. The room air samples were otherwise handled in the same way as breath samples.

Analysis was performed with a GCxGC-MS device consisting of GCMS-QP2010 Ultra and AOC-5000 Plus injection system (Shimadzu Scientific Instruments, Columbia, MD, USA), Optic-4 multi-mode inlet (GL Science, Eindhoven, The Netherlands) and ZX-1 thermal modulator (Zoex Corporation, Houston, TX, USA) (Figure 2d). The injection was done with automated injection of AOC-5000 Plus to the inlet of Optic injector. The temperature of the inlet was at the beginning 35 °C for 2 min and then rose to the 200 °C at the rate of 18 °C/min. The injection was done in high-pressure mode with split 5 allowing the pressure of the inlet decrease temporarily during the injection. The injected sample was preconcentrated to the cryotrap after the injector at −100 °C for 7 min and released rapidly at 200 °C (temperature rise 60 °C/s) to the GC. VOCs were separated on two serial capillary columns; polar Nukol (0.25 µm thick phase, 0.25 mm internal diameter, 30 m long, Supelco, Bellefonte, PA, USA) and non-polar Zebron ZB-35HT Inferno (0.8 µm/0.18 mm/1 m, Phenomenex Torrance, CA, USA), separated by a cryogenic Zoex-modulator. The modulation was done with 8 s modulation time and 10–30% filling of the 5 L dewar of the liquid nitrogen. Carrier gas was Helium 4.6 (AGA, Espoo, Finland) with column pressure 150 kPa, column flow 2.14 mL/min, and linear velocity 45.5. The GC oven was programmed to be 35 °C for 10 min, then raised by 3 °C/min to 200 °C. The duration of the GC program was 70 min. The detection was done with MS SCAN 35–300 *m/z*, event time 0.02 s, and scan speed 20,000 unit/s. Temperature of the ion source was 200 °C and for MS interface 220 °C.

The data were analyzed using ChromSquare 2.2 data analysis software (Chromaleont, Messina, Italy). All the visible blobs in two-dimensional chromatograph were manually selected for identification. The tentative identification was performed by comparing their mass spectra with data from NIST 11 Mass Spectral library (The National Institute of Standards and Technology, Gaithersburg, MD, USA), Wiley Registry 10th Edition (John Wiley & Sons, Hoboken, NJ, USA), and Flavour & Fragrance Natural & Synthetic Compounds GCMS library FFNSC 2 (Shimadzu Corp., Kyoto, Japan). The identification was checked precisely for each blob by the researcher, but it was not confirmed with analytical standards or retention indices. Tentatively identified VOCs were reported only if they were found in more than a single exhaled breath sample. Four exhaled breath samples and five background room air samples were excluded because of technical problems in the GC-MS analysis. Siloxanes and polyethylene glycol compounds were excluded from analysis because they likely originate from column phases.

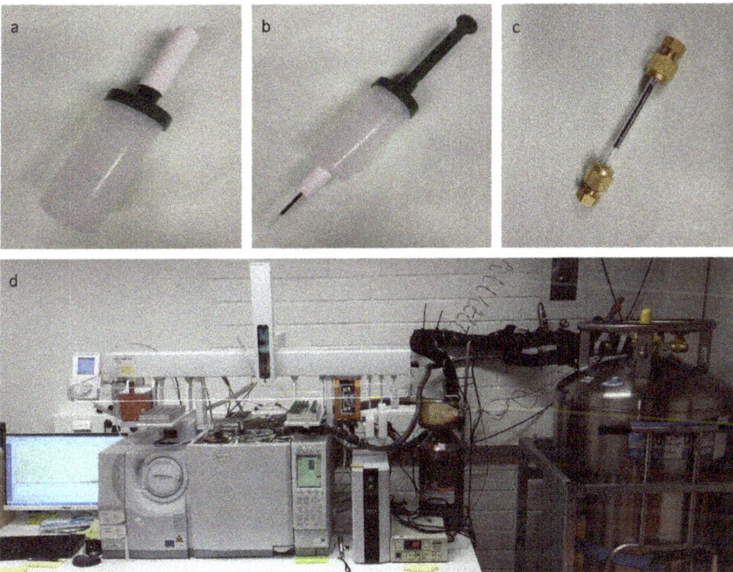

Figure 2. Exhaled breath analysis. End-tidal exhaled breath was sampled with the Bio-VOC® sampler (**a**), injected into a glass liner containing Tenax GR absorbent (**b**), removed to laboratory in a sealed liner (**c**), and analyzed with comprehensive two-dimensional gas chromatography–mass spectrometry (**d**).

5. Conclusions

Exhaled breath VOCs reflect metabolism and lifestyle, thus having large interindividual variation and a lot of so far unidentified molecules. Since diet affects exhaled breath VOCs, they could be utilized in studying the metabolic effects of diets. The GCxGC-MS technology offers some advantage in making the detection of human VOCs sensitive and selective. Exhaled breath benzoic acid derivatives, phenolic compounds, and furanones are interesting compounds to study further when exploring the metabolic effects of whole grains.

Supplementary Materials: The following are available online, Table S1: The detected VOCs (86) with tentative identification [1] from exhaled breath samples (n = 32) of nine study participants.

Author Contributions: Conceptualization, K.R., K.P., H.M. and O.R.; methodology, K.R., E.J.-R., H.M. and O.R.; validation, K.R. and E.J.-R.; formal analysis, K.R. and R.N.; investigation, K.R. and R.N.; resources, K.R., R.N., H.M., K.P. and O.R.; data curation, K.R.; writing—original draft preparation, K.R. and R.N.; writing—review and editing, K.R, E.J.-R., K.P., H.M. and O.R.; visualization, K.R.; supervision, K.P., H.M. and O.R.; project administration, K.P., H.M. and O.R.; funding acquisition, K.R., K.P., H.M. and O.R. All authors have read and agreed to the published version of the manuscript.

Funding: This research was funded by the Doctoral School of the University of Eastern Finland and with a personal grant by Raisio Plc Reseach Foundation to K.R.

Institutional Review Board Statement: The study was conducted according to the guidelines of the Declaration of Helsinki and approved by the Ethics Committee of the Hospital District of Northern Savo (University of Eastern Finland, Hannu Mykkänen, 40/2015, 27 January 2015).

Informed Consent Statement: Informed consent was obtained from all subjects involved in the study.

Data Availability Statement: The data presented in this study are available on request from the corresponding author. The original data are not publicly available due to privacy of participants.

Acknowledgments: We would like to thank The Doctoral School of the University of Eastern Finland and the Raisio Plc Research Foundation financing the PhD work of K.R. The SIB Labs unit enabling GCxGC-MS analysis is greatly appreciated, special thanks to Arto Koistinen and Teemu Vilppo for your support. We also warmly thank Marjukka Kolehmainen for cheering us on to finish this work.

Conflicts of Interest: The authors declare no conflict of interest.

Sample Availability: Samples of the compounds are not available from the authors.

References

1. Cena, H.; Calder, P.C. Defining a Healthy Diet: Evidence for the role of contemporary dietary patterns in health and disease. *Nutrients* **2020**, *12*, 334. [CrossRef] [PubMed]
2. Lappi, J.; Mykkänen, H.; Kolehmainen, M.; Poutanen, K. Wholegrain foods and health. In *Fibre-Rich and Wholegrain Foods*; Delcour, J.A., Poutanen, K., Eds.; Woodhead Publishing Limited: Cambridge, UK, 2013; pp. 76–95.
3. Reynolds, A.; Mann, J.; Cummings, J.; Winter, N.; Mete, E.; Te Morenga, L. Carbohydrate quality and human health: A series of systematic reviews and meta-analyses. *Lancet* **2019**, *393*, 434–445. [CrossRef]
4. Zhang, B.; Zhao, Q.; Guo, W.; Bao, W.; Wang, X. Association of whole grain intake with all-cause, cardiovascular, and cancer mortality: A systematic review and dose-response meta-analysis from prospective cohort studies. *Eur. J. Clin. Nutr.* **2018**, *72*, 57–65. [CrossRef] [PubMed]
5. Aune, D.; Keum, N.; Giovannucci, E.; Fadnes, L.T.; Boffetta, P.; Greenwood, D.C.; Tonstad, S.; Vatten, L.J.; Riboli, E.; Norat, T. Whole grain consumption and risk of cardiovascular disease, cancer, and all cause and cause specific mortality: Systematic review and dose-response meta-analysis of prospective studies. *BMJ* **2016**, *353*, i2716. [CrossRef]
6. Maki, K.C.; Palacios, O.M.; Koecher, K.; Sawicki, C.M.; Livingston, K.A.; Bell, M.; Nelson Cortes, H.; McKeown, N.M. The relationship between whole grain intake and body weight: Results of meta-analyses of observational studies and randomized controlled trials. *Nutrients* **2019**, *11*, 1245. [CrossRef] [PubMed]
7. Schlesinger, S.; Neuenschwander, M.; Schwedhelm, C.; Hoffmann, G.; Bechthold, A.; Boeing, H.; Schwingshackl, L. Food groups and risk of overweight, obesity, and weight gain: A systematic review and dose-response meta-analysis of prospective studies. *Adv. Nutr.* **2019**, *10*, 205–218. [CrossRef] [PubMed]
8. Chanson-Rolle, A.; Meynier, A.; Aubin, F.; Lappi, J.; Poutanen, K.; Vinoy, S.; Braesco, V. Systematic review and meta-analysis of human studies to support a quantitative recommendation for whole grain intake in relation to type 2 diabetes. *PLoS ONE* **2015**, *10*, e0131377. [CrossRef] [PubMed]
9. Aune, D.; Norat, T.; Romundstad, P.; Vatten, L.J. Whole grain and refined grain consumption and the risk of type 2 diabetes: A systematic review and dose-response meta-analysis of cohort studies. *Eur. J. Epidemiol.* **2013**, *28*, 845–858. [CrossRef] [PubMed]
10. Ye, E.Q.; Chacko, S.A.; Chou, E.L.; Kugizaki, M.; Liu, S. Greater whole-grain intake is associated with lower risk of type 2 diabetes, cardiovascular disease, and weight gain. *J. Nutr.* **2012**, *142*, 1304–1313. [CrossRef] [PubMed]
11. Kelly, S.A.; Hartley, L.; Loveman, E.; Colquitt, J.L.; Jones, H.M.; Al-Khudairy, L.; Clar, C.; Germanò, R.; Lunn, H.R.; Frost, G.; et al. Whole grain cereals for the primary or secondary prevention of cardiovascular disease. *Cochrane Database Syst. Rev.* **2017**, *8*, CD005051. [CrossRef]
12. Benisi-Kohansal, S.; Saneei, P.; Salehi-Marzijarani, M.; Larijani, B.; Esmaillzadeh, A. Whole-grain intake and mortality from all causes, cardiovascular disease, and cancer: A systematic review and dose-response meta-analysis of prospective cohort studies. *Adv. Nutr.* **2016**, *7*, 1052–1065. [CrossRef] [PubMed]
13. Jacobs, D.R.; Gallaher, D.D. Whole grain intake and cardiovascular disease: A review. *Curr. Atheroscler. Rep.* **2004**, *6*, 415–423. [CrossRef] [PubMed]
14. Xie, M.; Liu, J.; Tsao, R.; Wang, Z.; Sun, B.; Wang, J. Whole grain consumption for the prevention and treatment of breast cancer. *Nutrients* **2019**, *11*, 1769. [CrossRef] [PubMed]
15. Vieira, A.R.; Abar, L.; Chan, D.S.M.; Vingeliene, S.; Polemiti, E.; Stevens, C.; Greenwood, D.; Norat, T. Foods and beverages and colorectal cancer risk: A systematic review and meta-analysis of cohort studies, an update of the evidence of the WCRF-AICR continuous update project. *Ann. Oncol.* **2017**, *28*, 1788–1802. [CrossRef] [PubMed]
16. Slavin, J. Why whole grains are protective: Biological mechanisms. *Proc. Nutr. Soc.* **2003**, *62*, 129–134. [CrossRef] [PubMed]
17. Ross, A.B.; Kamal-Eldin, A.; Aman, P. Dietary alkylresorcinols: Absorption, bioactivities, and possible use as biomarkers of whole-grain wheat- and rye-rich foods. *Nutr. Rev.* **2004**, *62*, 81–95. [CrossRef] [PubMed]
18. Di Renzo, L.; Gualtieri, P.; Romano, L.; Marrone, G.; Noce, A.; Pujia, A.; Perrone, M.A.; Aiello, V.; Colica, C.; De Lorenzo, A. Role of personalized nutrition in chronic-degenerative diseases. *Nutrients* **2019**, *11*, 1707. [CrossRef]
19. Amann, A.; Costello Bde, L.; Miekisch, W.; Schubert, J.; Buszewski, B.; Pleil, J.; Ratcliffe, N.; Risby, T. The human volatilome: Volatile organic compounds (VOCs) in exhaled breath, skin emanations, urine, feces and saliva. *J. Breath Res.* **2014**, *8*, 034001. [CrossRef] [PubMed]
20. Van Malderen, K.; De Winter, B.Y.; De Man, J.G.; De Schepper, H.U.; Lamote, K. Volatomics in inflammatory bowel disease and irritable bowel syndrome. *EBioMed.* **2020**, *54*, 102725. [CrossRef] [PubMed]

21. Sinha, R.; Lockman, K.A.; Homer, N.Z.M.; Bower, E.; Brinkman, P.; Knobel, H.H.; Fallowfield, J.A.; Jaap, A.J.; Hayes, P.C.; Plevris, J.N. Volatomic analysis identifies compounds that can stratify non-alcoholic fatty liver disease. *JHEP Rep.* **2020**, *2*, 100137. [CrossRef] [PubMed]
22. Rondanelli, M.; Perdoni, F.; Infantino, V.; Faliva, M.A.; Peroni, G.; Iannello, G.; Nichetti, M.; Alalwan, T.A.; Perna, S.; Cocuzza, C. Volatile organic compounds as biomarkers of gastrointestinal diseases and nutritional status. *J. Anal. Methods Chem.* **2019**, *2019*, 7247802. [CrossRef]
23. Ajibola, O.A.; Smith, D.; Spaněl, P.; Ferns, G.A. Effects of dietary nutrients on volatile breath metabolites. *J. Nutr. Sci.* **2013**, *2*, e34. [CrossRef]
24. Rattray, N.J.W.; Hamrang, Z.; Trivedi, D.K.; Goodacre, R.; Fowler, S.J. Taking your breath away: Metabolomics breathes life in to personalized medicine. *Trends Biotechnol.* **2014**, *32*, 538–548. [CrossRef]
25. de Lacy Costello, B.; Amann, A.; Al-Kateb, H.; Flynn, C.; Filipiak, W.; Khalid, T.; Osborne, D.; Ratcliffe, N.M. A review of the volatiles from the healthy human body. *J. Breath Res.* **2014**, *8*, 014001. [CrossRef]
26. Phillips, M.; Herrera, J.; Krishnan, S.; Zain, M.; Greenberg, J.; Cataneo, R.N. Variation in volatile organic compounds in the breath of normal humans. *J. Chromatogr. B Biomed. Sci. Appl.* **1999**, *729*, 75–88. [CrossRef]
27. Pleil, J.D.; Stiegel, M.A.; Risby, T.H. Clinical breath analysis: Discriminating between human endogenous compounds and exogenous (environmental) chemical confounders. *J. Breath Res.* **2013**, *7*, 017107. [CrossRef] [PubMed]
28. Smith, D.; Španěl, P.; Davies, S. Trace gases in breath of healthy volunteers when fasting and after a protein-calorie meal: A preliminary study. *J. Appl. Physiol.* **1999**, *87*, 1584–1588. [CrossRef] [PubMed]
29. Turner, C.; Parekh, B.; Walton, C.; Spaněl, P.; Smith, D.; Evans, M. An exploratory comparative study of volatile compounds in exhaled breath and emitted by skin using selected ion flow tube mass spectrometry. *Rapid Commun. Mass Spectrom.* **2008**, *22*, 526–532. [CrossRef] [PubMed]
30. Spaněl, P.; Dryahina, K.; Rejskova, A.; Chippendale, T.W.; Smith, D. Breath Acetone Concentration; Biological Variability and the Influence of Diet. *Physiol. Meas.* **2011**, *32*, N23–N31. [CrossRef] [PubMed]
31. Dowlaty, N.; Yoon, A.; Galassetti, P. Monitoring states of altered carbohydrate metabolism via breath analysis: Are times ripe for transition from potential to reality? *Curr. Opin. Clin. Nutr. Metab. Care* **2013**, *16*, 466–472. [CrossRef] [PubMed]
32. Minh, T.D.; Blake, D.R.; Galassetti, P.R. The clinical potential of exhaled breath analysis for diabetes mellitus. *Diabetes Res. Clin. Pract.* **2012**, *97*, 195–205. [CrossRef]
33. Novak, B.J.; Blake, D.R.; Meinardi, S.; Rowland, F.S.; Pontello, A.; Cooper, D.M.; Galassetti, P.R. Exhaled Methyl nitrate as a noninvasive marker of hyperglycemia in type 1 diabetes. *Proc. Natl. Acad. Sci. USA* **2007**, *104*, 15613–15618. [CrossRef]
34. Minh, T.D.; Oliver, S.R.; Ngo, J.; Flores, R.; Midyett, J.; Meinardi, S.; Carlson, M.K.; Rowland, F.S.; Blake, D.R.; Galassetti, P.R. Noninvasive measurement of plasma glucose from exhaled breath in healthy and type 1 diabetic subjects. *Am. J. Physiol. Endocrinol. Metab.* **2011**, *300*, E1166–E1175. [CrossRef]
35. Lee, J.; Ngo, J.; Blake, D.; Meinardi, S.; Pontello, A.M.; Newcomb, R.; Galassetti, P.R. Improved predictive models for plasma glucose estimation from multi-linear regression analysis of exhaled volatile organic compounds. *J. Appl. Physiol.* **2009**, *107*, 155–160. [CrossRef] [PubMed]
36. Galassetti, P.R.; Novak, B.; Nemet, D.; Rose-Gottron, C.; Cooper, D.M.; Meinardi, S.; Newcomb, R.; Zaldivar, F.; Blake, D.R. Breath ethanol and acetone as indicators of serum glucose levels: An initial report. *Diabetes Technol. Ther.* **2005**, *7*, 115–123. [CrossRef] [PubMed]
37. Minh, T.D.; Oliver, S.R.; Flores, R.L.; Ngo, J.; Meinardi, S.; Carlson, M.K.; Midyett, J.; Rowland, F.S.; Blake, D.R.; Galassetti, P.R. Noninvasive measurement of plasma triglycerides and free fatty acids from exhaled breath. *J. Diabetes Sci. Technol.* **2012**, *6*, 86–101. [CrossRef]
38. Baranska, A.; Mujagic, Z.; Smolinska, A.; Dallinga, J.W.; Jonkers, D.M.; Tigchelaar, E.F.; Dekens, J.; Zhernakova, A.; Ludwig, T.; Masclee, A.A.; et al. Volatile organic compounds in breath as markers for irritable bowel syndrome: A metabolomic approach. *Aliment. Pharmacol. Ther.* **2016**, *44*, 45–56. [CrossRef] [PubMed]
39. Bodelier, A.G.; Smolinska, A.; Baranska, A.; Dallinga, J.W.; Mujagic, Z.; Vanhees, K.; van den Heuvel, T.; Masclee, A.A.; Jonkers, D.; Pierik, M.J.; et al. Volatile organic compounds in exhaled air as novel marker for disease activity in crohn's disease: A metabolomic approach. *Inflamm. Bowel Dis.* **2015**, *21*, 1776–1785. [CrossRef]
40. Van Berkel, J.J.; Dallinga, J.W.; Moller, G.M.; Godschalk, R.W.; Moonen, E.J.; Wouters, E.F.; Van Schooten, F.J. A profile of volatile organic compounds in breath discriminates COPD patients from controls. *Respir. Med.* **2010**, *104*, 557–563. [CrossRef] [PubMed]
41. Baranska, A.; Tigchelaar, E.; Smolinska, A.; Dallinga, J.W.; Moonen, E.J.; Dekens, J.A.; Wijmenga, C.; Zhernakova, A.; van Schooten, F.J. Profile of volatile organic compounds in exhaled breath changes as a result of gluten-free diet. *J. Breath Res.* **2013**, *7*, 037104. [CrossRef]
42. Raninen, K.; Lappi, J.; Kolehmainen, M.; Kolehmainen, M.; Mykkänen, H.; Poutanen, K.; Raatikainen, O. Diet-derived changes by sourdough-fermented rye bread in exhaled breath aspiration ion mobility spectrometry profiles in individuals with mild gastrointestinal symptoms. *Int. J. Food Sci. Nutr.* **2017**, *68*, 987–996. [CrossRef] [PubMed]
43. Raninen, K.J.; Kolehmainen, M.; Tuomainen, T.; Mykkänen, H.; Poutanen, K.; Raatikainen, O. Exhaled breath aspiration ion mobility spectrometry profiles reflect metabolic changes induced by diet. *J. Physiobiochem. Metab.* **2015**, *3*. [CrossRef]
44. Mathew, T.L.; Pownraj, P.; Abdulla, S.; Pullithadathil, B. Technologies for Clinical Diagnosis using Expired Human Breath Analysis. *Diagnostics* **2015**, *5*, 27–60. [CrossRef]

45. Phillips, M.; Cataneo, R.N.; Chaturvedi, A.; Kaplan, P.D.; Libardoni, M.; Mundada, M.; Patel, U.; Zhang, X. Detection of an extended human volatome with comprehensive two-dimensional gas chromatography time-of-flight mass spectrometry. *PLoS ONE* **2013**, *8*, e75274. [CrossRef]
46. Raninen, K.J.; Lappi, J.E.; Mukkala, M.L.; Tuomainen, T.; Mykkänen, H.M.; Poutanen, K.S.; Raatikainen, O.J. Fiber content of diet affects exhaled breath volatiles in fasting and postprandial state in a pilot crossover study. *Nutr. Res.* **2016**, *36*, 612–619. [CrossRef]
47. Fischer, S.; Bergmann, A.; Steffens, M.; Trefz, P.; Ziller, M.; Miekisch, W.; Schubert, J.S.; Kohler, H.; Reinhold, P. Impact of food intake on in vivo voc concentrations in exhaled breath assessed in a caprine animal model. *J. Breath Res.* **2015**, *9*, 047113. [CrossRef] [PubMed]
48. Mochalski, P.; King, J.; Haas, M.; Unterkofler, K.; Amann, A.; Mayer, G. Blood and breath profiles of volatile organic compounds in patients with end-stage renal disease. *BMC Nephrol.* **2014**, *15*, 43. [CrossRef]
49. Marco, E.; Grimalt, J.O. A rapid method for the chromatographic analysis of volatile organic compounds in exhaled breath of tobacco cigarette and electronic cigarette smokers. *J. Chromatogr. A* **2015**, *1410*, 51–59. [CrossRef]
50. Bajtarevic, A.; Ager, C.; Pienz, M.; Klieber, M.; Schwarz, K.; Ligor, M.; Ligor, T.; Filipiak, W.; Denz, H.; Fiegl, M.; et al. Noninvasive detection of lung cancer by analysis of exhaled breath. *BMC Cancer* **2009**, *9*, 348. [CrossRef]
51. Salerno-Kennedy, R.; Cashman, K.D. Potential applications of breath isoprene as a biomarker in modern medicine: A concise overview. *Wien. Klin. Wochenschr.* **2005**, *117*, 180–186. [CrossRef]
52. King, J.; Mochalski, P.; Unterkofler, K.; Teschl, G.; Klieber, M.; Stein, M.; Amann, A.; Baumann, M. Breath isoprene: Muscle dystrophy patients support the concept of a pool of isoprene in the periphery of the human body. *Biochem. Biophys. Res. Commun.* **2012**, *423*, 526–530. [CrossRef]
53. Hibbard, T.; Killard, A.J. Breath ammonia levels in a normal human population study as determined by photoacoustic laser spectroscopy. *J. Breath Res.* **2011**, *5*, 037101. [CrossRef] [PubMed]
54. Hwang, L.; Low, K.; Khoshini, R.; Melmed, G.; Sahakian, A.; Makhani, M.; Pokkunuri, V.; Pimentel, M. Evaluating breath methane as a diagnostic test for constipation-predominant IBS. *Dig. Dis. Sci.* **2010**, *55*, 398–403. [CrossRef]
55. Wong, J.M.; de Souza, R.; Kendall, C.W.; Emam, A.; Jenkins, D.J. Colonic health: Fermentation and short chain fatty acids. *J. Clin. Gastroenterol.* **2006**, *40*, 235–243. [CrossRef]
56. Koistinen, V.M.; Hanhineva, K. Microbial and endogenous metabolic conversions of rye phytochemicals. *Mol. Nutr. Food Res.* **2017**, *61*, 1600627. [CrossRef]
57. Nordlund, E.; Aura, A.M.; Mattila, I.; Kosso, T.; Rouau, X.; Poutanen, K. Formation of phenolic microbial metabolites and short-chain fatty acids from rye, wheat, and oat bran and their fractions in the metabolical in vitro colon model. *J. Agric. Food Chem.* **2012**, *60*, 8134–8145. [CrossRef] [PubMed]
58. Di Gilio, A.; Palmisani, J.; Ventrella, G.; Facchini, L.; Catino, A.; Varesano, N.; Pizzutilo, P.; Galetta, D.; Borelli, M.; Barbieri, P.; et al. Breath analysis: Comparison among methodological approaches for breath sampling. *Molecules* **2020**, *25*, 5823. [CrossRef] [PubMed]
59. Braden, B. Methods and functions: Breath tests. *Best Pract. Res. Clin. Gastroenterol.* **2009**, *23*, 337–352. [CrossRef] [PubMed]
60. Schauer, J.J.; Kleeman, M.J.; Cass, G.R.; Simoneit, B.R.T. Measurement of emissions from air pollution sources. 1. C_1 through C_{29} organic compounds from meat charbroiling. *Environ. Sci. Technol.* **1999**, *33*, 1566–1577. [CrossRef]
61. Schauer, J.J.; Kleeman, M.J.; Cass, G.R.; Simoneit, B.R.T. Measurement of emissions from air pollution sources. 4. C_1-C_{27} organic compounds from cooking with seed oils. *Environ. Sci. Technol.* **2002**, *36*, 567–575. [CrossRef] [PubMed]
62. Heinonen, S.; Nurmi, T.; Liukkonen, K.; Poutanen, K.; Wahala, K.; Deyama, T.; Nishibe, S.; Adlercreutz, H. In vitro metabolism of plant lignans: New precursors of mammalian lignans enterolactone and enterodiol. *J. Agric. Food Chem.* **2001**, *49*, 3178–3186. [CrossRef] [PubMed]
63. Hallmans, G.; Zhang, J.X.; Lundin, E.; Stattin, P.; Johansson, A.; Johansson, I.; Hulten, K.; Winkvist, A.; Aman, P.; Lenner, P.; et al. Rye, lignans and human health. *Proc. Nutr. Soc.* **2003**, *62*, 193–199. [CrossRef] [PubMed]
64. Hålldin, E.; Eriksen, A.K.; Brunius, C.; da Silva, A.B.; Bronze, M.; Hanhineva, K.; Aura, A.M.; Landberg, R. Factors explaining interpersonal variation in plasma enterolactone concentrations in humans. *Mol. Nutr. Food Res.* **2019**, *63*, e1801159. [CrossRef] [PubMed]

Article

Comparison of Targeted and Untargeted Approaches in Breath Analysis for the Discrimination of Lung Cancer from Benign Pulmonary Diseases and Healthy Persons

Michalis Koureas [1], Dimitrios Kalompatsios [1], Grigoris D. Amoutzias [2], Christos Hadjichristodoulou [1], Konstantinos Gourgoulianis [3] and Andreas Tsakalof [1,4,*]

[1] Department of Hygiene and Epidemiology, University Hospital of Larissa, Faculty of Medicine, University of Thessaly, 22 Papakyriazi Street, 41222 Larissa, Greece; mkoureas@med.uth.gr (M.K.); thejimkal@gmail.com (D.K.); xhatzi@med.uth.gr (C.H.)

[2] Bioinformatics Laboratory, Department of Biochemistry and Biotechnology, University of Thessaly, 41500 Larissa, Greece; amoutzias@bio.uth.gr

[3] Respiratory Medicine Department, University Hospital of Larissa, Faculty of Medicine, University of Thessaly, 41110 Larissa, Greece; kgourg@uth.gr

[4] Department of Biochemistry, Faculty of Medicine, University of Thessaly, 41500 Larissa, Greece

* Correspondence: atsakal@med.uth.gr; Tel.: +30-2410685580

Citation: Koureas, M.; Kalompatsios, D.; Amoutzias, G.D.; Hadjichristodoulou, C.; Gourgoulianis, K.; Tsakalof, A. Comparison of Targeted and Untargeted Approaches in Breath Analysis for the Discrimination of Lung Cancer from Benign Pulmonary Diseases and Healthy Persons. *Molecules* **2021**, *26*, 2609. https://doi.org/10.3390/molecules26092609

Academic Editors: Natalia Drabińska and Ben de Lacy Costello

Received: 28 March 2021
Accepted: 27 April 2021
Published: 29 April 2021

Publisher's Note: MDPI stays neutral with regard to jurisdictional claims in published maps and institutional affiliations.

Abstract: The aim of the present study was to compare the efficiency of targeted and untargeted breath analysis in the discrimination of lung cancer (Ca+) patients from healthy people (HC) and patients with benign pulmonary diseases (Ca−). Exhaled breath samples from 49 Ca+ patients, 36 Ca− patients and 52 healthy controls (HC) were analyzed by an SPME–GC–MS method. Untargeted treatment of the acquired data was performed with the use of the web-based platform XCMS Online combined with manual reprocessing of raw chromatographic data. Machine learning methods were applied to estimate the efficiency of breath analysis in the classification of the participants. Results: Untargeted analysis revealed 29 informative VOCs, from which 17 were identified by mass spectra and retention time/retention index evaluation. The untargeted analysis yielded slightly better results in discriminating Ca+ patients from HC (accuracy: 91.0%, AUC: 0.96 and accuracy 89.1%, AUC: 0.97 for untargeted and targeted analysis, respectively) but significantly improved the efficiency of discrimination between Ca+ and Ca− patients, increasing the accuracy of the classification from 52.9 to 75.3% and the AUC from 0.55 to 0.82. Conclusions: The untargeted breath analysis through the inclusion and utilization of newly identified compounds that were not considered in targeted analysis allowed the discrimination of the Ca+ from Ca− patients, which was not achieved by the targeted approach.

Keywords: lung cancer; exhaled breath; volatile organic compounds; untargeted analysis; breath analysis; cancer biomarkers; volatolomics

1. Introduction

Human breath contains volatile organic compounds (VOCs) either originating from endogenous biochemical processes and thus distinguished as endogenous VOCs or environmental exposures (inhalation, ingestion, dermal absorption) and therefore pertaining to exogeneous VOCs. In case of disease, the biochemical pathways can be dysregulated or altered [1], and this will change the composition of exhaled breath in endogenous VOCs. Moreover, disease can also affect the absorption, distribution metabolism and excretion of the exogenous compounds. These alterations can be detected and used for disease detection and diagnosis. The analysis of exhaled breath is currently an area of intensive research aiming at the development of new non-invasive tests for preliminary screening and diagnosis of various pathological conditions. Particular attention is given to cancer, where early diagnosis is critical for successful disease treatment and which today is often diagnosed at late stages, and diagnosis procedures are invasive, time consuming or costly. Mass spectrometry (MS)-based breath analysis for disease diagnosis research is currently the mainstream choice that can be accomplished using two strategies, which are classified as targeted or non-targeted (also referred to as untargeted). The former is based on quantification of an a priori defined set of VOCs known or hypothesized as disease biomarkers and is thus a hypothesis-driven approach. In contrast, the non-targeted strategy is a (qualitative) hypothesis-generating approach that investigates the whole VOC profile in a breath sample without any a priori information about the chemical composition of the sample and aims to identify a maximum number of VOCs. By non-targeted breath analysis, novel biomarkers and disturbed metabolic pathways can be discovered or characteristic breath VOC profile of the disease can be defined and further used for disease detection and diagnosis. However, the non-targeted approach yields a huge amount of complex data and its application would be impossible without the development of bioinformatics software designed for the treatment and statistical analysis of raw chromatography–mass spectrometry data, and identification of detected unknown compounds. This has been done mostly in the last decade and currently there is a variety of commercial or open source software for the treatment and analysis of chromatography–mass spectrometry data and extraction of the relative biological information [2]. That has given great impetus for the development of non-targeted analysis in metabolomics in general [3] and opens new perspectives in breath research in particular [4]. One of the most widely used metabolomic software is XCMS Online, which is freely available [5].

However, the non-targeted approach has long-standing reproducibility issues [6,7] and is never truly unbiased since the acquired data are significantly affected by experimental design and instrumental parameters. In contrast to the targeted strategy, the lack of absolute quantification makes it difficult to assess variations in metabolite levels between groups, to normalize the acquired data and even to make interlaboratory comparisons of the results [7,8]. These weaknesses of the non-targeted approach are, at the same time, the strengths of the targeted approach and, recently, hybrid approaches bridging them have been developed [8,9]. In this study, we make a retrospective non-targeted analysis of full scan data previously acquired [10] in targeted analysis of the breath samples from lung cancer (Ca+) and benign pulmonary disease (Ca−) patients and healthy controls (HC). The targeted analysis was based on the quantitation of 19 pre-determined VOCs [10]. While Ca+ patients were satisfactorily discriminated from healthy controls, the analysis failed to discriminate Ca+ patients from Ca− patients (without LC but with pathological computed tomography findings). The aim of the present study is to compare the efficiency of the targeted and untargeted approaches in lung cancer discrimination with healthy people and patients with other pulmonary diseases and record the strengths and limitations of each approach on the same raw GC–MS data pool. Additionally, by merging (combining) targeted and untargeted approaches, we sought to improve the discrimination ability of the breath analysis.

2. Results

2.1. Characteristics of Study Participants

From the 85 patients with pathological computed tomography (CT) findings who underwent bronchoscopy, lung cancer was diagnosed in 49 patients (43 males/6 females). The mean age of Ca+ patients was 71.1 years (SD: 8.2). The majority of LC patients (n = 40) were diagnosed with non-small cell lung carcinoma, while 8 were diagnosed with small cell lung carcinoma (for one patient, the type was not available). Thirty-six patients (30 males/6 females, mean age 66.8 (SD: 10.8)) were not diagnosed with LC by histological/cytological examination. The possible pathological origins for this group include sarcoidosis, hypersensitivity pneumonitis, interstitial lung diseases or pulmonary infections such as tuberculosis. The control group consisted of 52 persons (35 males/17 females) with a mean age of 66.8 (SD: 10.8).

In regard to smoking habit, most of the LC patients (81.6%) were former smokers with a mean time from cessation of 9.4 years, while 12.2% were active smokers and 6.1% reported that they had never smoked. Patients that were not diagnosed with LC had slightly different frequencies of smoking habit, with 55.6% being former smokers (mean time from cessation: 10.6 years), 27.7% being active smokers and 16.7% never smokers. In the HC group, the percentage of active smokers was significantly higher (38.4%), as was the percentage of individuals that had never smoked (28.9%). The percentage of former smokers was 32.7%, with mean time from cessation of 20.1 years. Mean pack/years were 69.43 (SD: 48.47) for the Ca+ group, 48.70 (SD: 35.41) for the Ca− group and 32.74 (SD: 33.39) for the HC group.

Concerning self-reported co-morbidities derived from personal interviews with the use of questionnaires, the most common were hypertension (Ca+ group: 44.9%, Ca− group 47.22%, HC group: 42.31%), diabetes (Ca+ group: 24.49%, Ca− group 27.78%, HC group: 22.45%) and hypercholesterolemia (Ca+ group: 38.78%, Ca− group 30.56%, HC group: 26.53%).

2.2. Data Pre-Processing, Selection and Identification of Candidate Features

The processing of raw files with the use of the XCMS Online platform identified 358 informative features (ions) meeting the criteria defined in the Materials and Methods (Section 4.3) after peak identification, alignment, retention time correction and preliminary online statistical analysis. Figure 1 presents the metabolomic cloud plots obtained from XCMS Online, concerning the pairwise analysis of Ca+ vs. HC and Ca+ vs. Ca− groups. Features identified as differentiated between subgroups by XCMS Online were automatically grouped into 110 corresponding chromatographic peaks. These peaks were manually evaluated and verified in the acquired chromatograms. This process resulted in the exclusion of 28 peaks from further analysis due to unacceptable chromatographic characteristics such as low signal to noise ratio and co-elution with other substances. The mass spectra corresponding to the 82 remaining peaks were compared with those stored in the NIST library after subtracting mass spectra corresponding to noise. These procedures lead to the exclusion of additional peaks with spectra indicating silanes and silicon compounds that were considered interferences from SPME fiber, the chromatography column or septum materials. In addition, peaks with mass spectra corresponding to known contaminants from Tedlar® bag materials (phenol, *N,N*-dimethylacetamide) were also excluded [11]. In total, 53 compounds were not considered for further analysis. Thus, the remaining 29 peaks were considered for further investigation. For these, comparisons of mass spectra with those contained in the NIST library identified 12 compounds with a probability higher than 75%. Four monoaromatic compounds (benzene, styrene, ethylbenzene and toluene) were also verified with analytical standards. In addition, seven compounds were verified by retention time (RT) by comparing actual RTs with simulated RTs determined with the use of the Pro EZGC Chromatogram Modeler (Restek Corporation, Bellefonte, PA, USA). For 5 peaks, the NIST probability was 50–75%, indicating a considerable degree of uncertainty in compound identification, while 12 compounds (probability < 50%) were designated as unknowns. Moreover, experimentally determined retention indices (RIs) were compared with those stored in the NIST library. Small deviations were observed (<10%) for most compounds, while the RI values were in agreement with the order of elution of identified VOCs, with the exceptions of propionic acid and methylacetamide. Figure 2 presents the flow chart of the process applied for selecting and identifying informative compounds. In Table 1, the compounds are presented along with NIST probability and spectra match scores, actual and simulated retention times, experimentally determined RIs and RIs derived from the NIST workbook. The 17 identified compounds were further investigated by searching for their presence in the KEGG pathway database [12] and in the scientific literature to determine their putative origins and the involved metabolic pathways. For twelve compounds, no evidence of endogenous origin was found. These include monoaromatic hydrocarbons and furans, which are carcinogens contained in tobacco smoke, and produced by industrial sources and commercial uses, sulfur-containing compounds (methyl propyl sulfide, 1-methylthio-(E)-1-propene) used as flavor agents and contained in garlic and onion and eucalyptol, which is used as an asthma/COPD drug. Eight substances could be of both endogenous and exogenous origin. Most of the identified metabolic pathways concerned the degradation/metabolism of xenobiotic substances such as ethylbenzene, benzene and dimethyacetamide. Propionic acid is involved in multiple pathways of lipid biosynthesis, propanoate metabolism and vitamin K metabolism. P-benzoquinone can be formed from benzene metabolism [13], but also participates in other pathways, and acetic acid is involved in the formation of glycogen, cholesterol synthesis, fatty acid degradation and acetylation of amines [14].

Table 1. Identification of compounds based on spectra comparison with NIST library and retention time criteria.

Candidate Compound	Probability (NIST), %	Match Score (NIST)	Retention Time, min	Retention Time Simulated [1], min	Deviations in Retention Time, %	Experimentally Determined Retention Index	NIST Retention Index [2]	Deviations in Retention Index, %
3-methyl-furane	86	892	5.03	NA		615	602	2.16
acetaldoxime	53	753	5.38	NA		625	606	3.14
Benzene *	72	923	7.17	7.73	−7.81	677	647	4.64
acetic acid	59	912	7.86	NA		698	650	7.38
1-methoxy-2-propanol	69	891	8.27	7.83	5.32	711	658	8.05
dimethyl furane	78	852	8.33	8.66	−3.96	714	694	2.88
methyl propyl sulfide	89	840	8.79	NA		729	714	2.10
1-methylthio-(E)-1-propene **	90	877	9.57	NA		756	722	4.71
Toluene *	34	868	10.31	10.95	−6.21	782	750	4.27
propionic acid	78	702	10.59	NA		792	712	11.24
p- xylene **	81	845	12.00	12.1	−0.83	859	833	3.12
ethyl benzene *	61	877	12.10	12.38	−2.31	890	858	3.73
Styrene *	37	869	12.3	12.79	−3.98	906	876	3.42
methylacetamide	78	831	12.79	NA		959	825	16.24
p-benzoquinone	90	817	13.00	NA		982	888	10.59
N-2-Aminoethyl acetamide	62	800	13.2	NA		1005	NA	
eucalyptol	52	848	13.51	NA		1060	1017	4.23

* Verified by analytical standard. ** NIST probability is given for all isomer compounds. Mass spectra were very similar for isomers of these compounds, compounds were identified based on RI similarities. [1] Retention time was simulated with Pro EZGC Chromatogram Modeler, Restek Corporation. [2] Retention indices were derived from NIST database related to a fully non-polar column (100% polydimethylsiloxane). NA: not available with equivalent column.

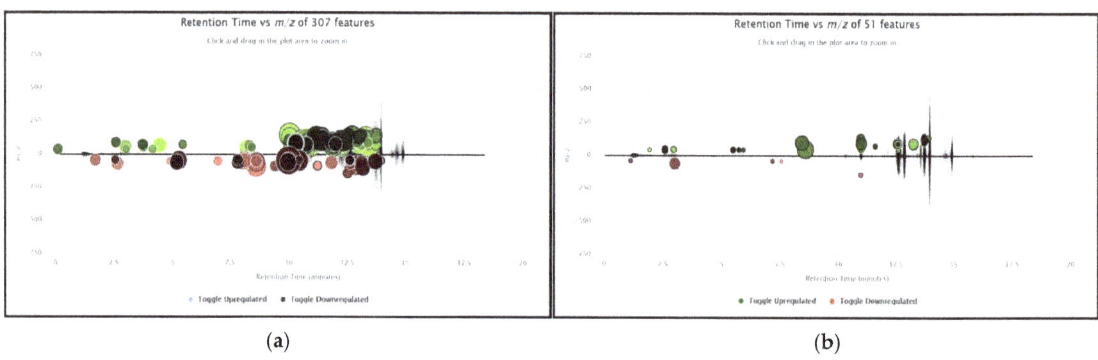

(a) (b)

Figure 1. Cloud plots with results of pairwise XCMS analysis between (**a**) Ca+ vs. HC. Detection settings: *p*-value < 0.01, fold change > 2, *m/z* range: 0–140, retention range: 0–14 min, max intensity > 10,000 and (**b**) Ca+ vs. Ca− characteristics (ion). Detection settings: *p*-value < 0.05, fold change > 1.1, *m/z* range: 0–140, retention range: 0–14 min, max intensity > 10,000.

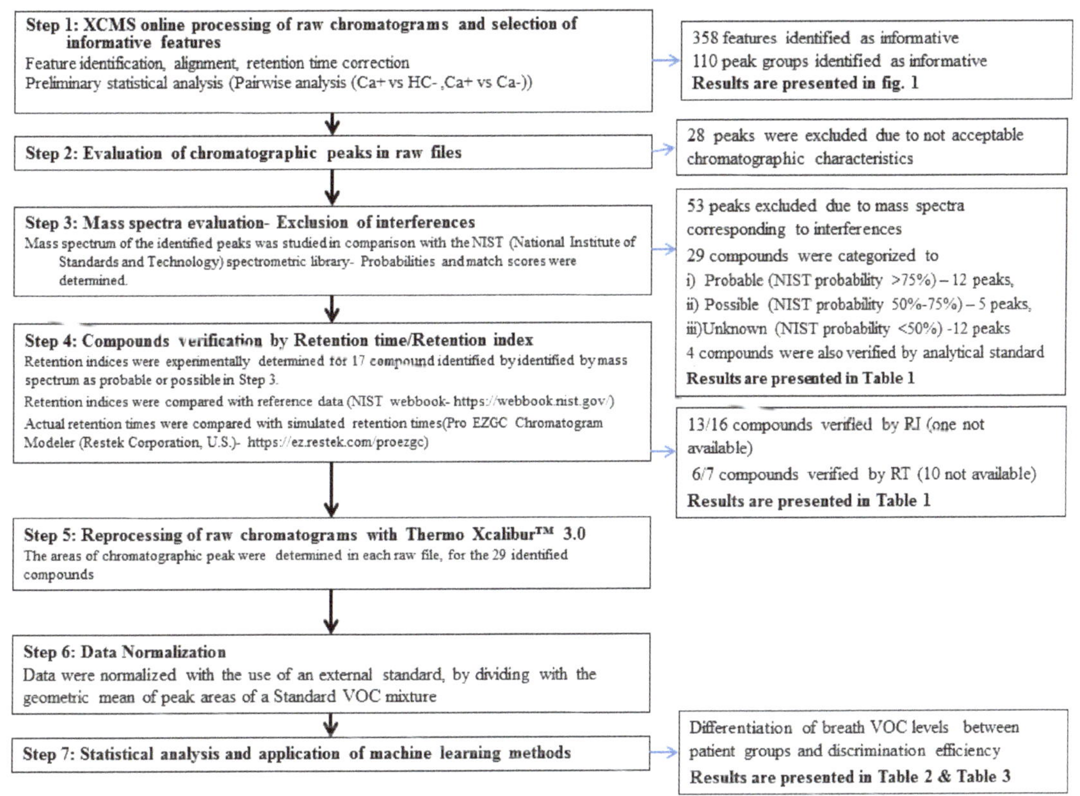

Figure 2. Flow chart of the process applied for selecting, identifying and processing informative compounds.

2.3. Reprocessing of Raw Chromatographs and Statistical Analysis of Identified/Verified Associations

Following the identification of the compounds, all raw files were reprocessed with Thermo Xcalibur™ software to obtain more valid data. This procedure allowed manual retention time correction, more accurate integration of chromatographic peaks and exclusion of false (noise) peaks. The areas of the chromatographic peaks were determined for each compound in exhaled breath samples but also in ambient air samples. Chromatographic peak areas were normalized with the use of an external standard mixture (see Section 4.4). Regarding ambient air levels, for six out of 29 compounds, the relative levels of ambient air were considered insignificant, for 5 compounds low, for 14 compounds moderate and for 4 compounds high (Table 2). Comparative statistical analysis confirmed the significant difference in breath levels between Ca+ patients and healthy controls for 18 out 29 compounds, while two were found to differ between Ca+ and Ca− patients. Lung cancer patients had significantly elevated levels of ethylbenzene, styrene, toluene, xylene, eucalyptol and four unknown compounds compared to healthy controls. Lower levels were observed for acetaldoxime, methyl propyl sulfide, 1-methylthio-(*E*)-1-propene, propionic acid, methylacetamide and three unknown compounds. Results concerning the comparative analysis of areas of chromatographic peaks between patient groups are summarized in Table 2.

Table 2. Comparative analysis of the areas of the 29 chromatographic peaks between patient groups and relative presence in ambient air.

Compound	Relative Presence in Ambient Air [1]	Ca+/HC		Ca+/Ca−	
		Trend in LC Patients	Significance *	Trend in LC Patients	Significance *
unknown	insignificant	↑	0.052	↑	0.311
unknown	moderate	↓	0.071	↓	0.056
3-methyl-furan *	low	↓	0.514	↓	0.482
acetaldoxime	high	↓↓↓	<0.001	↓	0.341
unknown	moderate	↓↓↓	<0.001	↓	0.689
unknown	low	↑↑	0.01	↓↓	0.013
benzene	moderate	↓↓↓	<0.001	↓	0.089
unknown	moderate	↓↓↓	<0.001	↓	0.756
acetic acid	low	↓↓↓	<0.001	↓	0.979
1-methoxy-2-propanol	high	↓↓↓	<0.001	↓	0.272
dimethyl furan	low	↓	0.125	↓	0.286
unknown	moderate	↑↑↑	0.002	↓	0.396
unknown	moderate	↓	0.902	↓	0.082
methyl propyl sulfide	insignificant	↓↓↓	<0.001	↓↓	0.035
1-methylthio-(E)-1-propene	insignificant	↓↓↓	<0.001	↓	0.239
unknown	insignificant	↓↓↓	<0.001	↓	0.185
toluene	moderate	↑↑↑	0.001	↑	0.986
propionic acid	insignificant	↓↓↓	<0.001	↑	0.384
unknown	high	↑	0.053	↑	0.752
unknown	moderate	↓	0.124	↓	0.175
ethylbenzene	moderate	↑↑↑	<0.001	↑	0.618
xylene(p,o,m)	moderate	↑↑↑	<0.001	↑	0.434
styrene	moderate	↑↑↑	<0.001	↑	0.423
methylacetamide	high	↓	0.178	↑	0.539
p-benzoquinone	insignificant	↓	0.076	↓	0.388
N-2-Aminoethyl acetamide	moderate	↓↓↓	<0.001	↓	0.824
unknown	moderate	↓↓↓	<0.001	↓	0.104
eucalyptol	low	↑	0.066	↑	0.511
unknown	moderate	↑	0.092	↑	0.463

[1] Determined from mean breath/mean air ratio. Insignificant: >20, low: 5–20, moderate: 0.5–5, high: <0.5. * Significance determined by Mann–Whitney test. ↑, ↓: $p > 0.05$, ↑↑, ↓↓: $p = 0.01$–0.05, ↓↓↓, ↑↑↑: $p < 0.01$.

2.4. Application of Machine Learning Methods to Estimate the Diagnostic Efficiency of the Breath Analysis

In our previous work, based on 19 selected VOCs, we identified subsets of features (VOCs) that were capable of efficiently discriminating healthy individuals from cancer patients, but not Ca+ from Ca− patients. In this section, we present the results of machine learning methods based on combinations of the 29 features, identified as differentiated between population subgroups by the untargeted approach. When all 29 features were included, correct classification of Ca+ and HC was 86% (AUC: 0.94) (Table 3, Analysis no. 9). After the two steps of feature selection, using a subset of eight features, the correct classification improved to 91% (AUC: 0.96) (Table 3, Analysis no. 10), which was higher than that of targeted analysis. Similarly, discrimination between Ca− patients and HC was also very efficient. The correct classification of datapoints ranged from 90% (AUC: 0.94), when using all 29 features (Table 3, Analysis no. 11), to 94% (AUC: 0.97) after the two steps of feature selection, using a subset of seven compounds (Table 3, Analysis no. 12). Not surprisingly, discrimination between pooled cancer-positive and non-cancer patients (Ca+ and Ca−) and HC was again very efficient. Overall, machine learning models based on compounds identified as differentiated by the untargeted approach achieved a very comparable if not marginally better accuracy than the targeted approach, when trying to discriminate healthy individuals from any of the three types of patients (cancer, non-cancer, pooled).

Table 3. Results of machine learning methods (random forest) to estimate the discrimination efficiency of the breath analysis.

Analysis no.	Approach	Variable	Comparison Groups	Smoking Habit	Features Used	Accuracy	AUC
1	targeted	Br	Ca+ vs. HC	All	t1–t19	85.14	0.95
2	targeted	Br	Ca+ vs. HC	All	t4, t5, t7–t11,t13–t15,t18	89.10	0.97
3	targeted	Br	Ca− vs. HC	All	t1–t19	86.36	0.91
4	targeted	Br	Ca− vs. HC	All	t4,t5, t7–t17	88.63	0.94
5	targeted	Br	Ca+ & Ca− vs. HC	All	t1–t19	86.70	0.96
6	targeted	Br	Ca+ & Ca− vs. HC	All	t1, t4,t5,t7–t15,t17	90.50	0.96
7	targeted	Br	Ca+ vs. Ca−	All	t1–t19	43.50	0.39
8	targeted	Br	Ca+ vs. Ca−	All	t4,t9, t17	52.90	0.55
9	untargeted	Br	Ca+ vs. HC	All	u1–u29	86.14	0.94
10	untargeted	Br	Ca+ vs. HC	All	u4,u8,u12,u14,u16,u19,u28,u29	91.08	0.96
11	untargeted	Br	Ca− vs. HC	All	u1–u29	89.77	0.94
12	untargeted	Br	Ca− vs. HC	All	u4,u6, u8, u12,u26,u27,u29	94.3	0.97
13	untargeted	Br	Ca+ & Ca− vs. HC	All	u1–u29	86.9	0.95
14	untargeted	Br	Ca+ & Ca− vs. HC	All	u4, u8, u11,u12,u19,u22,u26,u27, u29	92	0.97
15	untargeted	Br	Ca+ vs. Ca−	All	u1–u29	52.9	0.54
16	untargeted	Br	Ca+ vs. Ca−	All	u4, u20,u26	75.3	0.82
17	untargeted	Sbtr	Ca+ vs. Ca−	All	t1–t19, u1–u29	57.6	0.54
18	untargeted	Sbtr	Ca+ vs. Ca−	All	u2, u4,u6, u11,u14, u25, u28,u29	71.76	0.78
19	merged	Br	Ca+ vs. Ca−	All	u1–u29, t1–t19	44.7	0.44
20	merged	Br	Ca+ vs. Ca−	All	t9, u4, u26	72.9	0.72
21	untargeted	Br	Ca+ vs. Ca−	Non-smokers	u1–u29	59.4	0.57
22	untargeted	Br	Ca+ vs. Ca−	Non-smokers	u4, u20,u 26	72.5	0.68
23	untargeted	Br	Ca+ vs. Ca−	Non-smokers	u4, u11, u13,u20,u26	76.8	0.85

Br: corresponds to breath compound levels, Sbtr: corresponds to breath subtract levels, Ca+: patients diagnosed with lung cancer, Ca−: patients with pathological CT findings not diagnosed with lung cancer by histological/cytological examination, HC: healthy controls. Features from targeted analysis: t1: isoprene, t2: acetone, t3: 2-propanol, t4: hexane, t5: 1-propanol, t6: 2-butanone, t7: cyclohexane, t8: benzene, t9: thiophene, t10: 1-butanol, t11: toluene, t12: octane, t13: ethyl butyrate, t14: hexanal, t15: ethyl benzene, t16: styrene, t17: cyclohexanone, t18: octanal, t19: nonanal. Features from untargeted analysis: u1: unknown, u2: unknown, u3: 3-methyl-furan, u4: acetaldoxime, u5: unknown, u6: unknown, u7: benzene, u8: unknown, u9: acetic acid, u10: 1-methoxy-2-propanol, u11: dimethyl furan, u12: unknown, u13: unknown, u14: 1-methylthio-(E)-1-propene, u15: allyl methyl sulfide, u16: unknown, u17: toluene, u18: propionic acid, u19: unknown, u20: unknown, u21: ethylbenzene, u22: p-xylene, u23: styrene, u24: methylacetamide, u25: p-benzoquinone, u26: N-2-aminoacetyl acetamide, u27: unknown, u28: eucalyptol, u29: unknown.

Subsequently, we tested the potential for discrimination between Ca+ and Ca− patients, with the three machine learning algorithms, by using normalized peak areas of compounds from breath. The set of 29 VOCs was not capable of efficiently discriminating between cancer and non-cancer patients, irrespective of the machine learning algorithm applied. The best-performing algorithm (random forest) correctly predicted only 53% of datapoints (AUC: 0.54) (Table 3, Analysis no. 15) when using all 29 VOCs. However, when two successive steps of feature selection were implemented, the random forest's accuracy significantly increased to 75% (AUC: 0.82), by using a set of only three metabolites (Table 3, Analysis no. 16). We repeated the analysis to discriminate Ca+ from Ca− patients, by incorporating normalized levels after subtracting ambient air levels, in the hope that removal of any noise from the air would increase the discriminatory power of the random forests. However, the performance did not increase as much as it did when we used only normalized concentrations of breath. More specifically, by using all 29 VOCs, random forests achieved an accuracy of 58% (AUC: 0.54) (Table 3, Analysis no. 17), whereas, after two steps of feature selection, the performance was increased to an accuracy of 72% (AUC: 0.78) by using eight features (Table 3, Analysis no. 18).

We also examined whether the combination of the 19 VOCs measured by the targeted approach together with the 29 VOCs identified as differentiated by the untargeted approach would increase the discriminatory power of the machine learning models in Ca+ vs. Ca− patients. In this set, the concentrations of 19 VOCs in breath were used together with 29 VOCs selected as informative by the untargeted approach. By using all 48 variables, random forests achieved an accuracy of 45% (AUC: 0.44) (Table 3, Analysis no. 19), whereas, after two steps of feature selection, the performance was increased to an accuracy of 73% (AUC: 0.72) (Table 3, Analysis no. 20), using three features (thiophene from the targeted approach and acetaldoxime and N-methyl acetamide from the untargeted approach). Thus, the inclusion of the 19 targeted metabolites did not increase the discriminatory performance of random forests that were based only on targeted metabolites.

Finally, we tested if smoking was a confounding factor for the discrimination (with random forests) of cancer vs. non-cancer patients, using normalized breath measurements of VOCs selected as informative by the untargeted approach. In these analyses, we retained 43 cancer patients and 26 non-cancer patients that never smoked or had quit smoking. The best-performing algorithm (random forest) correctly predicted only 59% of datapoints (AUC: 0.57) when using all 29 untargeted VOCs (Table 3, Analysis no. 21). When we used the three untargeted VOCs that had yielded the best performance in the previous cancer vs. non-cancer patients analysis, random forests of the non-smokers achieved an accuracy of 72.5%, but with a significantly lower AUC of 0.68 (Table 3, Analysis no. 22). Thus, we also performed two rounds of feature selection specifically for the non-smokers and, this time, random forests achieved an accuracy of 77%, with an AUC of 0.85, by using five VOCs (Table 3, Analysis no. 23).

In summary, based on all the above analyses, we conclude that the best-performing algorithm is again random forests, whereas the normalized breath data from the untargeted approach are sufficient to help the algorithm achieve a very high performance, in all comparisons. Furthermore, the two successive rounds of feature selection significantly improved the performance of the random forests, especially in the case of Ca + vs. Ca− patients. This was not possible in a previous study that had used a limited set of 19 selected VOCs. Furthermore, smoking was not a confounding factor for the untargeted analysis, an observation that is in agreement with the results of targeted analysis. It is very clear that the given untargeted approach, in combination with machine learning algorithms and feature selection, identified sets of compounds with sufficient discriminatory power (accuracy of 91–94%) to help us understand if a sample comes from a healthy person or from a person with a pulmonary disease. This was achievable with only seven to nine metabolites. Furthermore, it is also possible to discriminate, with satisfactory accuracy (75–77%), cancer from non-cancer patients, by using only three to five untargeted metabolites.

3. Discussion

In this study, we performed analyses based on non-targeted screening of the raw chromatographic data obtained from breath analysis, for three population groups (Ca+, Ca− and HC) and compared the discriminatory power of this approach to that achieved by targeted analysis. In the targeted analysis, 19 pre-selected compounds were measured, which were selected based on literature indicating that they might be potential biomarkers of lung cancer. Seven of these pre-selected compounds were found to differ significantly between Ca+ and HC, and between pooled patient (Ca+ and Ca−) and HC groups, and none differed significantly between Ca+ and Ca− groups [10].

The non-targeted analysis was performed with the use of the XCMS Online data processing platform combined with manual processing of the raw chromatograms to select the informative compounds and develop a dataset containing the areas of chromatographic peaks of differentiated compounds. Processing of the raw files with XCMS Online was conducted to determine the subset of chromatographic peaks and corresponding ions (m/z) to focus on, and narrow the investigated peaks to those only identified as significantly differentiated between population subgroups (Figure 2: Step 1). Next, we manually cross-checked (Figure 2: Step 2) and reprocessed (Figure 2: Step 5) the identified peaks in the raw data, by integrating extracted ion chromatograms (EICs). This task was performed to confirm and, when necessary, correct the results obtained from automated online data processing, and increase the reliability of the developed dataset, before proceeding to statistical analyses and the application of machine learning methods. We considered this stage necessary since peak misalignment or identification of "false peaks" by preprocessing software has been reported as a potential limitation of this approach due to the variance and complexity of raw chromatograms [15–17]. Indeed, a number of peaks identified by XCMS as informative could not be satisfactorily processed in the raw chromatograms and had to be excluded from the analysis, due to noise interferences or co-elution issues. It was interesting that two compounds (1-propanol and 2-propanol) identified as differentiated between population groups by the targeted approach were filtered out by the selection criteria applied in the untargeted workflow. By searching for 2-propanol and 1-propanol in the XCMS results, we observed that the corresponding peaks were correctly identified and their levels were found to differ between Ca+ and HC, while fold changes in LC patients were in agreement with those observed when concentrations determined by calibration curves (targeted analysis) were compared. However, the level of statistical significance of non-normalized values (determined by a *t*-test) was 0.0185 for 2-propanol and 0.0198 for 1 propanol, which was marginally higher than the selection criterion ($p < 0.01$) set for Ca+ vs. Ca− pairwise (online automated) analysis. It should also be mentioned that the *t*-test is not the appropriate significance criterion for non-normally distributed data.

It is also noteworthy that 53 compounds identified as informative by the analysis with XCMS Online were at a later stage excluded as they corresponded to silicon-based compounds and presumably derived from the SPME fiber and chromatographic column bleed (Figure 2: Step 3). The vast majority of these compounds were selected based on the Ca+ vs. HC pairwise analysis and the associations can be attributed to different experimental conditions during the time periods of the collection and analysis of the population subgroups. It is therefore assumed that these compounds were selected due to systematic variations in experimental conditions. This effect is often corrected through normalization processes where signal intensity is adjusted by the total intensity, the highest value or by an external or internal standard [18]. In untargeted metabolomics, the use of pooled samples as external standards is often applied [19] but this practice would be extremely complicated in exhaled air samples. In the present study, external standard normalization was conducted by incorporating spiked standard mixtures with known concentrations that were used in targeted analysis (Figure 2: Step 6). Moreover, after manual processing of the detected peaks and external standard normalization, a few associations that were determined as significant from XCMS Online analysis were not confirmed by offline statistical analysis of reprocessed data.

Some of the identified compounds have been reported previously to differ in the breath of LC patients and other pulmonary diseases. In particular, monoaromatics are reported by numerous publications. A very recent review by Ratiu identified 21 aromatic hydrocarbons differentiated in lung cancer [20]. Furans, such as 3-methylfuran and 2,5-dimethylfuran, have also been identified by previous studies but these compounds are considered biomarkers of both active and passive exposure to tobacco smoke [21]. Allyl methyl sulfide and methyl propyl sulfide (an isomer of 1-methylthio-(*E*)-1-propene), which were found in lower levels in LC patients, are known to suppress the proliferation of human lung tumor cells and possess anti-carcinogenic properties [22,23]. Moreover, similar structures, such as dimethyl sulfide and methionol, are involved in the metabolism of methionine [24]. Differences in the exhaled breath levels of acetic acid and propionic acid have also been reported by previous studies, albeit less frequently [25,26]. Exhaled p-benzoquinone has been proposed as a marker of malignant pleural mesothelioma [27]. For other identified substances (N-2-Aminoethyl acetamide, 1 methoxy propanol, methylacetamide, acetaldoxime, eucalyptol), we did not find any references in the scientific literature concerning the potential association of the exhaled breath concentrations with lung cancer. It should be noted that for some compounds (e.g., propionic acid, acetic acid), we report lower levels in the exhaled breath of LC patients, a finding which apparently contradicts existing evidence. The lack of reproducibility between independent research groups is a known obstacle in breath research. It should also be mentioned that for a few compounds, the identification is questionable. This statement is based on the observation that deviations in RTs and RIs (Figure 2: Step 4) for these compounds do not follow the trend established by known compounds. These include propionic acid, methylacetamide, acetaldoxime and 1-methxy-propanol. The utilization of retention indices in compound identification confirmation through the comparison with available retention data can be of great importance, especially when mass spectral matches are derived from multiple candidate compounds with similar spectra (e.g., isomer compounds) [28]. In our investigation, the use of RIs assisted in the confirmation of mass spectra matches and in distinguishing which isomer compound corresponds to the chromatographic peak (1-methylthio-(E)-1-propene, p-xylene). The small deviations between calculated and library-derived RIs were expected since RIs were experimentally determined with a DB-624 column (6% cyanopropyl/phenyl, 94% polydimethylsiloxane (PDMS)) and retrieved RIs were related to a 100% PDMS column. Naturally, the RI is dependent on the kind of stationary phase and different stationary phases give rise to different RIs of the same compound. However, the same trend in the abovementioned deviation was observed in the vast majority of the identified compounds. The combination of mass spectra and RI data has been proposed in both targeted and untargeted GC–MS data processing protocols [29].

By searching for the identified compounds in metabolic pathway databases and in the scientific literature, we found no direct evidence linking these VOCs to biochemical alterations that occur in cancer and therefore the biochemical interpretation of the results is not straightforward. While instrumental techniques, sampling methods and informatics approaches for studying diseases through the analysis of exhaled breath are constantly evolving [30–32], it is critical for future research to advance the knowledge concerning the understanding of underlying mechanisms that result in alteration of VOC breath composition. Current scientific knowledge provides some evidence and hypotheses concerning the biochemical background of endogenous VOCs [33], but the origin of the majority of these compounds is largely uncertain. Further research on endogenous products is of great importance not only for diagnostic purposes but also for targeting treatment [34].

It is evident that most of the compounds identified as differentiated in population groups in the present study are of exogenous origin or are produced endogenously during the metabolism of exogenous compounds. This observation enhances the findings of our previous publication, where it was hypothesized that alterations in pulmonary function and in the metabolism and excretion of exogenous compounds in disease can have an effect on the concentrations measured in exhaled breath. This hypothesis is also supported by several clinical tests and recent research that use exogenous VOCs (EVOCs) as probes to "measure the activity of metabolic enzymes in vivo, as well as the function of organs, through breath analysis" [35]. Future research should further elucidate the potential of the administration of harmless exogenous compounds as probes to study diseases.

In accordance with acquired data, the discrimination of LC patients from patients with abnormal CT findings was substantially increased by the untargeted approach and subsequent feature selection/machine learning in comparison to a previously conducted targeted approach. The correct classification was 75–77% for Ca+ vs. Ca− in the untargeted analysis compared to approximately 50% in the targeted analysis. Additionally, we report 91% accuracy for the discrimination of LC patients from healthy controls based on the investigation of 29 VOCs selected as informative by a non-targeted approach. The discriminatory power was slightly increased compared to the targeted analysis focusing on the quantification of a set 19 pre-determined VOCs. Although the targeted approach has the advantage of the absolute determination of VOC levels and is less prone to biases, untargeted screening allowed us to detect new distinctive features and incorporate a larger compound set into the classification analysis, thus resulting in better discrimination. Previous studies investigating VOC profiles by gas chromatography–mass spectrometry also reported high discriminatory power in distinguishing LC patients from healthy controls [36–47]. However, the major concerns are the limited reproducibility regarding the compounds identified by different research groups and the uncertainties regarding the origins of VOCs that differentiate lung cancer. The lower discriminant power between Ca+ and Ca− patients underlines the importance of evaluating the interference of other pulmonary diseases in the identification of LC biomarkers [46,47]. The combination of the datasets developed by the targeted and untargeted approaches did not significantly improve the discrimination, an observation that underlines that the information provided by targeted analysis is contained to a large extent in the data obtained by the untargeted approach. Untargeted VOC screening detected four (toluene, benzene, styrene, ethylbenzene) out of seven compounds that were found to differ significantly in targeted analysis, and exploited numerous features that could not be identified by the targeted approach. In agreement with targeted analysis, incorporating breath subtracts (ambient air was subtracted from breath measurements) slightly decreased the discriminatory power of the analysis. This can be explained by the fact that for some VOCs with high concentrations in ambient air, the information contained in breath measurements was not exploited. Including breath substrate (also referred to as alveolar gradient) in the analysis is a double-edged decision. On the one hand, not considering the ambient air chemical composition may introduce environmental interferences, while, in parallel, subtracting air levels from breath may result in the exclusion of valuable information.

Some further issues should be considered when interpreting the results of the present study. Although SPME has many advantages as a solvent-free and versatile pre-concentration method, it is not without limitations. During SPME, VOCs compete for the active sites of the fiber, and molecules with higher molecular weight may displace smaller ones. Thus, varying the composition of samples may influence the amounts of VOC extracted [48]. Moreover, different fiber coatings are suitable for different classes of analytes [49]. The fiber used in this study (CAR/PDMS) is suitable for VOCs with low molecular weight and a Kovats index of less than 980 [50]. According to a study conducted to evaluate the performance of different fiber coatings in the isolation of VOCs from feces, the particular fiber used isolated 60% of the total examined VOCs [51]. Concerning sampling, pre-concentration and instrumental procedures, we adopted a mixed expiratory breath sampling/SPME/GC–MS approach, but a variety of alternative methods are available. In brief, sampling can also focus on later or end-tidal expiratory breath, pre-concentration can be achieved with thermal desorption (TD) and needle trap devices (NTDs) [52] and instrumental analysis can also be performed with proton transfer reaction MS (PTR-MS) and selective ion flow tube MS (SIFT-MS) [18]. Cross-reactive sensors have also been developed and tested by numerous research groups [53].

Another limitation of this study is that the participants who formed the HC group did not undergo clinical examination or diagnostic tests to exclude the possibility of having undiagnosed cancer or serious pulmonary diseases, instead they were recruited based on personal interviews. Thus, the possibility that a few individuals were falsely classified as controls cannot be entirely excluded.

In summary, untargeted VOC profiling captured, to a large extent, the information provided by targeted analysis and performed more efficiently in discriminating lung cancer patients from patients with benign pulmonary diseases, through the utilization of new compounds that were not previously considered. However, uncertainties in compound identification and automated processing of raw data should be carefully addressed. Subsequence steps for the verification and manual correction of automatically identified peaks in the raw chromatographic files can increase the reliability of the acquired datasets.

4. Materials and Methods

4.1. Participant Recruitment and Breath Sampling

A detailed description concerning the procedures followed for participant recruitment and sampling of exhaled breath can be found in a previous publication [10]. In brief, the study population consisted of 85 patients from the General University Hospital of Larissa (Greece) who underwent bronchoscopy due to abnormal CT findings and a control group of 52 individuals of similar age were recruited from local health centers. Samples were collected from October 2018 to October 2019. After bronchoscopy, patients were categorized according to the presence of LC, according to results of the cytological/histological examination. The control group (referred to in the text as healthy controls (HC)) was selected on the basis of the absence of self-reported pulmonary diseases and cancer. The absence of these diseases was determined by self-report during the personal interviews conducted on the day of sampling.

Breath samples were collected in Tedlar® bags (Sigma-Aldrich, St. Louis, MO, USA). Participants were asked to inhale deeply and hold their breath for 30 s, then exhale through a disposable mouthpiece into the 1 L Tedlar® bag until filled. Two breath samples were collected with approximately two-minute intervals in between. Ambient air samples were also collected with the use of a portable Laboport® UN 86 KTP (KNF Neuberger GmbH, Freiburg, Germany) pump.

4.2. Materials, Solid Phase Microextraction and GC–MS Analysis

A detailed description of the materials and methods used in the present study can be found in our previous publication [10]. In brief, extraction and pre-concentration of the analytes from breath samples was achieved by solid phase microextraction (SPME) using a 75 μm carboxen-polydimethylsiloxane (CAR/PDMS)-coated fused silica fiber assembly (Sigma-Aldrich, St. Louis, MO, USA), and desorption of analytes from the fiber was performed for 5 min at 270 °C. Instrumental analysis was performed with a Finnigan Trace GC Ultra/Polaris Ion Trap GC/MSn system equipped with a DB-624 GC capillary column (inner diameter: 0.25 mm, length: 30 m, film: 1.4 μm, 6% cyanopropylphenyl/94% dimethylpolysiloxan, Agilent, Santa Clara, CA, USA). GC–MS chromatograms were acquired in total ion current (TIC) mode of the mass analyzer, and then extracted at one or two specific m/z values for analyte quantification. Data acquisition and processing were carried out using Xcalibur™ 3.0 software (ThermoFisher Scientific, San Francisco, CA, USA). Furthermore, for the determination of RIs, SAK-100-1 and SMA-200-1 (Agilent, Santa Clara, CA, USA) analytical standards containing C5 to C12 alkanes were used. Gas samples were prepared, spiked with methanolic solution of C5-C12 alkanes and retention times of each alkane were determined.

4.3. Data Pre-Processing and Analysis

After GC/MS analysis, all raw data were converted to mzml files using ProteoWizard, and subsequently the converted files were imported into XCMS Online software (XCMS Online version 3.7.1) (https://xcmsonline.scripps.edu) for feature detection, alignment and retention time correction. The raw data processing was carried out using the following parameters: general: Rt, format: minutes, polarity: positive, feature detection: centWave, ppm: 900, minimum peak width: 5, maximum peak width: 30, mzdiff: 0.1, signal/noise threshold: 3, integration method: 1, prefilter peaks: 3, prefilter intensity: 100, noise filter: 0, Rt. correction: obiwarp, profStep: 1, alignment: bw 1, minfrac: 0.2, mzwid: 0.25, minsamp: 1, max: 500, statistics: statistical test: t-test. All chromatograms were simultaneously analyzed with identical settings. Selection of the most informative variables (m/z) was based on statistical criteria (p-value < 0.01, fold change > 2, m/z < 140, Rt < 14.00 min for Ca+ vs. HC; p-value < 0.05, fold change > 1.1, m/z < 140, Rt < 14.00 min for Ca+ vs. Ca−) of differentiated peak intensity between patients and controls.

4.4. Identification of Candidate Features and Raw Data Reprocessing

All features identified as differentiated between population groups with the XCMS analysis were searched for in the raw chromatograms and the corresponding peaks were identified. The mass spectrum of the identified peaks was studied in comparison with the National Institute of Standards and Technology (NIST) spectrometric library. Peaks of compounds corresponding to technical interferences (siloxanes, Tedlar® bag compounds) were excluded from further analysis. Extracted ion chromatograms were obtained for the ions identified as significantly differentiated between population subgroups by XCMS analysis, and were reprocessed by calculation of the areas of the chromatographic peaks in SIM mode using Thermo Xcalibur™ software. The most discriminatory features were assigned based on mass spectral similarities to the NIST 2011 mass spectral library. Compounds were categorized as "probable" (probability > 75%), "possible" (probability 50–75%) and unknown (probability < 50%). To further confirm the identification of compounds, retention characteristics were examined. Retention times were simulated by using the Pro

EZGC Chromatogram Modeler (Restek Corporation, Bellefonte, PA, USA), introducing an equivalent chromatographic column and an identical temperature program. Simulated RTs were compared to actual RTs for substances contained in the Restek database. Retention indices of these compounds were retrieved from the NIST webbook and related to a fully non-polar column (100% polydimethylsiloxane). Moreover, retention indices for each compound were experimentally determined. SAK-100-1 and SMA-200-1 (Agilent) analytical standards with C5 to C12 alkanes were used to calculate the retention indices from the unknown compounds. Experimental retention indices of these compounds were calculated according to the following formula:

$$I = 100 \left[n + (t_i - t_n)/(t_{n+1} - t_n) \right]$$

I: retention index
n: number of carbons of heading *n*-alkane peak *i*
t_i: retention time of specific compound *i* (minutes)
t_n, t_{n+1}: retention times of heading and trailing n-alkanes

Normalization of chromatographic peak areas was performed with an external standard, by dividing instrument response by the geometric mean peak areas of three monoaromatic compounds (benzene, toluene and ethyl benzene) of a standard mixture ($\approx$20 ng/L air each) analyzed on the same day.

4.5. Machine Learning Methods

The machine learning analyses were performed with Waikato Environment for Knowledge Analysis (Weka). For each comparison, group 1 vs. group 2 or cases vs. controls were analyzed using naive Bayes, logistic regression and random forest methods, with 10-fold cross-validation. However, random forests consistently outperformed the other algorithms, therefore, all results are shown for this specific type of algorithm. Feature selection within the appropriate Weka module was also performed, in order to detect subsets of informative metabolites that could more efficiently separate the groups from each other. In particular, feature selection was performed in two steps with a wrapper that evaluates various subsets of the features (WrapperSubsetEval), using the Best_First method in order to maximize the performance of the random forest, based on the metric of the area under the curve (AUC). In the first step, the wrapper functions in a feature selection mode that performs 10-fold cross-validation. The output of this first feature selection step assesses how many times a feature has been selected in the 10-fold cross-validations. The features that are selected in at least 50% of the cross-validations form another subset that is fed into the second step. Thus, we repeat (in the second step) the feature selection, by starting with the abovementioned informative subset, and this time the wrapper runs in a feature selection mode that uses the full training set and selects only a certain final subset of features.

Author Contributions: Conceptualization, A.T.; Data curation, M.K. and D.K.; Formal analysis, M.K. and G.D.A.; Funding acquisition, C.H., K.G. and A.T.; Investigation, D.K.; Methodology, M.K., G.D.A. and A.T.; Project administration, A.T.; Resources, C.H., K.G. and A.T.; Supervision, A.T.; Validation, M.K.; Visualization, M.K.; Writing—original draft, M.K., G.D.A. and A.T.; Writing—review and editing, C.H., K.G. and A.T. All authors have read and agreed to the published version of the manuscript.

Funding: This work was supported by a postdoctoral scholarship program (Project No.: 5394.02.07) implemented by the University of Thessaly and funded by the Stavros Niarchos Foundation.

Institutional Review Board Statement: The study protocol was approved by the Scientific Council of the General University Hospital of Larissa with a 16/20/06-12-2018 decision.

Informed Consent Statement: Informed consent was obtained from all subjects involved in the study.

Data Availability Statement: The data are not publicly available because they contain sensitive information at an individual level.

Acknowledgments: We would like to thank Ioanna Chatzelli for her assistance in the investigation of the compounds' possible origin in the KEGG PATHWAY database.

Conflicts of Interest: The authors declare no conflict of interest.

Sample Availability: Air samples cannot be stored and thus they are not available.

References

1. Forsberg, E.M.; Huan, T.; Rinehart, D.; Benton, H.P.; Warth, B.; Hilmers, B.; Siuzdak, G. Data processing, multi-omic pathway mapping, and metabolite activity analysis using XCMS Online. *Nat. Protoc.* **2018**, *13*, 633–651. [CrossRef]
2. Tian, H.; Li, B.; Shui, G. Untargeted LC–MS Data Preprocessing in Metabolomics. *J. Anal. Test.* **2017**, *1*, 187–192. [CrossRef]
3. Meier, R.; Ruttkies, C.; Treutler, H.; Treutler, H.; Neumann, S. Bioinformatics can boost metabolomics research. *J. Biotechnol.* **2017**, *261*, 137–141. [CrossRef] [PubMed]
4. Giannoukos, S.; Agapiou, A.; Brkić, B.; Taylor, S. Volatolomics: A broad area of experimentation. *J. Chromatogr. B* **2019**, *1105*, 136–147. [CrossRef]
5. Patti, G.J.; Yanes, O.; Siuzdak, G. Innovation: Metabolomics: The apogee of the omics trilogy. *Nat. Rev. Mol. Cell. Biol.* **2012**, *13*, 263–269. [CrossRef] [PubMed]
6. Lin, Y.; Caldwell, G.W.; Li, Y.; Lang, W.; Masucci, J. Inter-laboratory reproducibility of an untargeted metabolomics GC-MS assay for analysis of human plasma. *Sci. Rep.* **2020**, *10*, 10918. [CrossRef] [PubMed]
7. Ribbenstedt, A.; Ziarrusta, H.; Benskin, J.P. Development, characterization and comparisons of targeted and non-targeted metabolomics methods. *PLoS ONE* **2018**, *13*, e0207082. [CrossRef] [PubMed]
8. Cajka, T.; Fiehn, O. Toward Merging Untargeted and Targeted Methods in Mass Spectrometry-Based Metabolomics and Lipidomics. *Anal. Chem.* **2016**, *88*, 524–545. [CrossRef]
9. Chen, L.; Zhong, F.; Zhu, J. Bridging Targeted and Untargeted Mass Spectrometry-Based Metabolomics via Hybrid Approaches. *Metabolites* **2020**, *10*, 348. [CrossRef] [PubMed]
10. Koureas, M.; Kirgou, P.; Amoutzias, G.; Hadjichristodoulou, C.; Gourgoulianis, K.; Tsakalof, A. Target Analysis of Volatile Organic Compounds in Exhaled Breath for Lung Cancer Discrimination from Other Pulmonary Diseases and Healthy Persons. *Metabolites* **2020**, *10*, 317. [CrossRef]
11. Beauchamp, J.; Herbig, J.; Gutmann, R.; Hansel, A. On the use of Tedlar® bags for breath-gas sampling and analysis. *J. Breath Res.* **2008**, *2*, 046001. [CrossRef]
12. Qiu, Y.-Q. KEGG Pathway Database. In *Encyclopedia of Systems Biology*; Dubitzky, W., Wolkenhauer, O., Cho, K.-H., Eds.; Springer: New York, NY, USA, 2013; pp. 1068–1069.
13. Kalf, G.F.; Renz, J.F.; Niculescu, R. p-Benzoquinone, a reactive metabolite of benzene, prevents the processing of pre-interleukins-1 alpha and -1 beta to active cytokines by inhibition of the processing enzymes, calpain, and interleukin-1 beta converting enzyme. *Environ. Health Perspect.* **1996**, *104* (Suppl. 6), 1251–1256. [CrossRef]
14. Pravasi, S.D. Acetic Acid. In *Encyclopedia of Toxicology*, 3rd ed.; Wexler, P., Ed.; Oxford University Press: New York, NY, USA, 2014; pp. 33–35.
15. Coble, J.B.; Fraga, C.G. Comparative evaluation of preprocessing freeware on chromatography/mass spectrometry data for signature discovery. *J. Chromatogr. A* **2014**, *1358*, 155–164. [CrossRef]
16. Myers, O.D.; Sumner, S.J.; Li, S.; Barnes, S.; Du, X. Detailed Investigation and Comparison of the XCMS and MZmine 2 Chromatogram Construction and Chromatographic Peak Detection Methods for Preprocessing Mass Spectrometry Metabolomics Data. *Anal. Chem.* **2017**, *89*, 8689–8695. [CrossRef]
17. Koh, Y.; Pasikanti, K.K.; Yap, C.W.; Chan, E.C.Y. Comparative evaluation of software for retention time alignment of gas chromatography/time-of-flight mass spectrometry-based metabonomic data. *J. Chromatogr. A* **2010**, *1217*, 8308–8316. [CrossRef] [PubMed]
18. Horváth, I.; Barnes, P.J.; Loukides, S.; Sterk, P.J.; Högman, M.; Olin, A.-C.; Amann, A.; Antus, B.; Baraldi, E.; Bikov, A.; et al. A European Respiratory Society technical standard: Exhaled biomarkers in lung disease. *Eur. Respir. J.* **2017**, *49*. [CrossRef]
19. Koek, M.M.; Jellema, R.H.; van der Greef, J.; Tas, A.C.; Hankemeier, T. Quantitative metabolomics based on gas chromatography mass spectrometry: Status and perspectives. *Metabolomics* **2011**, *7*, 307–328. [CrossRef] [PubMed]
20. Ratiu, I.A.; Ligor, T.; Bocos-Bintintan, V.; Mayhew, C.A.; Buszewski, B. Volatile Organic Compounds in Exhaled Breath as Fingerprints of Lung Cancer, Asthma and COPD. *J. Clin. Med.* **2020**, *10*, 32. [CrossRef] [PubMed]
21. Buszewski, B.; Ulanowska, A.; Ligor, T.; Denderz, N.; Amann, A. Analysis of exhaled breath from smokers, passive smokers and non-smokers by solid-phase microextraction gas chromatography/mass spectrometry. *Biomed. Chromatogr. BMC* **2009**, *23*, 551–556. [CrossRef]
22. Sakamoto, K.; Lawson, L.D.; Milner, J.A. Allyl sulfides from garlic suppress the in vitro proliferation of human A549 lung tumor cells. *Nutr. Cancer* **1997**, *29*, 152–156. [CrossRef] [PubMed]
23. Fukushima, S.; Takada, N.; Hori, T.; Wanibuchi, H. Cancer prevention by organosulfur compounds from garlic and onion. *J. Cell. Biochem. Suppl.* **1997**, *27*, 100–105. [CrossRef]
24. Tangerman, A. Measurement and biological significance of the volatile sulfur compounds hydrogen sulfide, methanethiol and dimethyl sulfide in various biological matrices. *J. Chromatogr. B* **2009**, *877*, 3366–3377. [CrossRef] [PubMed]

25. Wang, C.; Dong, R.; Wang, X.; Lian, A.; Chi, C.; Ke, C.; Guo, L.; Liu, S.; Zhao, W.; Xu, G.; et al. Exhaled volatile organic compounds as lung cancer biomarkers during one-lung ventilation. *Sci. Rep.* **2014**, *4*, 7312. [CrossRef] [PubMed]
26. Itoh, T.; Miwa, T.; Tsuruta, A.; Akamatsu, T.; Izu, N.; Shin, W.; Park, J.; Hida, T.; Eda, T.; Setoguchi, Y. Development of an Exhaled Breath Monitoring System with Semiconductive Gas Sensors, a Gas Condenser Unit, and Gas Chromatograph Columns. *Sensors* **2016**, *16*, 1891. [CrossRef]
27. Di Gilio, A.; Catino, A.; Lombardi, A.; Palmisani, J.; Facchini, L.; Mongelli, T.; Varesano, N.; Belloti, R.; Galetta, D.; de Gennaro, G.; et al. Breath Analysis for Early Detection of Malignant Pleural Mesothelioma: Volatile Organic Compounds (VOCs) Determination and Possible Biochemical Pathways. *Cancers* **2020**, *12*, 1262. [CrossRef]
28. Babushok, V.I. Chromatographic retention indices in identification of chemical compounds. *Trends Analyt. Che.-TrAC* **2015**, *69*, 98–104. [CrossRef]
29. Fiehn, O. Metabolomics by Gas Chromatography-Mass Spectrometry: Combined Targeted and Untargeted Profiling. *Curr. Protoc. Mol. Biol.* **2016**, *114*, 30.4.1–30.4.32. [CrossRef]
30. Nidheesh, V.R.; Mohapatra, A.K.; Unnikrishnan, V.K.; Sinha, R.K.; Nayak, R.; Kartha, V.B.; Chidangil, S. Breath analysis for the screening and diagnosis of diseases. *Appl. Spectrosc. Rev.* **2020**. [CrossRef]
31. Ghosh, C.; Singh, V. Recent advances in breath analysis to track human health by new enrichment technologies. *J. Sep. Sci.* **2020**, *43*, 226–240. [CrossRef]
32. Smolinska, A.; Hauschild, A.C.; Fijten, R.R.; Dallinga, J.W.; Baumbach, J.; van Schooten, F.J. Current breathomics–a review on data pre-processing techniques and machine learning in metabolomics breath analysis. *J. Breath Res.* **2014**, *8*, 027105. [CrossRef]
33. Haick, H.; Broza, Y.Y.; Mochalski, P.; Ruzsanyi, V.; Amann, A. Assessment, origin, and implementation of breath volatile cancer markers. *Chem. Soc. Rev.* **2014**, *43*, 1423–1449. [CrossRef] [PubMed]
34. Boots, A.W.; Bos, L.D.; van der Schee, M.P.; van Schoten, F.-J.; Sterk, P.J. Exhaled Molecular Fingerprinting in Diagnosis and Monitoring: Validating Volatile Promises. *Trends Mol. Med.* **2015**, *21*, 633–644. [CrossRef]
35. Gaude, E.; Nakhleh, M.K.; Patassini, S.; Boschmans, J.; Allsworth, M.; Boyle, B.; van der Schee, M.P. Targeted breath analysis: Exogenous volatile organic compounds (EVOC) as metabolic pathway-specific probes. *J. Breath Res.* **2019**, *13*, 032001. [CrossRef] [PubMed]
36. Buszewski, B.; Ligor, T.; Jezierski, T.; Wenda-Piesik, A.; Walczak, M.; Rudnicka, J. Identification of volatile lung cancer markers by gas chromatography-mass spectrometry: Comparison with discrimination by canines. *Anal. Bioanal. Chem.* **2012**, *404*, 141–146. [CrossRef] [PubMed]
37. Kischkel, S.; Miekisch, W.; Sawacki, A.; Straker, E.M.; Trefz, P.; Amann, A.; Schubert, J.K. Breath biomarkers for lung cancer detection and assessment of smoking related effects—Confounding variables, influence of normalization and statistical algorithms. *Clin. Chim. Acta* **2010**, *411*, 1637–1644. [CrossRef]
38. Ma, W.; Gao, P.; Fan, J.; Hashi, Y.; Chen, Z. Determination of breath gas composition of lung cancer patients using gas chromatography/mass spectrometry with monolithic material sorptive extraction. *Biomed. Chromatogr.* **2015**, *29*, 961–965. [CrossRef]
39. Pesesse, R.; Stefanuto, P.H.; Schleich, F.; Louis, R.; Focant, J.-F. Multimodal chemometric approach for the analysis of human exhaled breath in lung cancer patients by TD-GC × GC-TOFMS. *J. Chromatogr. B* **2019**, *1114–1115*, 146–153. [CrossRef]
40. Rudnicka, J.; Kowalkowski, T.; Buszewski, B. Searching for selected VOCs in human breath samples as potential markers of lung cancer. *Lung Cancer* **2019**, *135*, 123–129. [CrossRef]
41. Rudnicka, J.; Kowalkowski, T.; Ligor, T.; Buszewski, B. Determination of volatile organic compounds as biomarkers of lung cancer by SPME-GC-TOF/MS and chemometrics. *J. Chromatogr. B* **2011**, *879*, 3360–3366. [CrossRef]
42. Sakumura, Y.; Koyama, Y.; Tokutake, H.; Hida, T.; Sato, K.; Itoh, T.; Akamatsu, T.; Shin, W. Diagnosis by volatile organic compounds in exhaled breath from lung cancer patients using support vector machine algorithm. *Sensors* **2017**, *17*, 287. [CrossRef]
43. Schallschmidt, K.; Becker, R.; Jung, C.; Bremser, W. Comparison of volatile organic compounds from lung cancer patients and healthy controls—Challenges and limitations of an observational study. *J. Breath Res.* **2016**, *10*. [CrossRef] [PubMed]
44. Song, G.; Qin, T.; Liu, H.; Xu, G.-B.; Pan, Y.-Y.; Xiong, F.-X.; Gu, K.-S.; Sun, G.-P.; Chen, Z.-D. Quantitative breath analysis of volatile organic compounds of lung cancer patients. *Lung Cancer* **2010**, *67*, 227–231. [CrossRef]
45. Ulanowska, A.; Kowalkowski, T.; Trawińska, E.; Buszewski, B. The application of statistical methods using VOCs to identify patients with lung cancer. *J. Breath Res.* **2011**, *5*. [CrossRef] [PubMed]
46. Wang, M.; Sheng, J.; Wu, Q.; Zou, Y.; Hu, Y.; Ying, K.; Wang, P. Confounding effect of benign pulmonary diseases in selecting volatile organic compounds as markers of lung cancer. *J. Breath Res.* **2018**, *12*. [CrossRef] [PubMed]
47. Zou, Y.; Zhang, X.; Chen, X.; Hu, Y.; Ying, K.; Wang, P. Optimization of volatile markers of lung cancer to exclude interferences of non-malignant diase. *Cancer Biomark.* **2014**, *14*, 371–379. [CrossRef] [PubMed]
48. Murray, R.A. Limitations to the Use of Solid-Phase Microextraction for Quantitation of Mixtures of Volatile Organic Sulfur Compounds. *Anal. Chem.* **2001**, *73*, 1646–1649. [CrossRef] [PubMed]
49. Yu, A.-N.; Sun, B.-G.; Tian, D.-T.; Qu, W.-Y. Analysis of volatile compounds in traditional smoke-cured bacon(CSCB) with different fiber coatings using SPME. *Food Chem.* **2008**, *110*, 233–238. [CrossRef]
50. Garcia-Esteban, M.; Ansorena, D.; Astiasarán, I.; Ruiz, J. Study of the effect of different fiber coatings and extraction conditions on dry cured ham volatile compounds extracted by solid-phase microextraction (SPME). *Talanta* **2004**, *64*, 458–466. [CrossRef]

51. Dixon, E.; Clubb, C.; Pittman, S.; Ammann, L.; Rasheed, Z.; Kazmi, N.; Keshavarzian, A.; Gillevet, P.; Rangwala, H.; Couch, R.D. Solid-phase microextraction and the human fecal VOC metabolome. *PLoS ONE* **2011**, *6*, e18471. [CrossRef]
52. Lawal, O.; Ahmed, W.M.; Nijsen, T.M.E.; Goodacre, R.; Fowler, S.J. Exhaled breath analysis: A review of 'breath-taking' methods for off-line analysis. *Metabolomics* **2017**, *13*, 110. [CrossRef]
53. Hashoul, D.; Haick, H. Sensors for detecting pulmonary diseases from exhaled breath. *Eur. Respir. Rev.* **2019**, *28*, 190011. [CrossRef] [PubMed]

Article

Needle Trap Device-GC-MS for Characterization of Lung Diseases Based on Breath VOC Profiles

Fernanda Monedeiro [1], Maciej Monedeiro-Milanowski [1], Ileana-Andreea Ratiu [1,2,3], Beata Brożek [4], Tomasz Ligor [1,3,*] and Bogusław Buszewski [1,3]

1   Interdisciplinary Centre of Modern Technologies, Nicolaus Copernicus University in Toruń, 4 Wileńska St., 87-100 Toruń, Poland; fmonedeiro@gmail.com (F.M.); milanowski.maciej@gmail.com (M.M.-M.); andreea_ratiu84@yahoo.com (I.-A.R.); bbusz@chem.umk.pl (B.B.)
2   "Raluca Ripan" Institute for Research in Chemistry, Babeş-Bolyai University, 30 Fântânele St., RO-400294 Cluj-Napoca, Romania
3   Department of Environmental Chemistry and Bioanalytics, Faculty of Chemistry, Nicolaus Copernicus University in Toruń, 7 Gagarina St., 87-100 Toruń, Poland
4   Department of Lung Diseases, Provincial Polyclinic Hospital in Toruń, 4 Krasińskiego St., 87-100 Toruń, Poland; bebro@wp.pl
*   Correspondence: Tomasz.Ligor@umk.pl; Tel.: +48-(56)-665-60-58

Abstract: Volatile organic compounds (VOCs) have been assessed in breath samples as possible indicators of diseases. The present study aimed to quantify 29 VOCs (previously reported as potential biomarkers of lung diseases) in breath samples collected from controls and individuals with lung cancer, chronic obstructive pulmonary disease and asthma. Besides that, global VOC profiles were investigated. A needle trap device (NTD) was used as pre-concentration technique, associated to gas chromatography-mass spectrometry (GC-MS) analysis. Univariate and multivariate approaches were applied to assess VOC distributions according to the studied diseases. Limits of quantitation ranged from 0.003 to 6.21 ppbv and calculated relative standard deviations did not exceed 10%. At least 15 of the quantified targets presented themselves as discriminating features. A random forest (RF) method was performed in order to classify enrolled conditions according to VOCs' latent patterns, considering VOCs responses in global profiles. The developed model was based on 12 discriminating features and provided overall balanced accuracy of 85.7%. Ultimately, multinomial logistic regression (MLR) analysis was conducted using the concentration of the nine most discriminative targets (2-propanol, 3-methylpentane, (E)-ocimene, limonene, m-cymene, benzonitrile, undecane, terpineol, phenol) as input and provided an average overall accuracy of 95.5% for multiclass prediction.

Keywords: VOCs; NTD-GC-MS; breath; lung cancer; COPD; asthma; biomarkers

Citation: Monedeiro, F.; Monedeiro-Milanowski, M.; Ratiu, I.-A.; Brożek, B.; Ligor, T.; Buszewski, B. Needle Trap Device-GC-MS for Characterization of Lung Diseases Based on Breath VOC Profiles. *Molecules* **2021**, *26*, 1789. https://doi.org/10.3390/molecules26061789

Academic Editors: Natalia Drabińska and Ben de Lacy Costello

Received: 24 February 2021
Accepted: 19 March 2021
Published: 22 March 2021

Publisher's Note: MDPI stays neutral with regard to jurisdictional claims in published maps and institutional affiliations.

1. Introduction

Respiratory diseases are conditions which affect the airways and other structures of the lungs and they are represented by lung cancer, asthma, tuberculosis, chronic obstructive pulmonary disease (COPD) and pneumonia, being the leading causes of mortality and morbidity globally. Smoking or exposure to secondhand smoke is the main risk factor associated to most of respiratory diseases, with current smokers 11 times more likely to develop lung cancer compared to non-smokers [1]. Globally, respiratory diseases affect 1 billion people and account for 7% of all deaths worldwide. Nevertheless, even considering that lung cancer is one of the leading causes of death worldwide, COPD and asthma are predominant lung diseases that represent a burden on society in terms of health care costs [2]. The diagnosis of asthma or COPD is usually made by non-invasive techniques based on spirometry, however lung cancer is often diagnosed in late stages, due to the lack of noticeable clinical manifestations, or because these can be easily associated with other symptoms. This fact may reduce the chance of applying a timely and effective treatment. Currently used diagnostic methods for respiratory diseases includes physical examination followed by a set of chemical, imaging, endoscopic and immunological procedures [3]. Because different lung diseases are characterized by inflammation and other correspondent symptoms, direct assessment of airways may be applied, by using invasive procedures such as: computer tomography, bronchoscopy, bronchoalveolar lavage or biopsy. These are costly, time consuming and/or invasive procedures [2]. Consequently, a simple, reliable, low-cost and non-invasive test, able to achieve the diagnosis in real time (minutes up to hours), using a mere sample of exhaled breath in highly desired.

Therefore, fast detection and characterization of volatile organic compounds (VOCs) emitted from different biological matrices (breath, sweat, saliva, plasma, tissues, exudates, urine, etc.) as a tool for diagnosis was approached [4–11]. Breath tests are minimally invasive procedures, which are more easily accepted by the patients. An exhaled breath sample consists of VOCs and the breath aerosol [12]. Breath consists of almost 3000 compounds which are present in different combinations and quantities. Consequently, not only specific biomarkers, but the global VOC profile can be potentially associated to a characteristic fingerprint for each disease [2]. Exhaled breath is largely composed of nitrogen, oxygen, carbon dioxide, water, and inert gases. Trace components—volatile substances that are generated in the body or absorbed from the environment—present in the nmol/L–pmol/L (ppb volume—ppt volume) range make up the rest of the breath. The endogenous VOCs are generated by the cellular biochemical processes of the body, hence VOCs existent in human breath can reflect endogenous metabolic processes which occur in the tissues. VOCs-patterns in exhaled breath have been associated with various respiratory diseases such as cancer, asthma, COPD, cystic fibrosis, tuberculosis, etc. [13,14]. Breath samples are probably the most adequate to reach the rapid diagnosis of respiratory conditions, once substances from surrounding blood vessels and tissue can be exchanged in the alveoli and be available in the exhaled air. A large number of VOCs has been reported in scientific literature as markers of various diseases, as well as bacterial infections. These compounds can be divided into different chemical groups [15–17]: saturated hydrocarbons (stable end products of lipid peroxidation) and unsaturated hydrocarbons (e.g., from mevalonic pathway of cholesterol synthesis) [6,16], alcohols (which can be addressed as oxidized products of hydrocarbons and their precursors) [16], aldehydes (associated with inflammatory processes, resulting from lipid peroxidation) [5,18], ketones (products of fatty acid decarboxylation processes in the liver, associated to a diet rich in proteins and fat) [16], aromatic VOC–typically related to exogenous sources such as tobacco smoke and pollution [19], sulfur-containing compounds generated by incomplete metabolism of methionine in the transamination pathway and also associated with bacterial activity [20,21]), and nitrogen-containing compounds (such as ammonia, dimethylamine and trimethylamine, derived e.g., when conversion to urea is limited due to an impairment of liver function) [17].

Nowadays, gas chromatography–mass spectrometry (GC-MS) is considered a gold standard for VOC analysis [22]. Solid phase microextraction (SPME) or sorption tubes followed by thermal desorption are the most frequently used pre-concentration techniques in breath analyses. A prominent sampling tool is the needle trap device (NTD), which consists of a sorbent material packed inside a needle, working as an extraction trap [23]. This solventless technique provides exhaustive extraction and has potential for laboratory automation [24,25]. In the present work, NTD was used as extraction technique, followed by GC-MS analysis. VOCs were analyzed in breath samples belonging to healthy controls and patients with lung cancer, asthma and COPD, in an attempt to develop a classification model able to discriminate between these lung diseases, which have in common inflammatory processes in the lungs. In this sense, besides the assessment of global VOC profiles, 29 target compounds previously reported as potential biomarkers of the referred respiratory diseases were also investigated and quantified in breath samples.

The present study describes the non-invasive assessment of asthma, COPD and lung cancer, based on breath analysis of VOCs. Once all of these are lung diseases involving inflammatory mechanisms, the applied design of data analysis intended to find specific VOC patterns able to provide discrimination between these illnesses. The comparison between self-annotated discriminating features and compounds reported by literature as indicators of lung diseases represents an original approach for the validation of candidate biomarkers. The outline of the work presents the application of NTD for the determination of VOCs in breath. The found results aim to support the implementation of breath analysis to the clinical practice, as an accurate and reliable diagnostic tool.

2. Results and Discussion

2.1. Calibration Method and Quantitation of Analytes

Table S1 presents information regarding calibration method, while Table 1 displays data concerning the quantitation of analytes in breath samples. Obtained limits of quantitation (LOQs) ranged from 0.003 (3-methylpentane, 2-butanone, toluene, isododecane, 1,2,4-trimethylbenzene, (*E*)-ocimene, limonene, *m*-cymene and benzonitrile) to 6.21 ppbv (tridecane). Higher limits were obtained for heavier and more polar analytes, which also displayed wider linearity ranges. Lower limits were associated to compounds with higher volatility, a factor that seemed to contribute for their more efficient recovery, besides their expected greater stability in samples. Relative standard deviation (RSD%) did not exceed 10%, demonstrating that the proposed method provided adequate reproducibility. In general, suitability of NTD for preconcentration of analytes in gas mixtures could be inferred. Among the targets, isoprene and 1-propanol were found in each breath sample. Styrene, decane and phenol were observed in lowest frequency of appearance. Ethanol, isoprene and acetoin were the targets which occurred in higher concentrations in all sample's cohorts. Carry-over effect was not observed, indicating that there is no influence of previously analyzed samples on the current ones.

Table 1. Data regarding quantitation of the targets in breath samples (H = healthy, CA = lung cancer, COPD = chronic obstructive pulmonary disease, AS = asthma, SD = standard deviation, nd = not detected, (−) = SD not calculated because analyte was quantified just in a single sample, nd = not detected).

Analyte	Average Concentration (ppbv)									Frequency in Samples (%)				
	H (SD)		CA (SD)		COPD (SD)		AS (SD)		Total	H	CA	COPD	AS
2-Methylbutane	1.52	(1.32)	3.73	(6.15)	1.63	(0.50)	1.72	(1.85)	37.5	25.0	50.0	33.3	50.0
Pentane	1.66	(0.67)	2.21	(1.09)	1.87	(0.53)	2.11	(2.26)	51.8	45.0	62.5	41.7	50.0
Ethanol	70.60	(95.14)	179.08	(132.87)	218.64	(216.1)	100.89	(108.96)	98.2	95.0	93.8	100.0	100.0
Isoprene	32.85	(34.01)	34.19	(30.08)	34.61	(20.9)	48.70	(16.95)	100.0	95.0	100.0	100.0	100.0
2-Propanol	10.55	(9.30)	230.66	(190.62)	258.37	(255.01)	123.42	(67.37)	85.7	55.0	100.0	100.0	100.0
2-Methylpentane	1.24	(0.30)	3.44	(2.41)	2.61	(2.07)	4.59	(5.57)	55.4	25.0	75.0	75.0	50.0
3-Methylpentane	0.24	(0.12)	0.93	(0.72)	1.27	(0.49)	1.07	(1.25)	35.7	10.0	68.8	33.3	25.0
1-Propanol	14.59	(14.63)	34.10	(37.73)	28.15	(38.54)	9.94	(5.77)	100.0	95.0	100.0	100.0	100.0
Methylcyclopentane	1.80	(0.53)	2.49	(1.11)	2.20	(0.47)	2.20	(0.27)	87.5	75.0	93.8	83.3	100.0
2-Butanone	1.74	(1.15)	1.93	(1.30)	1.45	(1.00)	1.26	(0.80)	80.4	55.0	100.0	83.3	87.5
Benzene	1.13	(0.83)	0.29	(−)	0.57	(−)	0.60	(0.09)	16.1	25.0	6.3	8.3	25.0
Acetoin	44.02	(19.8)	60.39	(51.63)	55.22	(28.95)	41.72	(17.93)	53.6	45.0	56.3	50.0	75.0
Toluene	6.23	(8.38)	0.98	(1.40)	0.63	(0.42)	0.89	(0.60)	55.4	40.0	75.0	58.3	50.0
Ethylbenzene	0.650	(0.65)	2.73	(2.50)	0.34	(0.36)	1.41	(−)	17.9	5.0	25.0	33.3	12.5
p-Xylene	1.15	(0.92)	1.62	(1.86)	1.40	(1.11)	1.97	(0.95)	41.1	25.0	50.0	58.3	37.5
Styrene	0.27	(0.26)	3.78	(6.26)	1.61	(1.30)	0.73	(0.59)	53.6	5.0	81.3	75.0	87.5
Decane	nd	(−)	nd	(−)	0.23	(−)	nd	(−)	1.8	0.0	0.0	8.3	0.0
6-Methyl-2-heptanone	1.65	(−)	4.42	(2.85)	1.72	(−)	6.46	(−)	12.5	5.0	25.0	8.3	12.5
Isododecane	0.69	(0.49)	1.59	(1.57)	0.98	(0.48)	0.52	(0.35)	76.8	45.0	93.8	83.3	100.0
1,2,4-Trimethylbenzene	0.83	(0.61)	2.55	(1.94)	2.60	(2.46)	1.42	(0.76)	82.1	50.0	93.8	100.0	100.0
(*E*)-Ocimene	1.16	(0.80)	4.64	(4.03)	2.95	(1.99)	2.98	(1.54)	82.1	50.0	100.0	91.7	100.0
Limonene	1.57	(1.20)	1.87	(1.80)	1.71	(1.75)	5.05	(2.15)	89.3	75.0	93.8	91.7	100.0
m-Cymene	0.61	(0.21)	0.41	(0.35)	0.38	(0.23)	0.32	(0.21)	46.4	10.0	62.5	50.0	87.5
Benzonitrile	1.44	(1.23)	3.57	(3.95)	4.57	(2.64)	2.12	(1.67)	78.6	50.0	93.8	91.7	87.5
Phenol	nd	(−)	52.78	(47.13)	75.02	(72.16)	nd	(−)	16.1	0.0	37.5	25.0	0.0
Undecane	0.80	(0.11)	3.83	(3.09)	2.44	(1.38)	1.78	(0.30)	41.1	20.0	75.0	41.7	25.0
Dodecane	5.18	(0.72)	10.58	(8.66)	9.51	(7.18)	6.27	(3.19)	73.2	45.0	87.5	91.7	87.5
Terpineol	3.57	(0.30)	17.36	(21.72)	26.53	(36.34)	6.87	(6.38)	71.4	15.0	100.0	100.0	100.0
Tridecane	3.43	(1.69)	42.16	(38.89)	28.36	(21.59)	8.59	(7.87)	51.8	10.0	75.0	75.0	75.0

2.2. VOCs Detected in Breath

Regarding the obtained VOC global profiles, a total number of 112 different VOCs were detected. The VOCs most frequently observed in the samples were hydrocarbons, alcohols, aldehydes and ketones. A graph displaying the distribution of VOCs according to the functional groups in profiles belonging to the different studied groups is presented in Figure 1a. In general, the number of compounds belonging to each of the chemical classes seems to be proportional when evaluating the different studied conditions, however, some particularities of the qualitative composition of each group of profiles can be evidenced. Lung cancer and COPD profiles appear to be associated to a greater variety of compounds (103 and 95 detected VOCs, respectively), while asthma profiles are composed by smaller number of compounds (84 detected VOCs). An increased number of hydrocarbons is observed in the VOC composition in breath of lung cancer patients. Moreover, samples from patients with lung cancer and COPD appear related to a greater variety of aldehydes (12 and 11, respectively, against 9 found in healthy). This observation can be due to the fact that hydrocarbons and aldehydes are frequently reported as the most characteristic products of oxidative stress induced by inflammatory process [26–28].

A matrix displaying number and percentage of overlapping VOCs in the acquired profiles is presented in Figure 1b. By the content of coincident compounds, the level of similarity regarding the qualitative composition of breath samples can be inferred. In this sense, lung cancer and COPD profiles, display the greater similarity between each other, followed by the VOC profile of lung cancer and healthy individuals, while asthma breath samples present to be the most distinct in terms of composition.

(a)

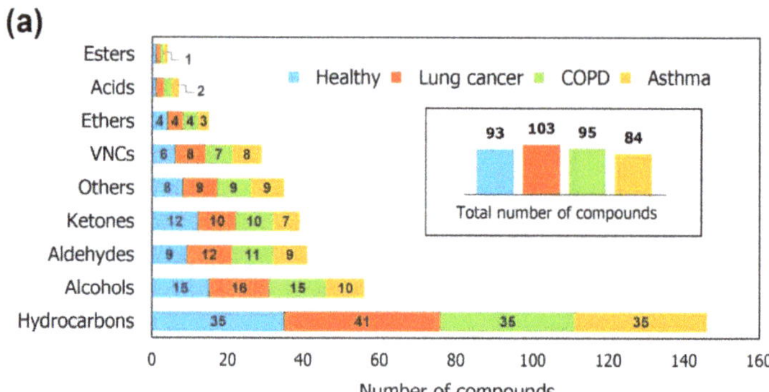

(b)

	Healthy	Lung cancer	COPD	Asthma
Healthy	**93** / 100%	86 / 92.47%	82 / 88.17%	69 / 74.19%
Lung cancer	86 / 83.50%	**103** / 100%	91 / 88.35%	82 / 79.61%
COPD	82 / 86.32%	91 / 95.79%	**95** / 100%	77 / 81.05%
Asthma	69 / 82.14%	82 / 97.62%	77 / 91.67%	**84** / 100%

Figure 1. (**a**) VOCs distribution according to main chemical classes, in profiles belonging to the different studied groups, the contoured box displays the total number of compounds found in each group; (**b**) Similarity matrix displaying number and percentage of overlapping VOCs in the acquired profiles.

2.3. Differential Distribution of VOCs

Principal component analysis (PCA) was performed intending to identify relationships and existing patterns within datasets. Peak area data regarding the global VOC profiles was used as input for generation of the score plot depicted in Figure 2a, in which 78.04% of variance was represented by the two first principal components. When using as input the calculated concentration values of the 29 preselected analytes in samples, the plot presented in Figure 2b is produced. In this case, 79.72% of total variance was described by the components **1** and **2**. In both cases, around 80% of the total variance can be assigned to the observed distribution. Although both score plots indicate a discrimination between control cases and remaining samples, a clearer grouping can be observed when considering the global profile, once in Figure 1b control samples appear confined to an isolated cluster. Still, in both situations, the lack of a distinct grouping according to each of the investigated conditions demonstrates that other factors play a relevant role in the observed pattern of distribution of VOCs. This can be mainly related to the variability in the nature and extension of the involved pathophysiological mechanisms, inherent to the different lung diseases. Therefore, the usage of supervised approaches is essential to achieve the classification of samples in agreement with the related diagnosis.

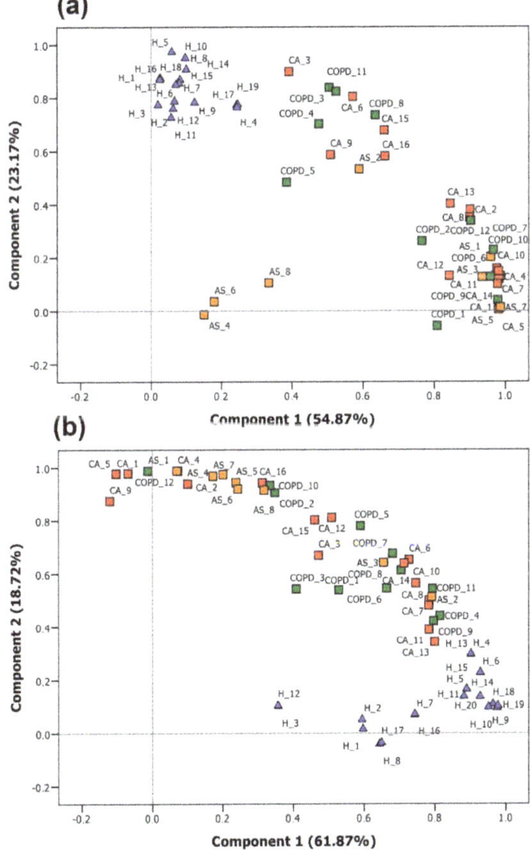

Figure 2. PCA plots using as input (**a**) VOCs' responses in global profile analysis, (**b**) responses of the targets quantified in the samples (triangles = control samples; squares = positive samples).

A volcano plot was built in order to present found discriminating features when considering obtained global VOCs profiles. In Figure 3a the overall trend of the detected VOCs (variables) is graphically represented. The variables located above the dashed line refer to the compounds which displayed greater statistically relevant changes in their responses when compared to the control group. The variables located along the y-axis correspond to VOCs absent in the healthy group and detected solely in positive samples. In the left part of the plot are displayed compounds with decreased responses in the positive samples, while in the right side of the plot are displayed VOCs presenting an increased response in samples of diseased. The VOCs located towards the top of the graph expressed the greatest statistical significance. The names of the most discriminative components are exhibited in the plot.

Figure 3b presents a bar graph showing the distribution of all compounds classified as discriminant features, considering as criteria $p \leq 0.05$. Most of the compounds which displayed significant alteration in their responses when compared to those presented in the healthy group belong to the class of hydrocarbons, followed by alcohols and aldehydes. In lung cancer profiles, a greater number of discriminating VOCs was verified (41 compounds). For asthma and COPD samples, 26 and 24 altered VOCs were indicated, respectively. As presented in Figure 1b, around 92, 88 and 74% of the compounds observed in lung cancer, COPD and asthma samples, respectively, were shown to be conserved in the healthy group profiles. This indicates that the differential abundance of VOCs in samples is determinant to discriminate between samples' group, once the similarity between the qualitative profiles belonging to the four studied groups is not so divergent. Such observation highlights the importance of validated quantitative assays' application regarding breath samples for diagnosis purposes.

Few compounds presented a more expressive incidence within the group of active smokers' individuals, thus possibly being ascribed as products of cigarette smoke. 1,3-Cyclopentadiene was identified solely in this group, in 40% of the samples; 2,5-Dimethylfuran was detected in 80% of samples from active smokers, which represented around 73% of its total incidence across samples. Other substances commonly related to tobacco smoke composition, such as benzene and toluene [29], did not present a specific distribution within samples of smokers, probably because these can be originated from other various sources.

With respect to the VOCs found altered, acetonitrile is typically present in cigarette smoke, although also present in automobile exhaust and other anthropogenic emissions [30]. Considering that most of the enrolled subjects were not smokers, differentiated levels of this substance would not be expected. However, together with the decreasing trend observed for p-xylene, the reduction in the abundances of such compounds in positive group can be an indicative of diminished ability of elimination of exogenous through exhaled air, or a consequence of the augmented activity of cytochrome P450 isoforms documented in lung cancer [31], which could be responsible for the rapid metabolization of inhaled compounds in the lungs.

The two main lung cancer types are small-cell lung carcinoma (SCLC) and non-small-cell lung carcinoma (NSCLC). Two hypothesis involve SCLC histogenesis: the first assumes that SCLC derives from cells of the diffuse endocrine system, i.e., the amine precursor uptake decarboxylation (APUD)-system, the second suggests this type of lung cancer originates from the endodermbronchial lining [32,33]. Adenocarcinoma (NSCLC subtype) arises from glandular cells of bronchial mucosa, whereas squamous lung cancer origins from the modified bronchial epithelial cells and adenosquamous carcinoma contains two types of cells: squamous cells (thin, flat cells that line certain organs) and gland-like cells. Finally, large cell (undifferentiated) carcinoma originates from epithelial cells of the lung [32]. The origin and nature of the malignant cells is crucial for different treatment strategies. Tumor tissue releases different protein biomarkers according to subtype of cancer. The same concerns different types or amounts of certain VOCs secreted by various malignant part of cell. The oxidation of fatty acids present in the cell membranes is pointed out as the source of VOCs associated to oxidative stress condition. The mentioned process is initiated by the reactive oxygen species (ROS) which are found in increased levels in inflamed tissues [17,28].

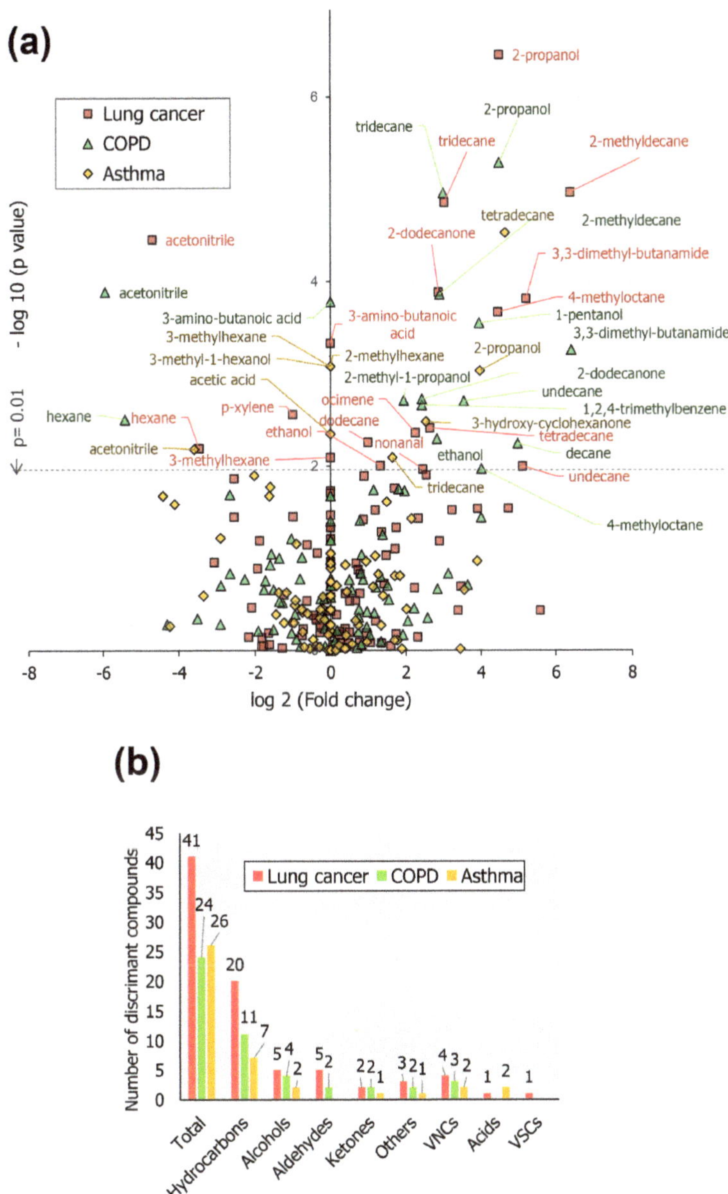

Figure 3. (**a**) Volcano plot displaying the most discriminating features, in terms of fold change (*x*-axis) and statistical relevance (*y*-axis), in which the dashed line represents the point of *y*-axis in which $p = 0.01$; (**b**) Graph of distribution of number of all compounds assigned as discriminating features, according to disease and chemical class (significance criteria: $p \leq 0.05$).

Due to the ROS activity, mechanism chain reactions occur, with radicals tending to be stabilized through alpha and beta scissions [34], leading to the formation of a variety of shorter chain fatty acids, alkanes, alkenes, alcohols and aldehydes. In addition, formed compounds can be subjected to other reactions, aiming their transformation into smaller and more polar molecules [35]. Cancer cells are characterized by their enhanced metabolism and altered functions in several biochemical pathways [36]. Therefore, metabolite profile consisting of a greater variety of compounds may be expected. Hexane can be possibly formed during the oxidation of oleic acid [34], while can be addressed as an exogenous substance as well. Hexane showed decreased abundance in cancer and COPD samples. This fact can be explained by three hypotheses: impaired excretion through exhalation [37], enhanced conversion of the specie into oxidized forms [38] and favoring of alternative mechanism, which gives rise to different products, during lipid oxidation associated to oxidative stress particular to the referred conditions.

1-Pentanol can be interpreted as a pentane oxidation product, caused by cytochrome P450, and recognized as a metabolite of reactive oxygen species reactions with omega-6 fatty acids [26]. Methyl ketones such as 2-dodecanone can be formed by the decarboxylation of β-keto acids during the metabolism of fatty acids [39]. Nonanal can be also formed by different mechanisms during ROS attack on oleic acid from cell membranes [34]. Medium-chain branched alkanes, such as 2-methyldecane and 4-methyloctane, were pointed out by previous works as oxidative stress indicators [40,41]. However, their generation by human organism due to the oxidation of lipids is questionable, as cell membranes contain only linear chain lipids [26].

Branched alkanes can be originated from microbial lipids, mostly produced in the fatty acid pathway of bacteria, by using amino acids as precursor molecules which are submitted to elongation in this biochemical path [42,43]. Considering this, the occurrence of methylated branched alkanes in breath could be connected with bacterial activity. Alternatively, these could be products of transformation/degradation of prenyl molecules in organism, a mechanism that also remains undescribed. Aromatic species, such as *p*-xylene (decreased in COPD) and 1,2,4-trimethylbenzene (increased in COPD), are frequently addressed as pollutants, although also possibly formed by bacterial shikimate and related pathways [44].

Regarding the 29 compounds belonging the set of selected targets, 15 of them presented themselves as discriminating features ($p < 0.05$) when assessing solely controls against positive samples, all of them displaying increased concentration in the positive group. They were 2-propanol, 2-methylpentane, 3-methylpentane, 1-propanol, 2-butanone, styrene, isododecane, 1,2,4-trimethylbenzene, (*E*)-ocimene, *m*-cymene, phenol, undecane, dodecane, terpineol and tridecane. However, as demonstrated in the next section, compounds other than these displayed usefulness in the characterization of studied groups, presenting themselves as discriminating variables related to disease type. A combination of mechanisms involved in carcinogenesis, inflammatory processes and microbiota activity—which develop important role in pathogenesis of several diseases, may play a part in the alterations observed for certain compounds in breath samples.

The propionic acid formed during microbial fermentation and the propionyl-CoA generated during amino acids degradation enters in the propanoate metabolism, which takes place in the mitochondria and comprehend a series of reactions coupled with other pathways related to cell energetics. In microorganisms, 1-propanol is a product of propanoyl-CoA transformation [45], while 2-propanol can be formed by the reduction of acetone produced during the synthesis of ketone bodies [46]. 2-Butanone is a secondary ketone, therefore its origin can be associated to the β-oxidation of fatty acids. The acetyl-CoA units generated in this process fuel the citric acid cycle, supplying energy generation [47]. Terpenoids are very diverse natural products synthetized by plants, but also by bacteria. These metabolites are associated to the mevalonate and deoxyxylulose phosphate pathways [48,49]. Although their biosynthesis in human so far remain unknown, studies have reported terpenoid derivatives as potential cancer indicators. Considering this, increased concentration of compounds such as (*E*)-ocimene, *m*-cymene and terpineol can either be a consequence of deficient metabolic function impairing proper elimination of these substances coming from diet [50], an indicative of specific bacterial activity, or even a product of transformation of isoprenoids derivatives due to the dysregulated mevalonate pathway in human during carcinogenesis [51].

Isododecane is known as a synthetic chemical with several applications in the industry [52], without any identified biosynthetic pathway so far. Styrene is a constituent of polymers, nevertheless, there is evidence that some microorganisms can produce styrene using phenylalanine as precursor molecule [53]. On the other hand, phenol is often reported as product of bacterial catabolism of aromatic amino acid species previously documented as elevated in gastroesophageal neoplasms [54].

Their formation of the *n*-alkanes undecane, dodecane and tridecane can be related to the oxidation of lipids, more precisely, a formed alkoxyl radical undergoes scission, generating an alkyl radical which abstracts a hydrogen atom, turning into a stable alkane [17,26]. 2-Methylpentane and 3-methylpentane are other branched species possibly derive from the oxidation of branched chain fatty acids generated by bacteria.

2.4. Diagnosis Prediction–Global Profiles

Most of the studies comprising the detection of diseases based on VOC analysis in biological samples compare paired data from healthy and diseased groups. Many of the compounds addressed as candidate biomarkers by literature are explained as produced by oxidative stress–a process promoted by typical inflammatory immune responses and thus non-specific. In this sense, illnesses sharing common etiological and pathological processes may play a part as confounding factors when a specific diagnosis is intended. For this reason, the present and following sections of the manuscript were dedicated to the development of statistical models able to identify and discriminate specific VOC patterns, allowing simultaneous differentiation of the studied lung diseases.

A random forest (RF) analysis was conducted on global profiles data, aiming to classify obtained VOC fingerprints into the four investigated categories. Variance importance was assessed based on the mean decrease Gini when one of the questioned variables is removed from a preliminary RF model. Gini impurity can be interpreted as the chance of a case sampled randomly to be incorrectly classified in relation to a given class, thus being related to the purity of cases within a tree node [55]. Therefore, greater decreases in this measurement indicate greater importance of a given variable. The resulting plot is presented in Figure 4a, the compounds are ranked from the most essential to those less relevant for the obtaining of homogenous classes. The 12 most important variables were assigned to compose the RF final model, the selected compounds appear depicted as the gray diamonds, in the upper part of the graph. The intention was to obtain the greater model overall accuracy as possible, including a minimum number of features. Predict probabilities of a case of the validation set to belong to a class were provided by RF modeling. The receiver operating characteristic (ROC) curves presenting the ability of the model to predict a certain condition are showed in Figure 4b, information on parameters regarding classification performance are presented in Table 2. It can be observed that class recognition was performed with at least 93% of sensitivity and 87.5% of specificity for lung cancer, asthma and healthy groups. Regarding the later mentioned groups, prediction with accuracy above 87% was achieved. The lower prediction capability obtained in case of COPD (67%). An exemplary decision tree, from the 1000 generated during modeling, is presented in Figure 4c.

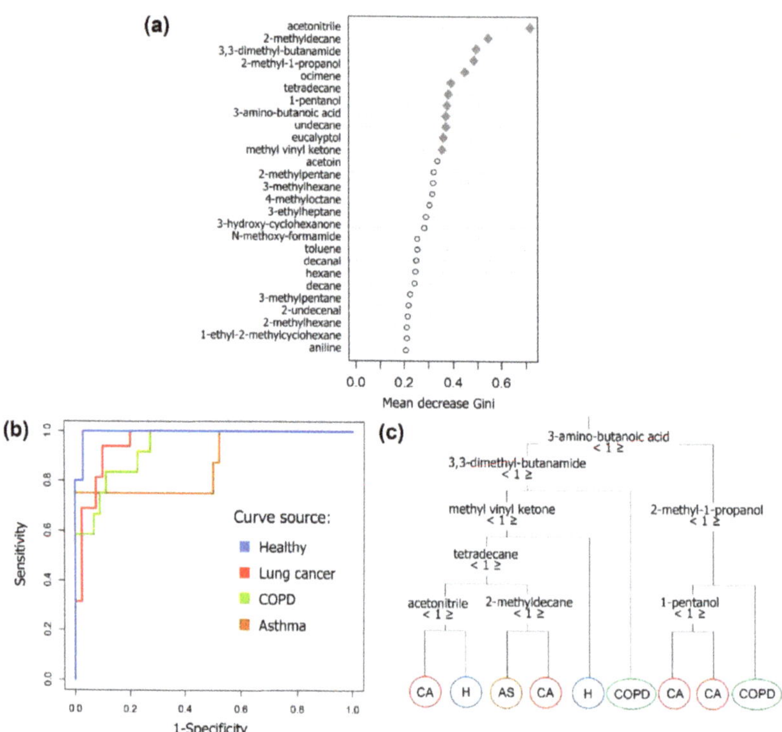

Figure 4. (**a**) Variable importance plot in terms of mean decrease Gini (node purity), obtained in the first training of RF model. Diamonds refer to VOCs selected for generation of the final classificatory model; (**b**) ROC curves based on RF's final model output regarding the test set, using a panel of 12 VOCs; (**c**) Example decision tree produced by RF analysis, in which obtained accuracy was 81% (AS = asthma, CA = lung cancer, COPD = chronic obstructive pulmonary disease, H = healthy).

Table 2. RF model performance (AUC = area under the curve, CI = confidence interval).

Statistics by Class	Sensitivity	Specificity	AUC	Balanced Accuracy
Asthma	75.0%	100%	0.872	87.5%
Lung cancer	93.8%	87.5%	0.956	90.6%
COPD	67.0%	97.7%	0.935	82.2%
Healthy	95.0%	94.5%	0.994	94.7%
RF overall accuracy (95% CI)	85.7% (73.7–93.6)			

2.5. Diagnosis Prediction–Target Analysis

In this section, in accordance with the criteria described in the Material and Methods section and empiric observations drawn from multinomial logistic regression (MLR) performance using different set of variables, 2-propanol, 3-methylpentane, (*E*)-ocimene, limonene, *m*-cymene, benzonitrile, undecane, terpineol, phenol were the compounds selected to build the MLR final model. A clearer depiction of variables distribution according to their importance can be observed in Figure 5. Table 3 presents information regarding the developed model, which, when applied to the train and test datasets provided 100% and 90.5% of accuracy, respectively (average overall accuracy = 95.3%).

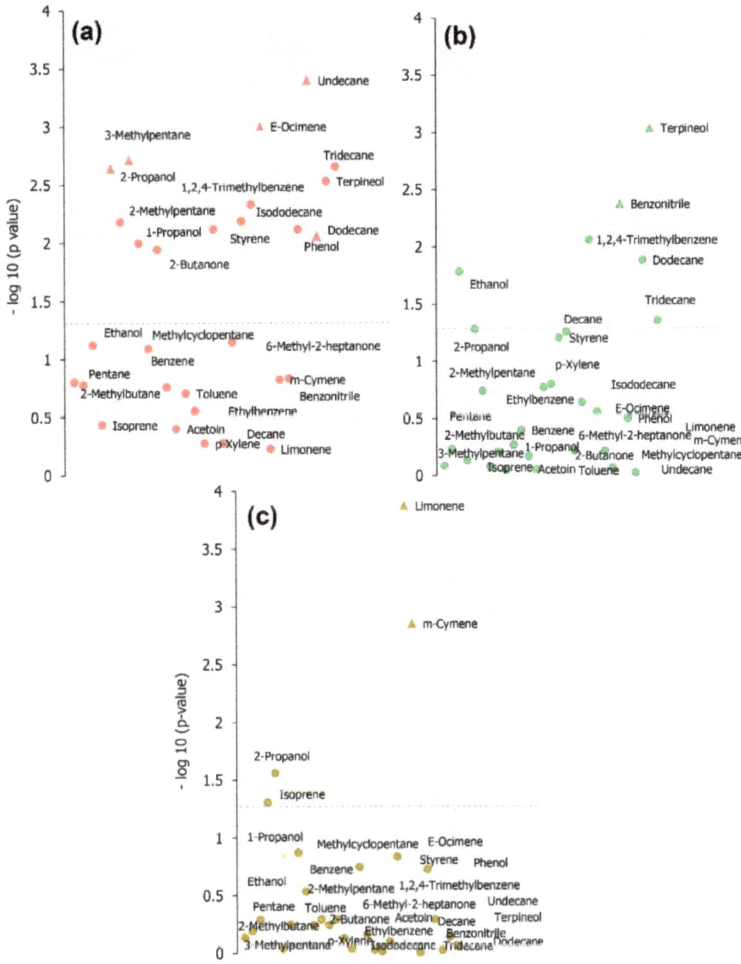

Figure 5. Plot of $-(\log_{10})$ of p values when applying Mann-Whitney test for specific classes: (**a**) lung cancer, (**b**) COPD or (**c**) asthma, against all other conditions. Dashed line represents where $p \leq 0.05$. Variables represented by triangle shape icon were those included in MLR final model.

Table 3. Description of MLR model (AS = asthma, CA = lung cancer, COPD = chronic obstructive pulmonary disease, SE = standard error).

Condition	Coefficients									
	Intercept	2-Propanol	3-Methyl-pentane	(*E*)-Ocimene	Limonene	*m*-Cymene	Benzonitrile	Undecane	Terpineol	Phenol
AS	−557.79	0.56	17.03	50.39	100.40	212.50	−22.69	−88.94	32.33	−12.95
(SE)	(42.62)	(0.37)	(2.55)	(2.48)	(10.34)	(11.52)	(1.32)	(3.64)	(1.31)	(5.31)
CA	−127.12	0.61	−129.32	32.92	32.33	31.11	−96.07	26.50	32.81	−1.89
(SE)	(29.25)	(0.38)	(17.43)	(8.11)	(9.90)	(6.47)	(16.91)	(11.17)	(8.40)	(0.86)
COPD	−23.49	0.49	−269.07	−64.27	11.09	−72.04	−7.72	−111.52	37.91	1.73
(SE)	(3.35)	(0.83)	(6.92)	(8.57)	(3.39)	(0.66)	(1.29)	(4.22)	(8.18)	(0.65)

In MLR, coefficients can be multiplied by the quantitative inputs for the calculation of probabilities of a case to belong to a specific condition. Equation (1) presents the model regression equation, where $\ln[P/(1-P)]$ represents the log-odds pertinent to a specific disease, β_0 is the intercept and $\beta_{1\ldots k}$ are the coefficients provided by the MLR model, referring to the variables X (in the case, the concentration of the selected targets). A case for which the calculated probabilities are greater than 50%, can be assigned as belonging to that class.

$$\ln[P/(1-P)] = \beta_0 + \beta_1 \cdot X_1 + \ldots + \beta_k \cdot X_k \tag{1}$$

The numerical coefficients provided by MLR can be interpreted as weights, or the contribution of these variables to the designated classes. Positive coefficients are related to compounds with increased response when comparing to the reference class ("Healthy"), while negative coefficients are associated to targets which were present in lower concentrations in positive samples. In a closer interpretation, the coefficients express multinomial log-odds. For example, assuming that all other variables remain constant, an increase of one unit in the concentration of 2-propanol multiplies the odds of a sample belonging to the asthma group instead of healthy group by 0.56. On the other hand, an increase in one unit of (*E*)-ocimene concentration in breath implies the log-odds of COPD to decrease by 64.27, in an assumption that the remaining variables are kept constant. Considering this, increased levels of limonene and *m*-cymene are characteristic from samples of asthma patients, while increased level of undecane and decreased concentrations of benzonitrile are observed for breath of individual with lung cancer; Moreover, greater concentrations of phenol and lower concentrations of *m*-cymene are particularly observed in samples from COPD patients. Values fitted for the train set and predictions performed by MLR method solely for the validation set were used as input to build ROC curves (Figures 6a–d and 6e, respectively). Values of area under the curve (AUC) presented in Figure 6a–d represent the probability of samples belonging to a given group to be classified as the state condition. For each class specified in the model, AUC was 1.0, meaning that 100% of sensitivity and specificity was obtained. On the other hand, cases not assigned as the state variable provided AUC ≤ 0.5 (curves below random guessing line). When considering the performance of the model on the test data, an overall accuracy of 91% was obtained, resulting in an average accuracy of 95.5% when both evaluated sets are considered. Detailed information regarding MLR performance is presented in Table 4.

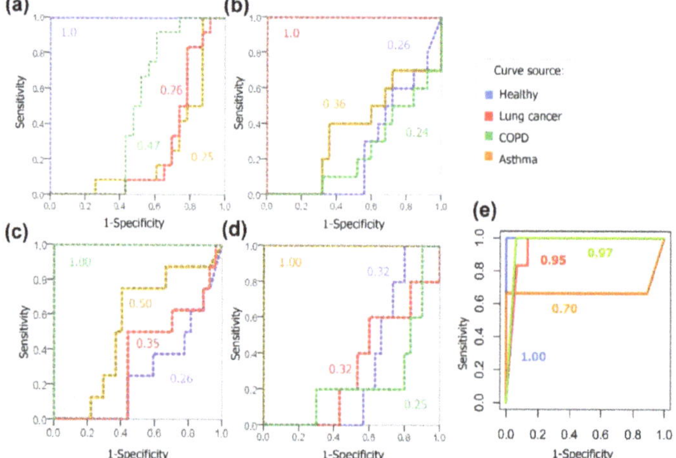

Figure 6. ROC curves generated from fitted values (train set) created by MLR model, labelling (**a**) healthy, (**b**) lung cancer, (**c**) COPD and (**d**) asthma as the state variables; (**e**) ROC curves generated from predictions computed by the MLR model for the test set. Colored numerals refer to values of AUC obtained for each depicted curve.

Table 4. MLR model performance (AUC = area under the curve, CI = confidence interval).

Statistics by Class	Sensitivity	Specificity	AUC	Balanced Accuracy
Asthma	68.0%	100%	0.700	83.4%
Lung cancer	83.4%	93.4%	0.950	88.4%
COPD	100%	94.1%	0.971	97.1%
Healthy	100%	100%	1.000	100%
MLR overall accuracy (95% CI)	91.0% (70.0–99.0)			

3. Materials and Methods

3.1. Apparatus and Standards

The analyses were conducted on a model 6890 A gas chromatograph coupled with a 5975 Inert XL MSD (Agilent Technologies, Waldbronn, Germany). Inlet temperature was kept at 260 °C and carrier gas (helium 6.0) flow was set at 2.2 mL min^{-1}. A DB-624 capillary column (Agilent) 60 m × 0.32 mm × 1.8 μm was used. The oven temperature program was as follows: initial temperature was 35 °C (held for 3 min), ramped to 50 °C, then 75 °C, 200 °C and finally 240 °C, at rates of 3 °C min^{-1}, 5 °C min^{-1}, 15 °C min^{-1} and 10 °C min^{-1}, respectively. The last temperature was kept for 15 min, resulting in a run time of 41.33 min. Full scan spectra were acquired within a range of 30–300 m/z, at electron ionization (EI) of 70 eV. The ion source and transfer line were set to 250 °C. Chromatographic data acquisition was performed using MSD ChemStation E.02.00.493 software (Agilent). Compounds identification was performed using NIST05 mass spectra library. Each peak was searched manually, including baseline subtraction and averaging over a peak. Forward match quality of at least 750/1000 was applied as the lower match threshold.

Needle trap device 700-60d-PXC (PDMS + Carbopack X + Carboxen 1000) was purchased from PAS Technology (Magdala, Germany). The air pump flow was conducted by a sampling case model SC-B (PAS Technology), designed for controlled air sampling. Prior first use, NTDs were conditioned in a heated conditioner (PAS Technology) at 300 °C under helium flow (1 bar), for 30 min, in order to remove VOC's contaminations from sorbent. One liter-Tedlar bags were obtained from SKC (Eighty Four, PA, USA).

Chemicals used as standards (2-methylbutane, pentane, ethanol, isoprene, 2-propanol, 2-methylpentane, 3-methylpentane, 1-propanol, methylcyclopentane, 2-butanone, benzene, acetoin, toluene, ethylbenzene, p-xylene, styrene, decane, 6-methyl-2-heptanone, isododecane, 1,2,4-trimethylbenzene, ocimene, D-limonene, m-cymene, benzonitrile, phenol, undecane, dodecane, terpineol and tridecane) were purchased from Sigma-Aldrich (St. Louis, MO, USA), all with purity > 98%.

3.2. Breath Collection

The study was approved by the local Ethics Committee of Collegium Medicum in Bydgoszcz (No. KB 621/2016–25.10.2016). Individuals aged over 18, with positive clinical diagnosis for lung cancer (non-small cell lung cancer, subtype: adenocarcinoma) (n = 16), chronic obstructive pulmonary disease (n = 12) and asthma (n = 8) were recruited at the Department of Lung Diseases of the Provincial Polyclinic Hospital in Toruń. Samples from enrolled cancer patients were obtained before any medical intervention (such as neoadjuvant therapies or surgery).

Individuals were refrained to eat, drink or smoke 2 h prior sample collection. No special dietary regimes were applied. All individuals gave informed consent to participation in the study. The patients completed a questionnaire describing their age, gender and current smoking status (active smokers, non-smokers). Samples of mixed alveolar breath gas (alveolar and dead space gas) were collected in Tedlar bags with parallel collection of ambient air at the same room. Breath samples were obtained after approximately after 10 min rest in the same ambient. Each subject provided breath samples using a disposable plastic straw connected to the Tedlar bag.

Control samples (n = 20) were collected from healthy individuals aged over 18 years, without any history of positive diagnosis for cancer or respiratory diseases, who were not suffering from any other inflammatory disease. All samples were analyzed within 2–3 h after collection–this timeframe was considered adequate to avoid the interference of gas composition losses [56]. In the total, 56 breath samples were collected. Information regarding enrolled volunteers is summarized in Table 5 (details regarding presented significance probabilities are described in the section "Data analysis and chemometrics approaches").

Tedlar bags involved in sample collection and calibration experiments were daily treated with several cycles of cleaning, each consisting of consecutively filling and evacuating argon 5.0 from the bag. Afterwards, the bags filled with argon were kept in an oven at 65 °C. The content bag was tested before breath sampling, by means of GC-MS, in order to verify the effectiveness of cleaning procedure.

Table 5. Main information regarding volunteers (SD = standard deviation, CA = lung cancer, COPD = chronic obstructive pulmonary disease, AS = asthma).

Group		Control		Positive		p
		n	%	n	%	
Total		20		36		0.367
Gender	Male	13	65.0%	27	75.0%	0.325
	Female	7	35.0%	9	25.0%	0.437
Age (SD)		41.2 (10.1)		66.8 (8.22)		0.078
Smoking status	Active smoker	2	10.0%	5 (2 COPD, 3 CA)	13.9%	0.287
	Ex-smoker	2	10.0%	22 (12 CA, 10 COPD)	61.1%	0.083
	Non-smoker	16	80.0%	9 (1 CA, 8 AS)	25.0%	0.640
Condition	Lung cancer	–		16	44.4%	
	COPD	–		12	33.3%	–
	Asthma	–		8	22.2%	

3.3. Selection of Targets

The compounds selected as targets were VOCs previously reported as potential breath biomarkers of lung cancer, COPD and asthma, in accordance with previous studies on this theme. A literature search was performed in the electronic database Web of Science Core Collection (from Clarivate Analytics; Philadelphia, PA, USA), as well as Google Scholar. The searched terms were: "volatile organic compounds", "gas chromatography", "biomarker", "lung cancer", "COPD" and "asthma", considering a time span from 1999 to 2016. The indexed literature is presented in the Supplementary Material (Table S2) [57–80].

3.4. Calibration Procedure

Gas mixtures of the analytes were prepared by injection of 1 µL of liquid standards into 1 L glass bulb (Supelco, Bellefonte, PA, USA) previously evacuated. Methanol HPLC was used for the preparation of 50:50 (v/v) dilution of acetoin, phenol and terpineol, which are solids at room temperature. After the complete vaporization of the liquids, the interior of the bulb was equilibrated with nitrogen, generating a gas mixture of the compounds of interest. Using a gas-tight syringe, different volumes of the stock gas solution were transferred to Tedlar bags filled with 1 L of nitrogen, in order to obtain the desired concentrations.

The concentrations were calculated in terms of part per billion per volume of analyte (ppbv), taking in consideration their molar volume. Six calibration levels were used in the construction of calibration curves, all analyzed in triplicates. The limit of detection (LOD) was defined as the lowest detectable concentration of analyte, considering a signal-to-noise (S/N) ratio of at least 3. LOQ was considered as the lowest concentration of analyte with imprecision of at least 15%, considering a minimum S/N value equal to 10. Calibration was conducted by linear regression analysis, using the obtained experimental data. Linearity was evaluated by the method of least squares and reported as the coefficient of determination (R^2). Linearity was confirmed for values of R^2 above 0.99. Inter-assay imprecision was assessed from the evaluation of assays in triplicate, these were expressed in terms of relative standard deviation (RSD%). Reported RSD% values are the average of imprecision calculated for lower (LOQs), medium (5.17–17.25 ppbv) and high calibration levels (9.52–3452.0 ppbv)—which concentrations varied depending on the linearity range displayed by the analyte.

3.5. Sample Extraction

Prior to sample extraction process, NTDs were conditioned for 10 min, at 300 °C, under helium 6.0 flow (Air Products, Warsaw, Poland). Samples in Tedlar bags were drawn through the air pump, at a flow rate of 30 mL min^{-1}. The fixed volume of 50 mL was sampled from each bag. Once extraction was complete, the loaded NTD was desorbed into GC inlet port for 2 min.

3.6. Data Analysis and Chemometrics Approaches

For the building of main dataset, area of peaks belonging to ambient air samples were subtracted from respective samples obtained from patients. Evaluation of normality of distributions was conducted using Kolmogorov-Smirnov test. Differences between volunteers' ages was assessed by t-test. Principal component analysis was performed in order to evaluate data patterns regarding sample's group. Mann-Whitney test was used to indicate VOCs which presented statistically relevant differences in their responses in the studied groups, $p \leq 0.05$ was considered as the relevance criteria. For the above cited tests, IBM SPSS Statistics v. 24 software (IBM Corp., Armonk, NY, USA) was used. The following approaches were executed in R environment, using RStudio console v. 1.1.463 (RStudio, PBC, Boston, MA, USA). Significant differences between the proportions of volunteers assigned to a certain group were assessed by the test of equal or given proportions, employing the R function "prop.test". For chemometrics approaches, the packages "gplots", "RandomForest", "caret", "ROCR" were employed. Random forest is a machine learning method based on recognition of latent patterns within global VOC profiles. RF was conducted in order to develop a multiclass model, able to distinguish between studied conditions. RF input consisted of peak table data converted into binary entries–once this algorithm was dedicated to non-quantified data, this format of dataset was considered as more appropriate than to express RF outcome in terms of peak area. Variable importance plots were assessed for selection of variables to be included in the model. Half of the data was randomly selected to compose the training set (bootstrap sampling method) and the remaining data was applied in the validation process. Receiver operating characteristic curves were built based on calculated probabilities output from RF modeling. Ultimately, the development of a classificatory model based solely on target compounds was aimed, for that, variables (targets) were selected according to their discriminative potential between all four investigated conditions. The criteria comprised most unique targets which presented higher discriminative relevance when considering a given condition against all others. MLR was performed using the package "nnet", employing the data regarding quantitation of the selected targets in analyzed samples. This multiclass categorical method performs a linear combination of features, allowing prediction through the calculated probabilities of an input (set of features' values) to belong to each class specified in the model. Sixty percent of the data regarding quantitation of targets in the samples was randomly addressed as the training set, while the remaining data was addressed to a testing set. "Healthy" group was defined as the reference class. ROC curves were prepared based on the predictions computed by developed MLR model for fitted values and test data.

4. Conclusions

The developed NTD-GC-MS method was demonstrated to be suitable for the determination of target VOCs in breath samples, providing considerably low limits of detection and quantitation, as well as appropriate reproducibility. From the 29 targets selected from literature, more than half of them presented significant differentiated responses among control and positive groups – found discriminating features were 2-propanol, 2-methylpentane, 3-methylpentane, 1-propanol, 2-butanone, styrene, isododecane, 1,2,4-trimethylbenzene, (*E*)-ocimene, *m*-cymene, phenol, undecane, dodecane, terpineol and tridecane, limonene and benzonitrile (which proved to serve for further differentiation between diseases). Built statistical models (using both self-annotated discriminating variables and quantified targets) aimed to simultaneously classify VOC profiles into lung cancer, COPD or asthma cases. Both classification models (RF and MLR), provided an overall accuracy above 80%. The distinction between VOC profiles related to clinical conditions involving concomitant molecular mechanisms is extremely relevant in order to assess cofounding aspects of breath analysis diagnosis. In this sense, machine learning tools and other mathematical models can be useful to identify disease-specific latent patterns. Cross-validated studies, comparing candidate biomarkers found by different research groups by means of different techniques, are essential for a future implementation of breath screening tests in a clinical setting. Such an approach can also enable a focused investigation of the pathways involved in the modulation of these potential biomarkers, as well as it can contribute to the establishment of optimized analysis protocols.

Supplementary Materials: The following are available online, Table S1: Data regarding calibration method of gas mixtures (LOD = limit of detection, LOQ = limit of quantitation, ppbv = part per billion per volume, R^2 = determination coefficient, RSD = relative standard deviation). Table S2: References which reported the targets selected in the present study as potential biomarker of lung diseases in breath samples, where: LC–lung cancer; COPD–chronic obstructive pulmonary disease.

Author Contributions: Conceptualization, F.M., T.L. and I.-A.R.; methodology, F.M. and T.L.; software, F.M.; validation, F.M. and M.M.-M.; formal analysis, F.M.; investigation, F.M. and M.M.-M.; resources, T.L. and Beata Brożek (B.B.); data curation, F.M., M.M.-M., I.-A.R. and Beata Brożek (B.B.); writing—original draft preparation, F.M. and I.-A.R.; writing—review and editing, F.M., M.M.-M. and T.L.; visualization, F.M.; supervision, T.L. and Bogusław Buszewski (B.B.); project administration, T.L. and Bogusław Buszewski (B.B.); funding acquisition, T.L. and Bogusław Buszewski (B.B.). All authors have read and agreed to the published version of the manuscript.

Funding: This research was financed by The National Centre for Research and Development (Warsaw, Poland) in frame of Polish-Turkish bilateral project "A comparative study of volatile organic compound biomarkers in breath and urine samples collected from Polish and Turkish communities for monitoring of various respiratory diseases" (POLTUR2/4/2018).

Institutional Review Board Statement: The study was conducted according to the guidelines of the Declaration of Helsinki, and approved by the local Ethics Committee of Collegium Medicum in Bydgoszcz (No. KB 621/2016–25.10.2016).

Informed Consent Statement: Informed consent was obtained from all subjects involved in the study.

Data Availability Statement: Data is contained within the article or Supplementary Material.

Conflicts of Interest: The authors declare no conflict of interest.

Sample Availability: Samples of the compounds are not available from the authors.

References

1. Varghese, C.; Troisi, G.; Schotte, K.; Prasad, V.M.; St Claire, S.M. World No Tobacco Day 2019 puts the spotlight on lung health. *J. Thorac. Dis.* **2019**, *11*, 2639–2642. [CrossRef] [PubMed]
2. Van de Kant, K.D.; Van der Sande, L.J.; Jöbsis, Q.; Van Schayck, O.C.; Dompeling, E. Clinical use of exhaled volatile organic compounds in pulmonary diseases: A systematic review. *Respir. Res.* **2012**, *13*, 117. [CrossRef] [PubMed]
3. Hashoul, D.; Haick, H. Sensors for detecting pulmonary diseases from exhaled breath. *Eur. Respir. Rev.* **2019**, *28*, 190011. [CrossRef]
4. Sun, X.; Shao, K.; Wang, T. Detection of volatile organic compounds (VOCs) from exhaled breath as noninvasive methods for cancer diagnosis. *Anal. Bioanal. Chem.* **2016**, *408*, 2759–2780. [CrossRef] [PubMed]
5. Monedeiro, F.; Dos Reis, R.B.; Peria, F.M.; Sares, C.T.G.; De Martinis, B.S. Investigation of sweat VOC profiles in assessment of cancer biomarkers using HS-GC-MS. *J. Breath Res.* **2020**, *14*, 026009. [CrossRef]

6. Monedeiro, F.; Milanowski, M.; Ratiu, I.-A.; Zmysłowski, H.; Ligor, T.; Buszewski, B. VOC Profiles of Saliva in Assessment of Halitosis and Submandibular Abscesses Using HS-SPME-GC/MS Technique. *Molecules* **2019**, *24*, 2977. [CrossRef]
7. Ratiu, I.-A.; Ligor, T.; Monedeiro, F.; Al-Suod, H.; Bocos-Bintintan, V.; Szeliga, J.; Jackowski, M.; Buszewski, B. "Features of infected versus uninfected chemical profiles released from human exudates ". *Stud. Univ. Babeş-Bolyai Chem.* **2019**, *64*, 207–216. [CrossRef]
8. Silva, C.L.; Passos, M.; Câmara, J.S. Investigation of urinary volatile organic metabolites as potential cancer biomarkers by solid-phase microextraction in combination with gas chromatography-mass spectrometry. *Br. J. Cancer* **2011**, *105*, 1894–1904. [CrossRef] [PubMed]
9. Buszewski, B.; Kęsy, M.; Ligor, T.; Amann, A. Human exhaled air analytics: Biomarkers of diseases. *Biomed. Chromatogr.* **2007**, *21*, 553–566. [CrossRef]
10. Filipiak, W.; Mochalski, P.; Filipiak, A.; Ager, C.; Cumeras, R.; E. Davis, C.; Agapiou, A.; Unterkofler, K.; Troppmair, J. A Compendium of Volatile Organic Compounds (VOCs) Released By Human Cell Lines. *Curr. Med. Chem.* **2016**, *23*, 2112–2131. [CrossRef]
11. Amann, A.; Miekisch, W.; Schubert, J.; Buszewski, B.; Ligor, T.; Jezierski, T.; Pleil, J.; Risby, T. Analysis of Exhaled Breath for Disease Detection. *Annu. Rev. Anal. Chem.* **2014**, *7*, 455–482. [CrossRef]
12. Ratiu, I.-A.; Bocos-Bintintan, V.; Monedeiro, F.; Milanowski, M.; Ligor, T.; Buszewski, B. An Optimistic Vision of Future: Diagnosis of Bacterial Infections by Sensing Their Associated Volatile Organic Compounds. *Crit. Rev. Anal. Chem.* **2019**, 1–12. [CrossRef]
13. Schmidt, K.; Podmore, I. Current Challenges in Volatile Organic Compounds Analysis as Potential Biomarkers of Cancer. *J. Biomarkers* **2015**, *2015*, 1–16. [CrossRef]
14. Mazzone, P.J. Analysis of Volatile Organic Compounds in the Exhaled Breath for the Diagnosis of Lung Cancer. *J. Thorac. Oncol.* **2008**, *3*, 774–780. [CrossRef]
15. Ratiu, I.-A.; Ligor, T.; Bocos-Bintintan, V.; Buszewski, B. Mass spectrometric techniques for the analysis of volatile organic compounds emitted from bacteria. *Bioanalysis* **2017**, *9*, 1069–1092. [CrossRef] [PubMed]
16. Nardi-Agmon, I.; Peled, N. Exhaled breath analysis for the early detection of lung cancer: Recent developments and future prospects. *Lung Cancer Targets Ther.* **2017**, *8*, 31–38. [CrossRef]
17. Miekisch, W.; Schubert, J.K.; Noeldge-Schomburg, G.F. Diagnostic potential of breath analysis—Focus on volatile organic compounds. *Clin. Chim. Acta* **2004**, *347*, 25–39. [CrossRef] [PubMed]
18. Milanowski, M.; Monedeiro, F.; Złoch, M.; Ratiu, I.-A.; Pomastowski, P.; Ligor, T.; De Martinis, B.S.; Buszewski, B. Profiling of VOCs released from different salivary bacteria treated with non-lethal concentrations of silver nitrate. *Anal. Biochem.* **2019**, *578*, 36–44. [CrossRef] [PubMed]
19. Vishinkin, R.; Haick, H. Nanoscale Sensor Technologies for Disease Detection via Volatolomics. *Small* **2015**, *11*, 6142–6164. [CrossRef]
20. Ratiu, I.A.; Bocos-Bintintan, V.; Patrut, A.; Moll, V.H.; Turner, M.; Thomas, C.L.P. Discrimination of bacteria by rapid sensing their metabolic volatiles using an aspiration-type ion mobility spectrometer (a-IMS) and gas chromatography-mass spectrometry GC-MS. *Anal. Chim. Acta* **2017**, *982*, 209–217. [CrossRef]
21. Ratiu, I.-A.; Ligor, T.; Bocos-Bintintan, V.; Al-Suod, H.; Kowalkowski, T.; Rafińska, K.; Buszewski, B. The effect of growth medium on an Escherichia coli pathway mirrored into GC/MS profiles. *J. Breath Res.* **2017**, *11*, 036012. [CrossRef]
22. Mametov, R.; Ratiu, I.-A.; Monedeiro, F.; Ligor, T.; Buszewski, B. Evolution and Evaluation of GC Columns. *Crit. Rev. Anal. Chem.* **2019**, 1–24. [CrossRef]
23. Dobrzyńska, E.; Buszewski, B. Needle-trap device for the sampling and determination of chlorinated volatile compounds. *J. Sep. Sci.* **2013**, *36*, 3372–3378. [CrossRef] [PubMed]
24. Lord, H.L.; Zhan, W.; Pawliszyn, J. Fundamentals and applications of needle trap devices. *Anal. Chim. Acta* **2010**, *677*, 3–18. [CrossRef] [PubMed]
25. Filipiak, W.; Filipiak, A.; Ager, C.; Wiesenhofer, H.; Amann, A. Optimization of sampling parameters for collection and preconcentration of alveolar air by needle traps. *J. Breath Res.* **2012**, *6*, 027107. [CrossRef]
26. Frank Kneepkens, C.M.; Lepage, G.; Roy, C.C. The potential of the hydrocarbon breath test as a measure of lipid peroxidation. *Free Radic. Biol. Med.* **1994**, *17*, 127–160. [CrossRef]
27. Fuchs, P.; Loeseken, C.; Schubert, J.K.; Miekisch, W. Breath gas aldehydes as biomarkers of lung cancer. *Int. J. Cancer* **2010**, *126*, 2663–2670. [CrossRef]
28. Hakim, M.; Broza, Y.Y.; Barash, O.; Peled, N.; Phillips, M.; Amann, A.; Haick, H. Volatile Organic Compounds of Lung Cancer and Possible Biochemical Pathways. *Chem. Rev.* **2012**, *112*, 5949–5966. [CrossRef] [PubMed]
29. Charles, S.M.; Batterman, S.A.; Jia, C. Composition and emissions of VOCs in main- and side-stream smoke of research cigarettes. *Atmos. Environ.* **2007**, *41*, 5371–5384. [CrossRef]
30. Kischkel, S.; Miekisch, W.; Sawacki, A.; Straker, E.M.; Trefz, P.; Amann, A.; Schubert, J.K. Breath biomarkers for lung cancer detection and assessment of smoking related effects—Confounding variables, influence of normalization and statistical algorithms. *Clin. Chim. Acta* **2010**, *411*, 1637–1644. [CrossRef]
31. Elfaki, I.; Mir, R.; Almutairi, F.M.; Abu Duhier, F.M. Cytochrome P450: Polymorphisms and roles in cancer, diabetes and atherosclerosis. *Asian Pacific J. Cancer Prev.* **2018**, *19*, 2057–2070. [CrossRef]

32. Zamay, T.; Zamay, G.; Kolovskaya, O.; Zukov, R.; Petrova, M.; Gargaun, A.; Berezovski, M.; Kichkailo, A. Current and Prospective Protein Biomarkers of Lung Cancer. *Cancers* **2017**, *9*, 155. [CrossRef]
33. Karachaliou, N.; Pilotto, S.; Lazzari, C.; Bria, E.; De Marinis, F.; Rosell, R. Cellular and molecular biology of small cell lung cancer: An overview. *Transl. Lung Cancer Res.* **2016**, *5*, 2–15. [CrossRef] [PubMed]
34. Schaich, K.M.; Shahidi, F.; Zhong, Y.; Eskin, N.A.M. Lipid Oxidation. In *Biochemistry of Foods*; Eskin, N.A., Shahidi, F., Eds.; Elsevier: Cambridge, UK, 2013; pp. 419–478. ISBN 9780122423529.
35. Altomare, D.F.; Di Lena, M.; Porcelli, F.; Trizio, L.; Travaglio, E.; Tutino, M.; Dragonieri, S.; Memeo, V.; De Gennaro, G. Exhaled volatile organic compounds identify patients with colorectal cancer. *Br. J. Surg.* **2013**, *100*, 144–150. [CrossRef] [PubMed]
36. Hanahan, D.; Weinberg, R.A. Hallmarks of Cancer: The Next Generation. *Cell* **2011**, *144*, 646–674. [CrossRef] [PubMed]
37. Phillips, M.; Herrera, J.; Krishnan, S.; Zain, M.; Greenberg, J.; Cataneo, R.N. Variation in volatile organic compounds in the breath of normal humans. *J. Chromatogr. B Biomed. Sci. Appl.* **1999**, *729*, 75–88. [CrossRef]
38. Mutti, A.; Falzoi, M.; Lucertini, S.; Arfini, G.; Zignani, M.; Lombardi, S.; Franchini, I. n-Hexane metabolism in occupationally exposed workers. *Occup. Environ. Med.* **1984**, *41*, 533–538. [CrossRef]
39. Forney, F.W.; Markovetz, A.J. The biology of methyl ketones. *J. Lipid Res.* **1971**, *12*, 383–395. [CrossRef]
40. Abaffy, T.; Duncan, R.; Riemer, D.D.; Tietje, O.; Elgart, G.; Milikowski, C.; DeFazio, R.A. Differential Volatile Signatures from Skin, Naevi and Melanoma: A Novel Approach to Detect a Pathological Process. *PLoS ONE* **2010**, *5*, e13813. [CrossRef]
41. Phillips, M.; Cataneo, R.N.; Cummin, A.R.C.; Gagliardi, A.J.; Gleeson, K.; Greenberg, J.; Maxfield, R.A.; Rom, W.N. Detection of Lung Cancer With Volatile Markers in the Breatha. *Chest* **2003**, *123*, 2115–2123. [CrossRef]
42. Bos, L.D.J.; Sterk, P.J.; Schultz, M.J. Volatile Metabolites of Pathogens: A Systematic Review. *PLoS Pathog.* **2013**, *9*, e1003311. [CrossRef]
43. Egge, H.; Murawski, U.; Ryhage, R.; György, P.; Chatranon, W.; Zilliken, F. Minor constitutents of human milk IV: Analysis of the branched chain fatty acids. *Chem. Phys. Lipids* **1972**, *8*, 42–55. [CrossRef]
44. Gosset, G. Production of aromatic compounds in bacteria. *Curr. Opin. Biotechnol.* **2009**, *20*, 651–658. [CrossRef] [PubMed]
45. Xue, J.; Murrieta, C.M.; Rule, D.C.; Miller, K.W. Exogenous or l-Rhamnose-Derived 1,2-Propanediol Is Metabolized via a pduD-Dependent Pathway in Listeria innocua. *Appl. Environ. Microbiol.* **2008**, *74*, 7073–7079. [CrossRef] [PubMed]
46. Ramachandriya, K.D.; Wilkins, M.R.; Delorme, M.J.M.; Zhu, X.; Kundiyana, D.K.; Atiyeh, H.K.; Huhnke, R.L. Reduction of acetone to isopropanol using producer gas fermenting microbes. *Biotechnol. Bioeng.* **2011**, *108*, 2330–2338. [CrossRef] [PubMed]
47. Houten, S.M.; Violante, S.; Ventura, F.V.; Wanders, R.J.A. The Biochemistry and Physiology of Mitochondrial Fatty Acid β-Oxidation and Its Genetic Disorders. *Annu. Rev. Physiol.* **2016**, *78*, 23–44. [CrossRef]
48. Ludwiczuk, A.; Skalicka-Woźniak, K.; Georgiev, M.I. Terpenoids. In *Pharmacognosy*; Badal, S., Delgoda, R., Eds.; Elsevier: London, UK, 2017; pp. 233–266. ISBN 9780128020999.
49. Audrain, B.; Farag, M.A.; Ryu, C.M.; Ghigo, J.M. Role of bacterial volatile compounds in bacterial biology. *FEMS Microbiol. Rev.* **2015**, *39*, 222–233. [CrossRef]
50. Friedman, M.I.; Preti, G.; Deems, R.O.; Friedman, L.S.; Munoz, S.J.; Maddrey, W.C. Limonene in expired lung air of patients with liver disease. *Dig. Dis. Sci.* **1994**, *39*, 1672–1676. [CrossRef]
51. Mullen, P.J.; Yu, R.; Longo, J.; Archer, M.C.; Penn, L.Z. The interplay between cell signalling and the mevalonate pathway in cancer. *Nat. Rev. Cancer* **2016**, *16*, 718–731. [CrossRef] [PubMed]
52. Johnson, W.; Bergfeld, W.F.; Belsito, D.V.; Hill, R.A.; Klaassen, C.D.; Liebler, D.; Marks, J.G.; Shank, R.C.; Slaga, T.J.; Snyder, P.W.; et al. Safety Assessment of Isoparaffins as Used in Cosmetics. *Int. J. Toxicol.* **2012**, *31*, 269S–295S. [CrossRef]
53. Fujiwara, R.; Noda, S.; Tanaka, T.; Kondo, A. Styrene production from a biomass-derived carbon source using a coculture system of phenylalanine ammonia lyase and phenylacrylic acid decarboxylase-expressing Streptomyces lividans transformants. *J. Biosci. Bioeng.* **2016**, *122*, 730–735. [CrossRef] [PubMed]
54. Wiggins, T.; Kumar, S.; Markar, S.R.; Antonowicz, S.; Hanna, G.B. Tyrosine, Phenylalanine, and Tryptophan in Gastroesophageal Malignancy: A Systematic Review. *Cancer Epidemiol. Biomarkers Prev.* **2015**, *24*, 32–38. [CrossRef] [PubMed]
55. Pal, M. Random forest classifier for remote sensing classification. *Int. J. Remote Sens.* **2005**, *26*, 217–222. [CrossRef]
56. Steeghs, M.M.L.; Cristescu, S.M.; Harren, F.J.M. The suitability of Tedlar bags for breath sampling in medical diagnostic research. *Physiol. Meas.* **2007**, *28*, 73–84. [CrossRef]
57. Bajtarevic, A.; Ager, C.; Pienz, M.; Klieber, M.; Schwarz, K.; Ligor, M.; Ligor, T.; Filipiak, W.; Denz, H.; Fiegl, M.; et al. Noninvasive detection of lung cancer by analysis of exhaled breath. *BMC Cancer* **2009**, *9*, 348. [CrossRef]
58. Poli, D.; Carbognani, P.; Corradi, M.; Goldoni, M.; Acampa, O.; Balbi, B.; Bianchi, L.; Rusca, M.; Mutti, A. Exhaled volatile organic compounds in patients with non-small cell lung cancer: Cross sectional and nested short-term follow-up study. *Respir. Res.* **2005**, *6*, 71. [CrossRef]
59. Fens, N.; Zwinderman, A.H.; Van der Schee, M.P.; De Nijs, S.B.; Dijkers, E.; Roldaan, A.C.; Cheung, D.; Bel, E.H.; Sterk, P.J. Exhaled Breath Profiling Enables Discrimination of Chronic Obstructive Pulmonary Disease and Asthma. *Am. J. Respir. Crit. Care Med.* **2009**, *180*, 1076–1082. [CrossRef]
60. Ligor, M.; Ligor, T.; Bajtarevic, A.; Ager, C.; Pienz, M.; Klieber, M.; Denz, H.; Fiegl, M.; Hilbe, W.; Weiss, W.; et al. Determination of volatile organic compounds in exhaled breath of patients with lung cancer using solid phase microextraction and gas chromatography mass spectrometry. *Clin. Chem. Lab. Med.* **2009**, *47*, 550–560. [CrossRef] [PubMed]

61. Ulanowska, A.; Kowalkowski, T.; Trawińska, E.; Buszewski, B. The application of statistical methods using VOCs to identify patients with lung cancer. *J. Breath Res.* **2011**, *5*, 046008. [CrossRef]
62. Phillips, C.O.; Syed, Y.; Mac Parthaláin, N.; Zwiggelaar, R.; Claypole, T.C.; Lewis, K.E. Machine learning methods on exhaled volatile organic compounds for distinguishing COPD patients from healthy controls. *J. Breath Res.* **2012**, *6*, 036003. [CrossRef] [PubMed]
63. Rudnicka, J.; Kowalkowski, T.; Ligor, T.; Buszewski, B. Determination of volatile organic compounds as biomarkers of lung cancer by SPME–GC–TOF/MS and chemometrics. *J. Chromatogr. B* **2011**, *879*, 3360–3366. [CrossRef]
64. Fens, N.; Roldaan, A.C.; Van der Schee, M.P.; Boksem, R.J.; Zwinderman, A.H.; Bel, E.H.; Sterk, P.J. External validation of exhaled breath profiling using an electronic nose in the discrimination of asthma with fixed airways obstruction and chronic obstructive pulmonary disease. *Clin. Exp. Allergy* **2011**, *41*, 1371–1378. [CrossRef]
65. Smolinska, A.; Klaassen, E.M.M.; Dallinga, J.W.; Van de Kant, K.D.G.; Jobsis, Q.; Moonen, E.J.C.; Van Schayck, O.C.P.; Dompeling, E.; van Schooten, F.J. Profiling of Volatile Organic Compounds in Exhaled Breath As a Strategy to Find Early Predictive Signatures of Asthma in Children. *PLoS ONE* **2014**, *9*, e95668. [CrossRef]
66. Phillips, M.; Gleeson, K.; Hughes, J.M.B.; Greenberg, J.; Cataneo, R.N.; Baker, L.; McVay, W.P. Volatile organic compounds in breath as markers of lung cancer: A cross-sectional study. *Lancet* **1999**, *353*, 1930–1933. [CrossRef]
67. Phillips, M.; Cataneo, R.N.; Ditkoff, B.A.; Fisher, P.; Greenberg, J.; Gunawardena, R.; Kwon, C.S.; Tietje, O.; Wong, C. Prediction of breast cancer using volatile biomarkers in the breath. *Breast Cancer Res Treat.* **2006**, *99*, 19–21. [CrossRef]
68. Gaspar, E.M.; Lucena, A.F.; Duro da Costa, J.; Chaves das Neves, H. Organic metabolites in exhaled human breath—A multivariate approach for identification of biomarkers in lung disorders. *J. Chromatogr. A* **2009**, *1216*, 2749–2756. [CrossRef] [PubMed]
69. Gaida, A.; Holz, O.; Nell, C.; Schuchardt, S.; Lavae-Mokhtari, B.; Kruse, L.; Boas, U.; Langejuergen, J.; Allers, M.; Zimmermann, S.; et al. A dual center study to compare breath volatile organic compounds from smokers and non-smokers with and without COPD. *J. Breath Res.* **2016**, *10*, 026006. [CrossRef]
70. Ibrahim, B.; Basanta, M.; Cadden, P.; Singh, D.; Douce, D.; Woodcock, A.; Fowler, S.J. Non-invasive phenotyping using exhaled volatile organic compounds in asthma. *Thorax* **2011**, *66*, 804–809. [CrossRef]
71. Martinez-Lozano Sinues, P.; Meier, L.; Berchtold, C.; Ivanov, M.; Sievi, N.; Camen, G.; Kohler, M.; Zenobi, R. Breath Analysis in Real Time by Mass Spectrometry in Chronic Obstructive Pulmonary Disease. *Respiration* **2014**, *87*, 301–310. [CrossRef] [PubMed]
72. Song, G.; Qin, T.; Liu, H.; Xu, G.-B.; Pan, Y.-Y.; Xiong, F.-X.; Gu, K.-S.; Sun, G.-P.; Chen, Z.-D. Quantitative breath analysis of volatile organic compounds of lung cancer patients. *Lung Cancer* **2010**, *67*, 227–231. [CrossRef] [PubMed]
73. Dragonieri, S.; Schot, R.; Mertens, B.J.A.; Le Cessie, S.; Gauw, S.A.; Spanevello, A.; Resta, O.; Willard, N.P.; Vink, T.J.; Rabe, K.F.; et al. An electronic nose in the discrimination of patients with asthma and controls. *J. Allergy Clin. Immunol.* **2007**, *120*, 856–862. [CrossRef]
74. Oguma, T.; Nagaoka, T.; Kurahashi, M.; Kobayashi, N.; Yamamori, S.; Tsuji, C.; Takiguchi, H.; Niimi, K.; Tomomatsu, H.; Tomomatsu, K.; et al. Clinical contributions of exhaled volatile organic compounds in the diagnosis of lung cancer. *PLoS ONE* **2017**, *12*, e0174802. [CrossRef]
75. Chen, X.; Xu, F.; Wang, Y.; Pan, Y.; Lu, D.; Wang, P.; Ying, K.; Chen, E.; Zhang, W. A study of the volatile organic compounds exhaled by lung cancer cells in vitro for breath diagnosis. *Cancer* **2007**, *110*, 835–844. [CrossRef]
76. Cazzola, M.; Segreti, A.; Capuano, R.; Bergamini, A.; Martinelli, E.; Calzetta, L.; Rogliani, P.; Ciaprini, C.; Ora, J.; Paolesse, R.; et al. Analysis of exhaled breath fingerprints and volatile organic compounds in COPD. *COPD Res. Pract.* **2015**, *1*, 7. [CrossRef]
77. Van de Kant, K.D.G.; Van Berkel, J.J.B.N.; Jobsis, Q.; Lima Passos, V.; Klaassen, E.M.M.; Van der Sande, L.; Van Schayck, O.C.P.; De Jongste, J.C.; Van Schooten, F.J.; Derks, E.; et al. Exhaled breath profiling in diagnosing wheezy preschool children. *Eur. Respir. J.* **2013**, *41*, 183–188. [CrossRef] [PubMed]
78. Van Berkel, J.J.B.N.; Dallinga, J.W.; Möller, G.M.; Godschalk, R.W.L.; Moonen, E.J.; Wouters, E.F.M.; Van Schooten, F.J. A profile of volatile organic compounds in breath discriminates COPD patients from controls. *Respir. Med.* **2010**, *104*, 557–563. [CrossRef] [PubMed]
79. Van Vliet, D.; Smolinska, A.; Jöbsis, Q.; Rosias, P.P.R.; Muris, J.W.M.; Dallinga, J.W.; Van Schooten, F.J.; Dompeling, E. Association between exhaled inflammatory markers and asthma control in children. *J. Breath Res.* **2016**, *10*, 016014. [CrossRef] [PubMed]
80. Peng, G.; Hakim, M.; Broza, Y.Y.; Billan, S.; Abdah-Bortnyak, R.; Kuten, A.; Tisch, U.; Haick, H. Detection of lung, breast, colorectal, and prostate cancers from exhaled breath using a single array of nanosensors. *Br. J. Cancer* **2010**, *103*, 542–551. [CrossRef] [PubMed]

molecules

Article

Volatile Organic Compounds, Bacterial Airway Microbiome, Spirometry and Exercise Performance of Patients after Surgical Repair of Congenital Diaphragmatic Hernia

Gert Warncke [1], Georg Singer [1,*], Jana Windhaber [1], Lukas Schabl [1], Elena Friehs [1], Wolfram Miekisch [2], Peter Gierschner [2], Ingeborg Klymiuk [3], Ernst Eber [4], Katarina Zeder [4], Andreas Pfleger [4], Beate Obermüller [1], Holger Till [1] and Christoph Castellani [1]

[1] Department of Paediatric and Adolescent Surgery, Medical University Graz, 8036 Graz, Austria; gert.warncke@medunigraz.at (G.W.); jana.windhaber@klinikum-graz.at (J.W.); schabl.lukas@gmail.com (L.S.); elena.friehs@stud.medunigraz.at (E.F.); beate.obermueller@medunigraz.at (B.O.); holger.till@medunigraz.at (H.T.); christoph.castellani@medunigraz.at (C.C.)

[2] Department of Anesthesiology and Intensive Care Medicine, Rostock Medical Breath Research Analytics and Technologies (ROMBAT), Rostock University Medical Centre, 18057 Rostock, Germany; wolfram.miekisch@uni-rostock.de (W.M.); peter.gierschner@gmx.net (P.G.)

[3] Core Facility Molecular Biology, Center for Medical Research, Medical University of Graz, 8036 Graz, Austria; ingeborg.klymiuk@medunigraz.at

[4] Department of Paediatrics and Adolescent Medicine, Division of Paediatric Pulmonology and Allergology, Medical University of Graz, 8036 Graz, Austria; ernst.eber@medunigraz.at (E.E.); katarina.zeder@medunigraz.at (K.Z.); andreas.pfleger@medunigraz.at (A.P.)

* Correspondence: georg.singer@medunigraz.at; Tel.: +43-316-385-83722

Citation: Warncke, G.; Singer, G.; Windhaber, J.; Schabl, L.; Friehs, E.; Miekisch, W.; Gierschner, P.; Klymiuk, I.; Eber, E.; Zeder, K.; et al. Volatile Organic Compounds, Bacterial Airway Microbiome, Spirometry and Exercise Performance of Patients after Surgical Repair of Congenital Diaphragmatic Hernia. *Molecules* **2021**, *26*, 645. https://doi.org/10.3390/molecules26030645

Academic Editors: Natalia Drabińska and Ben de Lacy Costello

Received: 1 January 2021
Accepted: 22 January 2021
Published: 26 January 2021

Publisher's Note: MDPI stays neutral with regard to jurisdictional claims in published maps and institutional affiliations.

Abstract: The aim of this study was to analyze the exhaled volatile organic compounds (VOCs) profile, airway microbiome, lung function and exercise performance in congenital diaphragmatic hernia (CDH) patients compared to healthy age and sex-matched controls. A total of nine patients (median age 9 years, range 6–13 years) treated for CDH were included. Exhaled VOCs were measured by GC–MS. Airway microbiome was determined from deep induced sputum by 16S rRNA gene sequencing. Patients underwent conventional spirometry and exhausting bicycle spiroergometry. The exhaled VOC profile showed significantly higher levels of cyclohexane and significantly lower levels of acetone and 2-methylbutane in CDH patients. Microbiome analysis revealed no significant differences for alpha-diversity, beta-diversity and LefSe analysis. CDH patients had significantly lower relative abundances of *Pasteurellales* and *Pasteurellaceae*. CDH patients exhibited a significantly reduced Tiffeneau Index. Spiroergometry showed no significant differences. This is the first study to report the VOCs profile and airway microbiome in patients with CDH. Elevations of cyclohexane observed in the CDH group have also been reported in cases of lung cancer and pneumonia. CDH patients had no signs of impaired physical performance capacity, fueling controversial reports in the literature.

Keywords: CDH; microbiome; VOCs; spiroergometry; outcome

1. Introduction

Congenital diaphragmatic hernia (CDH) is a rare disease occurring with an incidence of 1:2000–1:5000 live births [1]. CDH is caused by disturbances in the formation of the diaphragm in the eighth week of gestation, leaving a defect with persistent communication between the abdominal and thoracic cavity [2]. Typically, this defect is located in the dorsal aspect of the diaphragm (Bochdalek hernia, 95% of cases, mostly located on the left side). Ventral hernias (Morgagni hernia) are rarer and typically located on the right side [3].

In fetuses with CDH, abdominal organs herniate into the thorax, subsequently restricting pulmonary development on the affected but also on the contralateral side. Consequently, patients with CDH suffer from pulmonary hypoplasia and vascular malformation with thickened muscle layers causing pulmonary hypertension with right ventricular dysfunction and left ventricular hypoplasia, reduced mobility of the diaphragm and impaired alveolar growth [4,5].

Even after surgical repair, 30–50% of CDH patients show—amongst others—persistent respiratory morbidity with impaired lung function [6–8] and/or recurrent respiratory infections [9]. While reduced lung function may be attributed to the congenital defect with lung hypoplasia and alveolar growth disturbances, the underlying reason for the recurrent infections may only partly be explained by impaired lung function. However, alterations of bacterial colonization can be speculated. Over recent decades, scientists have postulated sterility of the respiratory tract. However, with the advent of DNA based sequencing methods, microbial colonization of the healthy respiratory tract has been demonstrated in the last years [10]. This has led to the term "pulmonary microbiome" describing the collective genome of bacteria, archaea, fungi and viruses inhabiting the respiratory tract. Overall, there are still very limited data focusing on the pulmonary microbiome in pediatrics. While there is some evidence of alterations of the pulmonary microbiome in cases of asthma and cystic fibrosis [11,12], there are currently no data published concerning CDH patients.

While some studies describe reduced exercise tolerance in addition to impaired lung function in patients after CDH repair [5,13,14], others report normal values compared to healthy peers [7,15]. All of these studies rely on exercise performance testing, but do not look in the depth of the patients' metabolism. The emerging field of volatile organic compounds (VOCs) analysis in patients may offer novel insights. Additionally to oxygen, nitrogen and carbon dioxide, human breath contains several hundred different VOCs [16]. Among others, the VOC profile contains carbohydrates, ketones, aldehydes, cyclic components and sulphur- or nitrogen containing compounds [16]. Some of these substances have been attributed to the metabolic and inflammatory processes of the host [17], others may also be related to the (pulmonary) microbiome.

Although the analysis of body odors, for instance the fruity smell of ketones in the breath of diabetic patients, goes back many thousand years in medical history, only recent technical developments have allowed a detailed VOC analysis. For instance, the concentration of exhaled VOC profiles differs between type I diabetes patients and healthy children [18], and metabolic adaptation through postprandial hyperglycemia and related oxidative stress is immediately reflected in exhaled breath VOC concentrations [19]. Breath VOC profiles may help to understand basic mechanisms and metabolic adaptation accompanying progression of chronic kidney disease in pediatric patients at an early stage [20]. Investigations of children with cystic fibrosis have revealed increased levels of pentane correlating to nutritional status and lung function [21]. While the pulmonary long-term sequelae of CDH have been described in several reports, examinations of exhaled VOC profiles of patients after surgical repair of CDH as potential noninvasive disease markers have not yet been published. Thus, it was the aim of this study to analyze the breath VOC profile, airway microbiome, lung function and exercise performance of patients after CDH repair compared to healthy age and sex-matched controls in order to gain more detailed information about the pathophysiology of this disease.

2. Results

Nine patients following surgical repair of a CDH were recruited for long-term follow-up examinations consisting of assessment of the exhaled VOC profile, airway microbiome, lung function and exercise performance. As a control group, nine age and sex-matched controls were enrolled. The median age at the examination was 9 years (IQR 5). Within each group, six patients were male and three were female. The median gestational age of the CDH patients was 39 weeks (IQR 3.8), median birth weight was 3.4 kg (IQR 0.7). CDH occurred on the left side in five, the right side in three and bilaterally in one patient. The liver was partially herniated into the thoracic cavity in two patients. CDH patients were ventilated conventionally for a median of 7.5 days (IQR 17). One patient required high frequency oscillation. Three patients were on inhalative nitric oxide because of pulmonary hypertension.

Surgical repair was performed on median day of life four (IQR 6). Eight patients underwent direct closure and one underwent a patch repair. There was no recurrence. In the post-operative medical history two of the nine CDH patients reported recurrent respiratory infections. Eight of the nine patients with CDH and all nine control patients reported feeling fit in daily life.

Not all of the patients were eligible for all examinations. Table 1 gives an overview of the data available for matched pair analysis.

Table 1. Overview of CDH patients and their age and sex-matched healthy controls.

CDH	Control	Age	Gender	Muscle Mass	Body Fat	VOCs	Pulmonary Microbiome	Spirometry Before Ex.	Spiroergometry	Spirometry After Ex.
CDH-1	CDHK-7	12	m	X	X	X	X	X	X	X
CDH-2	CDHK-8	9	m	X	X	X	X	X	01	01
CDH-3	CDHK-6	8	m	X	X	X	01	X	01	01
CDH-4	CDHK-9	13	m	X	X	X	X	X	X	X
CDH-5	CDHK-2	9	f	X	X	X	X	X	X	X
CDH-6	CDHK-4	12	f	X	X	X	X	X	X	X
CDH-7	CDHK-3	6	m	01	X	X	X	X	01	01
CDH-8	CDHK-1	13	m	X	X	X	X	X	X	X
CDH-9	CDHK-5	7	f	02	02	X	02	02	02	02

X: Examination performed and valid; 0: Examination of one or both of the matched pairs missing; Ex: Exercise; Underlined: Patients have been subjected to passive smoke in their familial surroundings; 1: Patient physically unable to perform test/donate sample; 2: Patient refused to perform test/donate sample. CDH: congenital diaphragmatic hernia; VOC: volatile organic compound.

2.1. Clinical Examination

There were no significant differences for height, body weight, BMI, muscle mass or body fat between the groups (Table 2).

Table 2. Anthropometric data of the patients in the CDH and control group. Data presented as medians (IQRs) and the Mann–Whitney U test was performed for group comparison.

Parameter	Control Group	CDH Group	p-Value
Height (cm)	137.0 (39.5)	142.5 (36.5)	0.673
Body weight (kg)	30.0 (28.4)	36.4 (30.5)	0.673
BMI (kg/m^2)	16.0 (5.0)	18.0 (4.7)	0.888
Appendicular muscle mass (kg/m^2)	5.6 (3.6)	5.3 (2.8)	0.805
Body fat (%)	5.3 (2.8)	6.0 (15.0)	0.442

BMI: Body Mass Index.

2.2. Breath VOC Profile

In the breath samples a total of 67 different VOCs could be identified. Levels of 35 VOCs were not consistently above the limit of quantification (LOQ) and therefore had to be excluded for further quantitative analysis.

The remaining 32 substances were further processed and used for group comparison. Heatmap and dendrogram analysis showed different unspecific clusters. Alterations in the following 20 substances were significantly affected by room air contamination (levels in room air >20% of exhaled concentration): 1-methylbenzene, 1-propanol, 2,3-butandione, 2-butanone, 2-phenoxyethanol, 3-methyl-2-butanone, α-pinene, benzene, ethanol, ethylbenzene, hexanal, n-hexane, isopropylalcohol, nonanal, nonanone, octane, pentan, pentanal, p-xylene and toluene and were thus excluded as potential biomarkers (Supplementary Figure S1).

Out of the remaining 12 substances, nine substances did not show significant differences between the groups (Supplementary Figure S2). Significant differences occurred for 2-methylbutane, acetone and cyclohexane. The VOCs 2-methylbutane (p = 0.038) and acetone ($p = 0.002$) were significantly decreased and cyclohexane was significantly increased ($p = 0.004$) in CDH patients compared to the healthy group (Figure 1).

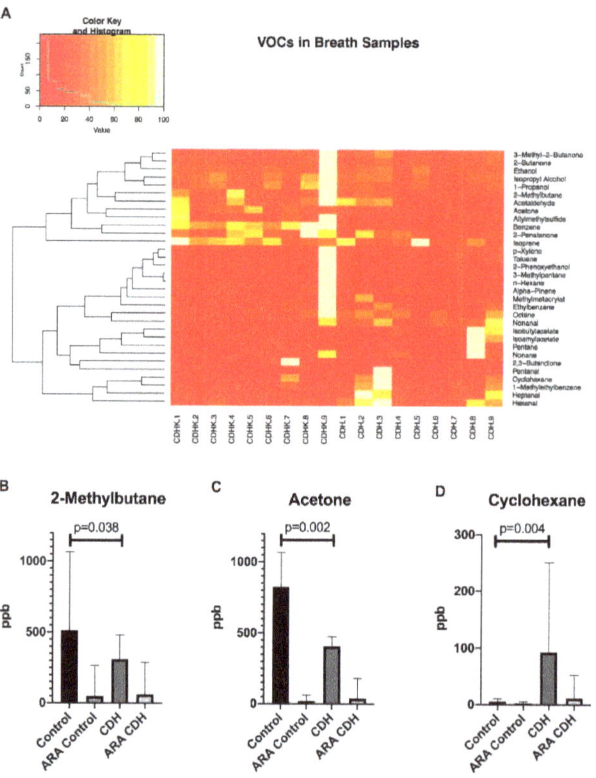

Figure 1. Concentrations of selected VOCs in breath samples of control patients and CDH patients (**A**). Concentrations of 2-methylbutane (**B**), acetone (**C**) and cyclohexane (**D**) were significantly different between the two groups; ARA: ambient room air.

2.3. 16S Based Airway Microbiome

The airway microbiome was measured with 16S based analysis of deep induced sputum samples. Alpha-diversity of the deep induced sputum samples did not differ significantly between CDH patients and controls (Shannon Index CDH median 6.94 ± IQR 0.577 vs. controls median 6.86 ± IQR 0.414; $p = 0.655$). Likewise, LefSe analysis over all hierarchical levels between the two groups did not result in significant different taxa. Beta-diversity analysis was not significantly different between the two groups (weighted unifrac $p = 0.97$, Bray–Curtis $p = 0.88$) (Figure 2). Analysis of the relative abundances revealed no statistically significant differences at the phylum, class and genus level between the two groups studied. On the order and family level, however, the relative abundances of *Pasteurellales* (controls median 0.022 ± IQR 0.01 vs. CDH median 0.016 ± IQR 0.01; $p = 0.038$) and *Pasteurellaceae* (controls median 0.022 ± IQR 0.01 vs. CDH median 0.016 ± IQR 0.01; $p = 0.038$) were significantly lower in CDH patients.

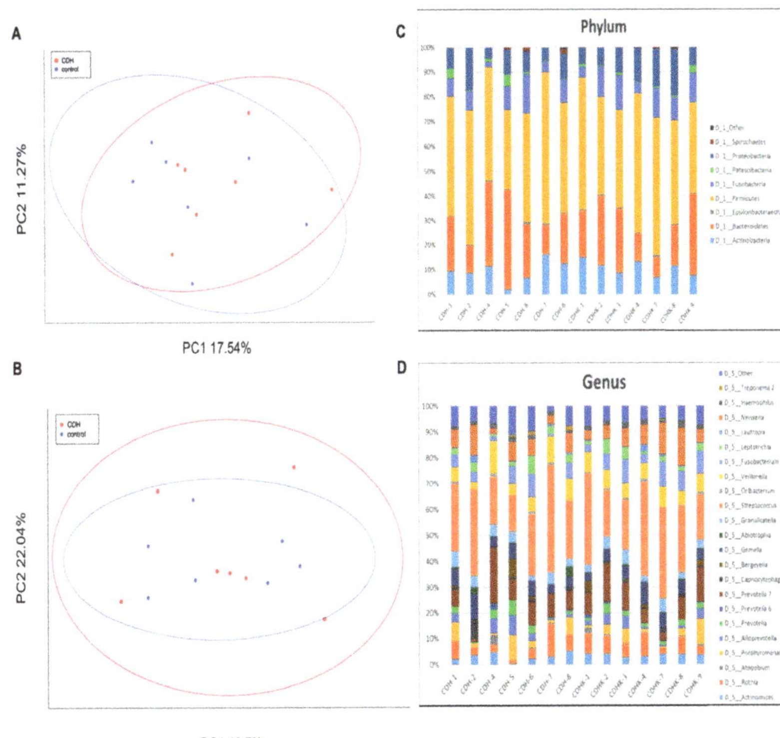

Figure 2. Principial coordinate analysis (PCoA) with weighted UniFrac comparison (**A**) and Bray–Curtis dissimilarity (**B**) tests; 95% confidence ellipses are indicated. The results revealed no obvious clustering of the14 deep induced sputum samples (CDH patients red dots, controls blue dots). PERMANOVA revealed no significant differences in both tests. Relative abundances of CDH patients and age and sex-matched controls at the phylum level (**C**) and genus level (**D**). Note that only bacteria with relative abundances of more than 1% are depicted.

2.4. Spirometry

Conventional spirometry was performed before and after exercise testing. CDH patients showed no differences in their forced vital capacity (FVC) before and after exercise in comparison to their healthy peers (Figure 3). The Tiffeneau index was significantly lower in CDH patients before ($p = 0.028$), but not after exercise ($p = 0.063$).

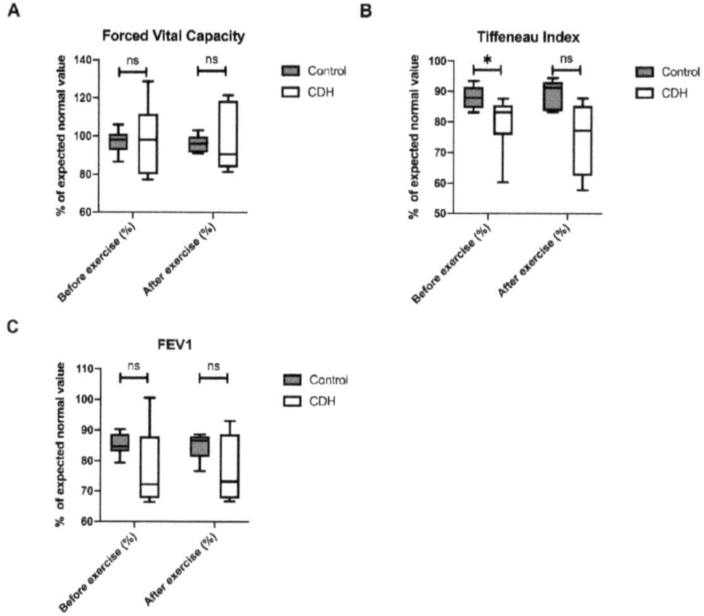

Figure 3. Forced vital capacity (FVC) (**A**), Tiffeneau Index (**B**) and forced expiratory volume in 1 s (FEV1) (**C**) before and after exercise of control patients and CDH patients; ns: not significant; * $p < 0.05$.

2.5. Spiroergometry

Bicycle spiroergometry was performed with a sex and body weight dependent protocol. There was no statistically significant difference between patients with CDH and their healthy peers (Table 3).

Table 3. Results of exercise performance testing (exhausting bicycle spiroergometry). Data displayed as medians (IQRs).

Parameter	Control Group	CDH Group	*p*-Value
Relative Performance Capacity (%)	118.0 (27.0)	108.0 (33.0)	0.095
VO$_2$max/kg (mL/kg/min)	46.7 (12.3)	42.3 (9.6)	0.222
Pmax/kg (W/kg)	3.4 (1.0)	3.3 (0.8)	0.310
O$_2$ Pulse (mL)	12.1 (7.3)	10.2 (7.6)	1.0
RER	1.2 (0.2)	1.2 (0.1)	0.841

VO$_2$max/kg: Maximum oxygen uptake per kilogram body weight; Pmax/kg: Maximum performance per kilogram body weight; RER: Respiratory exchange ratio.

3. Discussion

Our study gives a first insight into both the airway microbiome and volatile organic compounds in breath samples of patients 6 to 13 years following surgical CDH repair compared to healthy age and sex-matched peers. As a major result, it revealed no significant differences in the bacterial airway colonization but differences in the VOC profile.

We were not able to find statistically significant differences regarding anthropometric parameters such as height, body weight, BMI, body fat and muscle mass between the two groups. This is concordant with the literature describing no evidence for long-term growth impairment in patients following CDH repair [15].

Regarding the VOC profiles of exhaled breath samples, CDH patients exhibited significantly decreased levels of acetone and 2-methylbutane, in addition to significantly increased levels of cyclohexane. Acetone is formed by decarboxylation of acetoacetate generated by beta-oxidation of fatty acids and is thus linked to fat metabolism [22]. Under exercise, acetone levels have been shown to increase to the lactate threshold at about 45% of maximum exercise followed by a steady decrease. In this regard the acetone peak marks the switch between fat and carbohydrate metabolism [22]. Type I diabetes and fasting are medical conditions with increased ketone concentrations in breath and urine. Both are associated with predominant lipid metabolism. In type I diabetes, a lack of insulin prevents dextrose from entering the cells leading to impaired carbohydrate metabolism. In this condition, the body shifts to lipid oxidation as the energy source, resulting in increased ketone levels. Similarly, lipid oxidation is activated in response to a lack of carbohydrates under fasting conditions. In our collective, no patient was known to be diabetic. All patients fasted for 2 h prior to VOC sampling. Therefore, the conditions were similar in both studied groups. Since no metabolic parameters were determined in this study, possible unknown co-morbidities cannot be ruled out as reasons for the different acetone levels, especially as there are no known influences of CDH on carbohydrate or fat metabolism. There is currently no scientific information about the role of 2-methylbutane in in vivo experiments. The exact role of this VOC has to be elucidated in future studies. Cyclohexane is an organic solvent and part of raw oil. None of our participants reported increased exposure to organic solvents or gasoline vapors. The distribution of patients exposed to passive smoke did not differ between the groups (compare Table 1). There are no reports focusing on its origin in humans at present. However, cyclohexane has been mentioned in two studies in association with pulmonary diseases. First, Oguma et al. described increased levels of cyclohexane (and xylene) in patients with lung cancer compared to healthy controls (after ruling out possible confounders such as age, smoking status, gender and pulmonary function) [23]. Furthermore, the authors described an increase in cyclohexane (and xylene) levels in breath samples with progressing disease and a decrease in the healing process [23]. A second study revealed increased cyclohexane levels (among other VOCs) in cell cultures of human lung tissues infected with *E. coli*, *P. aeruginosa* and *S. aureus* [24]. Similarly, the authors could demonstrate increased cyclohexane levels in rabbits with pneumonia due to infection with the same pathogens [24]. While xylene showed no significant alterations in our study, cyclohexane was significantly increased as a possible sign for pulmonary impairment in CDH patients.

The airway microbiome analysis showed no significant differences of both α- and β-diversity between CDH and control patients. Regarding patient history, only two of the nine CDH patients reported recurrent respiratory infections. This low number makes a statistical comparison unfeasible. Additionally, the infections occurred before enrollment in this trial and it was therefore impossible to assess the nature of these infections (viral/bacterial). Taken together, the role of the airway microbiome and its role in a possible pre-disposition for respiratory infections remains unclear at present. Currently, there are no other reports in the literature, making comparisons to other patient groups impossible.

Spirometry revealed no significant differences in FVC between CDH and control patients. Similarly, Turchetta and coworkers reported no significant differences in lung function testing in CDH patients [25]. In contrast, however, Zaccara et al. and Marven et al. both have described a significant FVC reduction in CDH patients compared to healthy controls [15,26]. Regarding airway obstruction, our data confirm the findings of other authors who also reported a reduced FEV1 or FEV1/FVC in CDH patients [5,13,26].

While some authors have shown that CDH patients feel less fit than their healthy peers [15,25], the majority of our patients felt physically fit. The bicycle spiroergometry results underline this subjective impression showing no significant differences between CDH and control patients. While Marven and colleagues reported no significant impairment of exercise performance in CDH patients similar to our data [15], other authors have revealed evidence for reduced exercise performance (lower VO_2max, lower O_2-pulse or endurance time) in their CDH groups [5,13,14,26]. A possible reason for the discrepancies in this regard may be differences in the training status of the patients. Several studies could prove that CDH patients who exercised had a better performance in spiroergometry compared to those who did not [5,25]. Therefore, a different training status between the CDH and control group may explain some of the differences found in exercise performance testing. While we did not assess the training habits in our study, our collective showed no differences in BMI, muscle or fat mass as possible anthropometric signs influencing the exercise performance between the groups.

Study limitations include that, despite a high effort with repeated attempts to contact patients, the number of recruited patients is relatively low. This can be explained by the low prevalence of CDH of 2.6 out of 10,000 total births and a mortality rate of 37.7% [27] and the single-center setting of this investigation. A further fact decreasing the number of potential participants is the fact that we only recruited children between 6 and 16 years of age in order to assess changes of the examined parameters in children and adolescents. Including older patients might increase the number of confounders (smoking, exhaust, comorbidities, etc.) for the values investigated. Younger children, on the other side, would not have been able to sufficiently participate in controlled breath VOCs sampling. In case of low effect size, relevant group differences which would have been detected in a larger sample size might have been overlooked due to the low number of participants in this study. Therefore, our study can be interpreted as an observational pilot study. Nevertheless, we give a first thorough overview of the airway microbiome and VOC analysis of CDH patients. Moreover, all but one patient underwent direct closure of the diaphragmatic defect. As a potential consequence our patient group may present with better exercise performance and other parameters compared to patients with larger defects and possibly associated higher grad of pulmonary hypoplasia. Therefore, future multi-center trials including a larger group of patients will be required to expand this first data set. Another limitation is that when sampling deep induced sputum, the sample from the deep airways also passes the main bronchi, trachea, pharynx and mouth resulting in a possible contamination of the sample at these levels, probably masking potential biological differences in the deep airways. Therefore, the microbiome sample obtained can only be referred to as an airway microbiome. Harvesting the pulmonary microbiome is only possible by bronchoscopy with a broncho-alveolar lavage, which is ethically impossible in our setting. Further, analyzing the fungal airway microbiome might remain of potential interest. Regarding VOC measurements, effects of inhaled room air could be addressed by sampling room air at each measurement. We consistently excluded potential marker candidates with high room air concentrations in this study. However, there are many factors with possible influences on the breath VOC profile [28,29]. Despite a careful study design and patient questioning, influences of other factors (unknown co-morbidities, influences of diet, etc.) cannot be ruled out completely. In particular, acetone is known to be influenced by patient metabolism. As we did not expect differences in acetone levels, metabolic markers (urine ketone levels, blood dextrose levels or HbA1c) were not determined. Therefore, the reason for the different acetone levels remains unclear at present. Future studies in the CDH collective will have to include metabolic markers to assess the influence of CDH per se on this parameter.

4. Materials and Methods

After ethical approval (EK 28-528 ex 15/16) all patients who had undergone surgical CDH repair at our institution were contacted by letter and telephone and invited to participate in this investigation. For all patients, age and sex- matched healthy controls were recruited from families of the medical staff. In all cases written informed consent was obtained from patients and/or legal guardians. We excluded patients younger than 6 years due to potential difficulties with controlled breath gas sampling and exercise and lung function testing. Moreover, patients older than 16 years and children with acute (within 4 weeks before the examination) and chronic gastro-intestinal disease, acute urinary tract infection, antibiotic or probiotic treatment within 4 weeks before the examination were excluded. After inclusion, patients were invited to participate in the following examinations:

4.1. Clinical Examination

The clinical examination included investigation of the following anthropometric data: height, body weight (BW) and body mass index (BMI). The body fat in % was determined by the caliper method, as previously described [30]. Appendicular muscle mass was assessed by segmental multi-frequency impedance analysis as published before [31].

4.2. Breath VOC Sampling

Patients fasted 2 h before sampling. Alveolar breath gas sampling was performed by combining mainstream capnometry and needle trap microextraction (NTME) with an automated sampling device (PAS Technology Deutschland GmbH, Magdala, Germany), as previously reported [19]. Needle trap devices (NTDs) were pre-conditioned in a heating device (PAS Technology Deutschland GmbH, Magdala, Germany) at 200 °C for 30 min under permanent N_2-flow before each measurement. This device ensured alveolar sampling at a flow rate of 30 mL/min by means of a CO_2-triggered, fast responding valve with a CO_2 threshold of 25–30 mmHg. Sampling was repeated twice for every patient. Additionally, ambient room air was harvested after each patient measurement by automated NTME sampling. After sampling, NTDs were sealed by a Teflon cap (Shinawa LTD. Japan/PAS Technology Deutschland GmbH, Magdala, Germany) immediately. Specimens were measured within 48 h after sampling.

4.3. Breath VOC Analysis

An Agilent 7890A gas chromatograph connected to an Agilent 5975 inert XL mass selective detector (MSD) was used for GC-MS measurements, as previously described [19]. A total of 67 substances were identified by means of a mass spectral library (NIST 2005, Gatesburg, PA, USA). The total responses for each substance were recorded. A total of 32 potential marker candidate compounds were verified by pure reference substances. For that purpose, a mixture of gaseous standards (Gas-MIX, Ionicon Analytik GmbH, Innsbruck, Austria) and aqueous solutions of pure reference substances (Sigma Aldrich, Darmstadt, Germany) were evaporated by means of a liquid calibration unit (LCU, Ionicon Analytik GmbH, Innsbruck, Austria). Concentration levels of the gas standards were prepared from 1 ppb to 500 ppb by diluting the standards with nitrogen and water with a matrix adapted humidity of 25 g/m^3 as previously described [32,33]. Evaporated standard gas was pre-concentrated onto NTDs and analyzed by GC-MS.

For the calibration and determination of limit of detection (LOD, signal-to-noise ratio 3:1) and limit of quantification (LOQ, signal-to-noise ratio 10:1), different concentration levels of the reference substances were measured as previously described [34] (Supplementary Table S1). The signals of selected ions from the reference substances at different concentration levels were used to calculate a calibration curve for each substance. These curves allowed to derive the concentrations of marker substances in parts per billion per volume (ppbV). The median VOC concentrations for patients´ and room air derived substances were compared. If the room air concentration of a candidate substance exceeded 20% of the patients´ median the observed changes were considered as biased by room air contamination and therefore excluded as potential marker compounds.

4.4. 16S Based Airway Microbiome

Deep induced sputum samples were harvested as previously described in the literature [35]. Samples were stored at −80 °C until further measurement. Briefly, sputum samples were treated with 100 µg/mL DTT (Sigma), incubated at 37 °C for 20 min and centrifuged at 4000× *g* for 30 min. Supernatant was removed and the pellet was resuspended in 500 µL PBS (Roth). A total of 250 µL of the suspension were mixed with 250 µL bacterial lysis buffer (Roche, Mannheim, Germany) and total DNA was isolated according to manufacturer's instructions and as published [36] in a MagNA Pure LC 2.0 (Roche, Mannheim, Germany) with the MagNA Pure LC DNA Isolation Kit III (Bacteria, Fungi) (Roche, Mannheim, Germany). A total of 5 µL of total DNA was used in a 25 µL PCR reaction in triplicates using a Fast Start High Fidelity PCR system (Roche, Mannheim, Germany) according to Klymiuk et al. [36], with the target specific primers F27-AGAGTTTGATCCTGGCTCAG and R357-CTGCTGCCTYCCGTA [37]. The 6 pM library was sequenced on an Illumina MiSeq desktop sequencer (Eindhoven, The Netherlands) with 20% PhiX control DNA (Illumina, Eindhoven, The Netherlands) and v3 chemistry for 600 cycles in paired end mode according to manufacturer's instructions and FastQ raw reads were used for data analysis. A total of 2,711,449 (per sample minimum 94,894, maximum 235,181, mean 193,674) raw sequence reads were used for data analysis in an established Galaxy based workflow (Medical University of Graz, funded by the Austrian Federal Ministry of Education, Science and Research, Hochschulraum-Strukturmittel 2016 grant as part of BioTechMed Graz). Briefly, raw reads were quality-filtered, de-noised, de-replicated, merged and checked for chimeras using DADA2 pipeline [38] with standard settings in QIIME2.0 [39]. For taxonomic assignment SILVA rRNA database Release 132 at 97% identity was used.

4.5. Spirometry

Spirometry was performed before and after exercise testing (Oxycon Pro®, Carl Reiner GmbH, Vienna, Austria). Forced vital capacity (FVC) was assessed as the maximum amount of air exhaled after maximum inhalation and expressed as percent predicted values. Forced expiratory volume in 1 s (FEV1) was determined and used to calculate the Tiffeneau index (FEV1/FVC).

4.6. Spiroergometry

Bicycle spiroergometry (Excalibur Sport®, Lode B.V., Groningen, The Netherlands and spirometer Oxycon Pro®, Carl Reiner GmbH, Vienna, Austria) was performed with a sex and body weight dependent protocol [30]. The spiroergometry was continued to exhaustion followed by a 3-min recovery phase. The respiratory parameters included tidal volume, respiratory rate, minute volume (MV) and inspiratory (FiO_2) and expiratory (FeO_2) fraction of oxygen. The accuracy for FiO_2 and FeO_2 is given with ± 0.01 vol% by the manufacturer. Using these values and the minute volume the oxygen uptake was calculated. For each patient we recorded the maximum performance per kilogram body weight, the maximum oxygen uptake per kilogram body weight, the relative performance capacity, the respiratory exchange ratio and the oxygen pulse.

4.7. Statistics

All data were managed with Microsoft Excel 2016® (Microsoft Corporation, Redmond, WA, USA). For statistical analysis, data were transferred to SPSS 25.0® (IBM Austria, Vienna, Austria). Metric data are displayed as median (interquartile range, IQR). A Mann–Whitney U-Test was performed for group comparison. p-values <0.05 were regarded as statistically significant. Box plots were drawn with Prism 8.3.0® (GraphPad, San Diego, CA, USA) and heatmaps with R Studio 1.2.1335® (RStudio Inc., Boston, MA, USA) using the heatmap.2 library.

5. Conclusions

In conclusion, this is the first study to report on the airway microbiome and VOC profile in CDH. The alterations of the microbiome were minor and the clinical consequence of reduced Pasteurellaceae remains unclear at present. The elevations in cyclohexane levels that were observed in the CDH group have also been reported in cases of lung cancer and pneumonia. CDH patients showed signs of an obstructive pulmonary disease. CDH patients had no signs of impaired physical performance capacity mirroring controversial reports in the literature in this regard. Future larger scale multi-center studies will be required to confirm these first results.

Supplementary Materials: The following are available online, Figure S1: Exhaled and ambient room air concentrations (ARA) of VOC candidate substances regarded as affected by room air; Figure S2: Exhaled and ambient room air concentrations (ARA) of VOC candidate substance without significant differences between CDH patients and controls; Table S1: Limit of detection (LOD) and limit of quantification (LOQ) of 32 candidate VOCs detected in breath samples by needle trap micro-extraction (NTME).

Author Contributions: Conceptualization, G.W., G.S., E.E., H.T. and C.C.; methodology, J.W., W.M., P.G., I.K., K.Z., A.P. and B.O.; formal analysis, J.W., W.M., P.G., I.K., E.E., A.P., H.T. and C.C.; investigation, G.W., L.S., E.F., K.Z.; writing—original draft preparation, G.W., G.S and C.C.; writing—review and editing, J.W., L.S., E.F., W.M., P.G., I.K., E.E., K.Z., A.P., B.O. and H.T.; visualization, G.W., G.S., B.O. and C.C.; supervision, G.S., E.E., H.T. and C.C.; project administration, G.W., L.S. and E.F.; funding acquisition, G.W., G.S., H.T. and C.C. All authors have read and agreed to the published version of the manuscript.

Funding: The project was funded by the City of Graz, inVITA—Gesellschaft zur Foerderung der Gesundheit unserer Kinder and the Verein fuer Kinderchirurgie an der Medizinischen Universitaet Graz.

Institutional Review Board Statement: The study was conducted according to the guidelines of the Declaration of Helsinki, and approved by the Ethics Committee of the Medical University of Graz (EK 28-528 ex 15/16).

Informed Consent Statement: Informed consent was obtained from all subjects involved in the study.

Data Availability Statement: The data presented in this study are available on request from the corresponding author.

Conflicts of Interest: The authors declare no conflict of interest. The funders had no role in the design of the study; in the collection, analyses, or interpretation of data; in the writing of the manuscript, or in the decision to publish the results.

Registration: This trial was registered with www.clinicaltrials.gov (NCT 03787160).

Sample Availability: Samples of the compounds are not available from the authors.

References

1. Flemmer, A.W.; Jani, J.C.; Bergmann, F.; Muensterer, O.J.; Gallot, D.; Hajek, K.; Sugawara, J.; Till, H.; Deprest, J. Lung tissue mechanics predict lung hypoplasia in a rabbit model for congenital diaphragmatic hernia. *Pediatr. Pulmonol.* **2007**, *42*, 505–512. [CrossRef] [PubMed]
2. Zalla, J.M.; Stoddard, G.J.; Yoder, B.A. Improved mortality rate for congenital diaphragmatic hernia in the modern era of management: 15year experience in a single institution. *J. Pediatr. Surg.* **2015**, *50*, 524–527. [CrossRef] [PubMed]
3. Skari, H.; Bjornland, K.; Haugen, G.; Egeland, T.; Emblem, R. Congenital diaphragmatic hernia: A meta-analysis of mortality factors. *J. Pediatr. Surg.* **2000**, *35*, 1187–1197. [CrossRef] [PubMed]
4. Bohn, D. Congenital Diaphragmatic Hernia. *Am. J. Respir. Crit. Care Med.* **2002**, *166*, 911–915. [CrossRef]
5. Bojanić, K.; Grizelj, R.; Dilber, D.; Saric, D.; Vuković, J.; Pianosi, P.T.; Driscoll, D.J.; Weingarten, T.N.; Pritišanac, E.; Schroeder, D.R.; et al. Cardiopulmonary exercise performance is reduced in congenital diaphragmatic hernia survivors. *Pediatr. Pulmonol.* **2016**, *51*, 1320–1329. [CrossRef]
6. Bagolan, P.; Casaccia, G.; Crescenzi, F.; Nahom, A.; Trucchi, A.; Giorlandino, C. Impact of a current treatment protocol on outcome of high-risk congenital diaphragmatic hernia. *J. Pediatr. Surg.* **2004**, *39*, 313–318. [CrossRef]
7. Peetsold, M.G.; Heij, H.A.; Nagelkerke, A.F.; Ijsselstijn, H.; Tibboel, D.; Quanjer, P.H.; Gemke, R.J.B.J. Pulmonary function and exercise capacity in survivors of congenital diaphragmatic hernia. *Eur. Respir. J.* **2009**, *34*, 1140–1147. [CrossRef] [PubMed]
8. Trachsel, D.; Selvadurai, H.; Bohn, D.; Langer, J.C.; Coates, A.L. Long-term pulmonary morbidity in survivors of congenital diaphragmatic hernia. *Pediatr. Pulmonol.* **2005**, *39*, 433–439. [CrossRef]
9. Basek, P.; Bajrami, S.; Straub, D.; Moeller, A.; Baenziger, O.; Wildhaber, J.; Bernet, V. The pulmonary outcome of long-term survivors after congenital diaphragmatic hernia repair. *Swiss Med Wkly.* **2008**, *138*, 173–179.
10. Brar, T.; Nagaraj, S.; Mohapatra, S. Microbes and asthma. *Curr. Opin. Pulm. Med.* **2012**, *18*, 14–22. [CrossRef]
11. Boutin, S.; Graeber, S.Y.; Weitnauer, M.; Panitz, J.; Stahl, M.; Clausznitzer, D.; Kaderali, L.; Einarsson, G.; Tunney, M.M.; Elborn, J.S.; et al. Comparison of Microbiomes from Different Niches of Upper and Lower Airways in Children and Adolescents with Cystic Fibrosis. *PLoS ONE* **2015**, *10*, e0116029. [CrossRef] [PubMed]
12. Morris, A.; Beck, J.M.; Schloss, P.D.; Campbell, T.B.; Crothers, K.; Curtis, J.L.; Flores, S.C.; Fontenot, A.P.; Ghedin, E.; Huang, L.; et al. Comparison of the respiratory microbiome in healthy nonsmokers and smokers. *Am. J. Respir. Crit. Care Med.* **2013**, *187*, 1067–1075. [CrossRef] [PubMed]

13. Gischler, S.J.; van der Cammen-van Zijp, M.H.M.; Mazer, P.; Madern, G.C.; Bax, N.M.; De Jongste, J.C.; Van Dijk, M.; Tibboel, D.; Ijsselstijn, H. A prospective comparative evaluation of persistent respiratory morbidity in esophageal atresia and congenital diaphragmatic hernia survivors. *J. Pediatr. Surg.* **2009**, *44*, 1683–1690. [CrossRef]

14. van der Cammen-van Zijp, M.H.M.; Gischler, S.J.; Mazer, P.; Van Dijk, M.; Tibboel, D.; Ijsselstijn, H. Motor-function and exercise capacity in children with major anatomical congenital anomalies: An evaluation at 5years of age. *Early Hum. Dev.* **2010**, *86*, 523–528. [CrossRef] [PubMed]

15. Marven, S.S.; Smith, C.M.; Claxton, D.; Chapman, J.; Davies, H.A.; Primhak, R.A.; Powell, C.V.E. Pulmonary function, exercise performance, and growth in survivors of congenital diaphragmatic hernia. *Arch. Dis. Child.* **1998**, *78*, 137–142. [CrossRef]

16. Miekisch, W.; Schubert, J.; Nöldge-Schomburg, G. Diagnostic potential of breath analysis—focus on volatile organic compounds. *Clin. Chim. Acta* **2004**, *347*, 25–39. [CrossRef]

17. Van De Kant, K.D.; Van Der Sande, L.J.T.M.; Jöbsis, Q.; Van Schayck, O.C.P.; Dompeling, E. Clinical use of exhaled volatile organic compounds in pulmonary diseases: A systematic review. *Respir. Res.* **2012**, *13*, 117. [CrossRef]

18. Trefz, P.; Obermeier, J.; Lehbrink, R.; Schubert, J.K.; Miekisch, W.; Fischer, D.-C. Exhaled volatile substances in children suffering from type 1 diabetes mellitus: Results from a cross-sectional study. *Sci. Rep.* **2019**, *9*, 15707–15709. [CrossRef]

19. Trefz, P.; Rösner, L.; Hein, D.; Schubert, J.K.; Miekisch, W. Evaluation of needle trap micro-extraction and automatic alveolar sampling for point-of-care breath analysis. *Anal. Bioanal. Chem.* **2013**, *405*, 3105–3115. [CrossRef]

20. Obermeier, J.; Trefz, P.; Happ, J.; Schubert, J.K.; Staude, H.; Fischer, D.-C.; Miekisch, W. Exhaled volatile substances mirror clinical conditions in pediatric chronic kidney disease. *PLoS ONE* **2017**, *12*, e0178745. [CrossRef]

21. Barker, M.; Hengst, M.; Schmid, J.; Buers, H.-J.; Mittermaier, B.; Klemp, D.; Koppmann, R. Volatile organic compounds in the exhaled breath of young patients with cystic fibrosis. *Eur. Respir. J.* **2006**, *27*, 929–936. [CrossRef] [PubMed]

22. Schubert, R.; Schwoebel, H.; Mau-Moeller, A.; Behrens, M.; Fuchs, P.; Sklorz, M.; Schubert, J.K.; Bruhn, S.; Miekisch, W. Metabolic monitoring and assessment of anaerobic threshold by means of breath biomarkers. *Metabolomics* **2012**, *8*, 1069–1080. [CrossRef]

23. Oguma, T.; Nagaoka, T.; Kurahashi, M.; Kobayashi, N.; Yamamori, S.; Tsuji, C.; Takiguchi, H.; Niimi, K.; Tomomatsu, H.; Tomomatsu, K.; et al. Clinical contributions of exhaled volatile organic compounds in the diagnosis of lung cancer. *PLoS ONE* **2017**, *12*, e0174802. [CrossRef]

24. Zhou, Y.; Chen, E.; Wu, X.; Hu, Y.; Ge, H.; Xu, P.; Zou, Y.; Jin, J.; Wang, P.; Ying, K. Rational lung tissue and animal models for rapid breath tests to determine pneumonia and pathogens. *Am. J. Transl. Res.* **2017**, *9*, 5116–5126. [PubMed]

25. Turchetta, A.; Fintini, D.; Cafiero, G.; Calzolari, A.; Giordano, U.; Cutrera, R.; Morini, F.; Braguglia, A.; Bagolan, P. Physical activity, fitness, and dyspnea perception in children with congenital diaphragmatic hernia. *Pediatr. Pulmonol.* **2011**, *46*, 1000–1006. [CrossRef] [PubMed]

26. Zaccara, A.; Turchetta, A.; Calzolari, A.; Iacobelli, B.; Nahom, A.; Lucchetti, M.; Bagolan, P.; Rivosecchi, M.; Coran, A. Maximal oxygen consumption and stress performance in children operated on for congenital diaphragmatic hernia. *J. Pediatr. Surg.* **1996**, *31*, 1092–1095. [CrossRef]

27. Politis, M.D.; Bermejo-Sánchez, E.; Canfield, M.A.; Contiero, P.; Cragan, J.D.; Dastgiri, S.; De Walle, H.E.; Feldkamp, M.L.; Nance, A.; Groisman, B.; et al. Prevalence and mortality in children with congenital diaphragmatic hernia: A multicountry study. *Ann. Epidemiol.* **2020**. [CrossRef] [PubMed]

28. Fischer, S.; Bergmann, A.; Steffens, M.; Trefz, P.; Ziller, M.; Miekisch, W.; Schubert, J.S.; Köhler, H.; Reinhold, P. Impact of food intake on in vivo VOC concentrations in exhaled breath assessed in a caprine animal model. *J. Breath Res.* **2015**, *9*, 047113. [CrossRef]

29. Fischer, S.; Trefz, P.; Bergmann, A.; Steffens, M.; Ziller, M.; Miekisch, W.; Schubert, J.S.; Köhler, H.; Reinhold, P. Physiological variability in volatile organic compounds (VOCs) in exhaled breath and released from faeces due to nutrition and somatic growth in a standardized caprine animal model. *J. Breath Res.* **2015**, *9*, 027108. [CrossRef]

30. Windhaber, J.; Steinbauer, M.; Castellani, C.; Singer, G.; Till, H.; Schober, P. Do Anthropometric and Aerobic Parameters Predict a Professional Career for Adolescent Skiers? *Int. J. Sports Med.* **2019**, *40*, 409–415. [CrossRef]

31. Skrabal, F.; Pichler, G.P.; Penatzer, M.; Steinbichl, J.; Hanserl, A.-K.; Leis, A.; Loibner, H. The Combyn™ ECG: Adding haemodynamic and fluid leads for the ECG. Part II: Prediction of total body water (TBW), extracellular fluid (ECF), ECF overload, fat mass (FM) and "dry" appendicular muscle mass (AppMM). *Med. Eng. Phys.* **2017**, *44*, 44–52. [CrossRef] [PubMed]

32. Oertel, P.; Bergmann, A.; Fischer, S.; Trefz, P.; Küntzel, A.; Reinhold, P.; Köhler, H.; Schubert, J.; Miekisch, W. Evaluation of needle trap micro-extraction and solid-phase micro-extraction: Obtaining comprehensive information on volatile emissions from in vitro cultures. *Biomed. Chromatogr.* **2018**, *32*, e4285. [CrossRef] [PubMed]

33. Traxler, S.; Bischoff, A.-C.; Saß, R.; Trefz, P.; Gierschner, P.; Brock, B.; Schwaiger, T.; Karte, C.; Blohm, U.; Schröder, C.; et al. VOC breath profile in spontaneously breathing awake swine during Influenza A infection. *Sci. Rep.* **2018**, *8*, 1–10. [CrossRef] [PubMed]

34. Trefz, P.; Koehler, H.; Klepik, K.; Möbius, P.; Reinhold, P.; Schubert, J.K.; Miekisch, W. Volatile Emissions from Mycobacterium avium subsp. paratuberculosis Mirror Bacterial Growth and Enable Distinction of Different Strains. *PLoS ONE* **2013**, *8*, e76868. [CrossRef] [PubMed]

35. Pettigrew, M.M.; Gent, J.F.; Kong, Y.; Wade, M.; Gansebom, S.; Bramley, A.M.; Jain, S.; Arnold, S.L.R.; McCullers, J.A. Association of sputum microbiota profiles with severity of community-acquired pneumonia in children. *BMC Infect. Dis.* **2016**, *16*, 317. [CrossRef] [PubMed]

36. Klymiuk, I.; Bilgilier, C.; Stadlmann, A.; Thannesberger, J.; Kastner, M.-T.; Högenauer, C.; Püspök, A.; Biowski-Frotz, S.; Schrutka-Kölbl, C.; Thallinger, G.G.; et al. The Human Gastric Microbiome Is Predicated upon Infection with Helicobacter pylori. *Front. Microbiol.* **2017**, *8*, 2508. [CrossRef] [PubMed]
37. McKenna, P.; Hoffmann, C.; Minkah, N.; Aye, P.P.; Lackner, A.; Liu, Z.; Lozupone, C.A.; Hamady, M.; Knight, R.; Bushman, F.D. The Macaque Gut Microbiome in Health, Lentiviral Infection, and Chronic Enterocolitis. *PLoS Pathog.* **2008**, *4*, e20. [CrossRef]
38. Callahan, B.J.; Mcmurdie, P.J.; Rosen, M.J.; Han, A.W.; Johnson, A.J.A.; Holmes, S.P. DADA2: High-resolution sample inference from Illumina amplicon data. *Nat. Methods* **2016**, *13*, 581–583. [CrossRef]
39. Bolyen, E.; Rideout, J.R.; Dillon, M.R.; Bokulich, N.A.; Abnet, C.C.; Al-Ghalith, G.A.; Alexander, H.; Alm, E.J.; Arumugam, M.; Asnicar, F.; et al. Reproducible, interactive, scalable and extensible microbiome data science using QIIME 2. *Nat. Biotechnol.* **2019**, *37*, 852–857. [CrossRef]

 *molecules*

Article

The Influence of Smoking Status on Exhaled Breath Profiles in Asthma and COPD Patients

Stefania Principe [1,2], Job J.M.H. van Bragt [1], Cristina Longo [1], Rianne de Vries [1,3], Peter J. Sterk [1], Nicola Scichilone [2], Susanne J.H. Vijverberg [1] and Anke H. Maitland-van der Zee [1,*]

[1] Department of Respiratory Medicine, Amsterdam UMC, University of Amsterdam, 1105 AZ Amsterdam, The Netherlands; stefaniaprincipe90@gmail.com (S.P.); j.j.vanbragt@amsterdamumc.nl (J.J.M.H.v.B.); c.longo@amsterdamumc.nl (C.L.); riannedevries1@gmail.com (R.d.V.); p.j.sterk@amsterdamumc.nl (P.J.S.); s.j.vijverberg@amsterdamumc.nl (S.J.H.V.)
[2] Dipartimento Universitario di Promozione della Salute, Materno Infantile, University of Palermo, Medicina Interna e Specialistica di Eccellenza "G. D'Alessandro"(PROMISE) c/o Pneumologia, AOUP "Policlinico Paolo Giaccone", 90127 Palermo, Italy; nicola.scichilone@unipa.it
[3] Breathomix b.v., 2333 Leiden, The Netherlands
* Correspondence: a.h.maitland@amsterdamumc.nl

Abstract: Breath analysis using eNose technology can be used to discriminate between asthma and COPD patients, but it remains unclear whether results are influenced by smoking status. We aim to study whether eNose can discriminate between ever- vs. never-smokers and smoking <24 vs. >24 h before the exhaled breath, and if smoking can be considered a confounder that influences eNose results. We performed a cross-sectional analysis in adults with asthma or chronic obstructive pulmonary disease (COPD), and healthy controls. Ever-smokers were defined as patients with current or past smoking habits. eNose measurements were performed by using the SpiroNose. The principal component (PC) described the eNose signals, and linear discriminant analysis determined if PCs classified ever-smokers vs. never-smokers and smoking <24 vs. >24 h. The area under the receiver–operator characteristic curve (AUC) assessed the accuracy of the models. We selected 593 ever-smokers (167 smoked <24 h before measurement) and 303 never-smokers and measured the exhaled breath profiles of discriminated ever- and never-smokers (AUC: 0.74; 95% CI: 0.66–0.81), and no cigarette consumption <24h (AUC 0.54, 95% CI: 0.43–0.65). In healthy controls, the eNose did not discriminate between ever or never-smokers (AUC 0.54; 95% CI: 0.49–0.60) and recent cigarette consumption (AUC 0.60; 95% CI: 0.50–0.69). The eNose could distinguish between ever and never-smokers in asthma and COPD patients, but not recent smokers. Recent smoking is not a confounding factor of eNose breath profiles.

Keywords: exhaled breath; eNose; smoking; asthma; COPD

Citation: Principe, S.; van Bragt, J.J.M.H.; Longo, C.; de Vries, R.; Sterk, P.J.; Scichilone, N.; Vijverberg, S.J.H.; Maitland-van der Zee, A.H. The Influence of Smoking Status on Exhaled Breath Profiles in Asthma and COPD Patients. *Molecules* **2021**, *26*, 1357. https://doi.org/10.3390/molecules26051357

Academic Editor: Natalia Drabińska

Received: 27 January 2021
Accepted: 26 February 2021
Published: 4 March 2021

Publisher's Note: MDPI stays neutral with regard to jurisdictional claims in published maps and institutional affiliations.

1. Introduction

Asthma and chronic obstructive pulmonary disease (COPD) are complex and heterogeneous chronic airway diseases that include several phenotypes, characterized by different inflammatory pathways [1,2]. The complexity and the heterogeneity of these diseases is due to variability of clinical characteristics, environmental influences, and pathophysiology aspects that are different for each patient [3]. Therefore, there is still a clinical need for new biomarkers to characterize the underlying processes [4].

The study of exhaled breath composition ("breathomics"), could facilitate a phenotyping approach of chronic airway diseases [5]. Exhaled breath is partially composed of volatile organic compounds (VOCs), including exogenous VOCs (e.g., drinks, food, drugs, environment) and endogenous VOCs (e.g., microbiome and body (patho) physiological metabolic processes), which can directly originate from the metabolism of bacteria residing in alveoli, immune cells, etc., and can also diffuse into the bloodstream where they diffuse passively across the capillary/alveolar interface and are subsequently emanated in the exhaled breath with different configurations according to their origin [6–8]. The eNose, a non-invasive and rapid technique that is able to detect exhaled VOC patterns, has shown some promise in characterizing asthma and COPD based on inflammatory characteristics and discriminating between patients with asthma, COPD, and lung cancer [9,10].

Exhaled breath measurements are considered a promising diagnostic technique, being easy to perform and potentially giving additional information that may help phenotyping patients, there are still some limitations related to the fact that there are many factors (e.g., diet, smoking, co-morbidities, physical activities, age, gender, pregnancy, and medication use) which could influence the level of individual compounds present in the exhaled breath [11]. Since VOCs are the result of metabolic and inflammatory processes related to (patho) physiological changes that take place in the respiratory tract [12], and smoking contributes to altering these processes in asthma and COPD [13], it is critical to assess the sensitivity of the eNose for the smoking status of patients with chronic respiratory diseases. Therefore, we aimed to investigate whether the eNose is suitable as a non-invasive technique to identify how patients with different smoking habits may respond to smoke exposure and whether smoking has an influence on disease classification. To assess this, in this exploratory analysis, we hypothesized that the eNose is able to distinguish between ever- and never-smokers.

Among patients with asthma and COPD and healthy volunteers, we assessed whether the eNose can accurately discriminate between (1) ever- vs. never-smokers, and (2) smoking less than vs. greater than 24 h before the exhaled breath measurement.

2. Results

2.1. Baseline Characteristics and Study Design

The study subjects selected for the analysis were enrolled from December 2015 through May 2017 across six different sites. Of the included asthma and COPD patients (n = 896) 593 (60.4%) were ever-smokers (237 asthma and 356 COPD) and 303 (33.8%) were never-smokers (295 asthma and 8 COPD). Among the ever-smokers, 167 (28.2%) smoked their last cigarette <24 h before measurement. The healthy control group was composed of 199 ever-smokers (out of which 107 subjects smoked in the last 24 h) and 366 never smokers. Tables 1–3 display the baseline clinical (age, body mass index (BMI), gender, medications) and functional (FEV1, FVC, FEV1/FVC) characteristics of the overall population (Table 1), ever-smokers (Table 2), and the healthy controls (Table 3). Ever-smokers had more advanced age and worse lung function (FEV1/ FEV1/FVC). In the asthma and COPD training set, 469 patients were ever-smokers and 247 were never-smokers. In the validation set, 124 patients were ever-smokers and 56 were never-smokers. A flowchart of the study design is presented in Figure 1a. Among the healthy controls, 452 patients were in the training set (160 ever-smokers and 292 never-smokers), and 113 subjects were in the validation set (39 ever-smokers and 74 never-smokers) (Figure 1b).

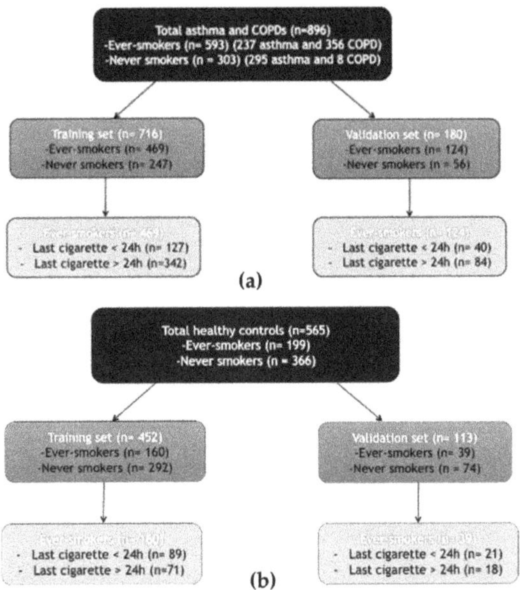

Figure 1. Flowchart of the study design of the asthma and COPD group (**a**), and healthy controls (**b**).

Table 1. Demographics of asthma and chronic obstructive pulmonary disease (COPD) patients stratified by smoking status. Data are expressed in number of patients, mean ± standard deviation or median and range for non-normal distributions.

Asthma and COPD Patients	Ever Smokers (n = 593)	Never Smokers (n = 303)	p-Value
Age (mean (SD))	60.99 (13.38)	48.46 (18.04)	<0.001
BMI (mean (SD))	27.78 (5.82)	26.95 (6.69)	0.055
Gender = M/F (%)	48.6/51.4	37.3/62.7	0.002
Allergy = Yes/No (%)	42.0/58.0	69.6/30.4	<0.001
FEV (mean (SD)) (l)	2.04 (0.89)	2.62 (0.93)	<0.001
FVC (mean (SD)) (l)	3.41 (1.07)	3.66 (1.13)	0.002
FEV1/FVC (mean (SD)) (%)	56 (16)	70 (14)	<0.001
FEV1 pred (mean (SD)) (%)	70.93 (23.81)	86.14 (21.74)	<0.001
ACQ (median [IQR])	1.60 [0.86, 2.50]	1.43 [0.71, 2.29]	0.208
CCQ (median [IQR])	1.00 [0.00, 2.30]	0.00 [0.00, 0.00]	<0.001
Current use of ICS = No/Yes (%)	27.8/72.2	15.5/84.5	<0.001
Oral corticosteroids (%)			0.780
Current use	2.4	1.7	
Previous use	10.3	10.6	
No	87.4	87.8	

BMI: body mass index; ACQ: asthma control questionnaire; CCQ: clinical COPD questionnaire; ICS: inhaled corticosteroids.

Table 2. Clinical characteristics of patients with recent cigarette consumption. Data are expressed in number of patients, mean ± standard deviation or median and range for non-normal distributions.

Ever Smokers (n = 593)	<24 h (n = 167)	>24 h (n = 426)	*p*-Value
Age (mean (SD))	56.98 (14.97)	60.83 (13.57)	<0.001
BMI (mean (SD))	26.63 (5.76)	27.75 (5.80)	0.114
Gender = M/F (%)	47.9/52.1	27.7/72.3	0.001
FEV1 (mean (SD)) (l)	2.03 (0.87)	2.05 (0.90)	<0.001
FVC (mean (SD)) (l)	3.37 (1.05)	3.41 (1.08)	<0.001
FEV1/FVC (mean (SD)) (%)	56 (16)	56 (16)	<0.001
FEV1 pred (mean (SD)) (%)	68.82 (21.79)	70.99 (23.83)	<0.001
Pack/year (median [IQR])	30.00 [17.00, 48.00]	25.00 [10.95, 41.25]	<0.001
ACQ (median [IQR])	1.86 [1.14, 2.86]	1.60 [0.88, 2.50]	0.010
CCQ (median [IQR])	1.40 [0.00, 2.55]	1.00 [0.00, 2.30]	<0.001

BMI: body mass index; ACQ: asthma control questionnaire; CCQ: clinical COPD questionnaire.

Table 3. Demographics of healthy subjects stratified by smoking status. Data are expressed in number of patients, mean ± standard deviation.

	Ever Smokers (n = 199)	Never Smokers (n = 366)	*p*-Value
Age (mean (SD))	46.95 (15.29)	35.67 (14.05)	<0.001
BMI (mean (SD))	26.02 (4.86)	23.72 (3.65)	<0.001
Gender = M/F (%)	72/127 (36.2/63.8)	127/239 (34.7/ 65.3)	0.795
FEV1(%) (mean (SD))	89.17(12.43)	92.08 (15.39)	<0.001
FEV1/VC (%) (mean (SD))	91.96 (12.26)	94.88(14.53)	<0.001
Last cigarette (%)			<0.001
<24 h	107 (53.7)	0	
>24 h	92 (46.2)	0	
Pack/years (mean (SD))	15.77 (19.20)	0	<0.001

2.2. The Ability of the eNose to Discriminate a History of Smoking in Asthma and COPD Patients

Out of 13 sensors, three principal components (PCs) were selected that captured 64% of the variance within the dataset of asthma and COPD patients (PC1 37%, PC2 16%, PC3 9%). There was no significant correlation between relevant PCs and ever- or never-smoker patients with asthma or COPD (see Supplementary Material, Figure S1). The ability to classify a history of smoking in patients with asthma or COPD showed reasonable accuracy in the training set (area under the receiver–operator characteristic curve (ROC–AUC) = 0.74, 95% CI = 0.70–0.77), and this accuracy was further confirmed in the validation set (ROC–AUC = 0.75, 95% CI = 0.68–0.82), with 67% of cross-validated grouped cases correctly classified after Leave One Out Cross-Validation (LOO-CV) in both groups (Figure 2a). This was confirmed with a higher accuracy using the number of pack-years among ever-smoker patients in both the training and validation sets (see Supplementary Material, Figure S2a).

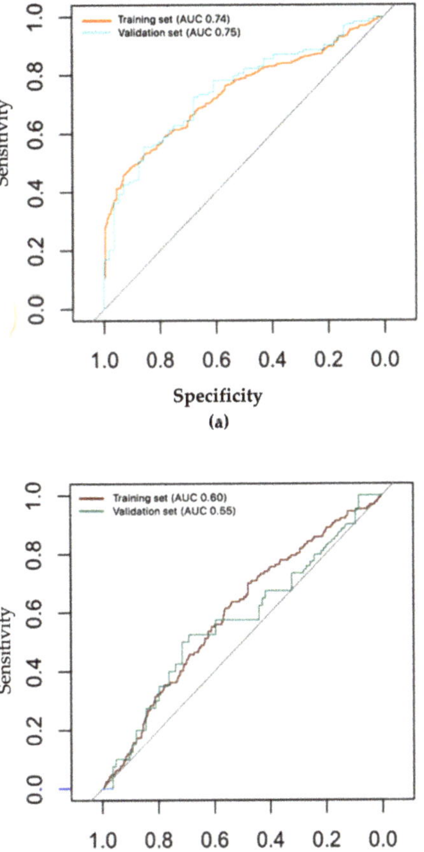

Figure 2. ROC analyses showing the accuracy of the linear discriminant model based on principal component (PC) reduction in the training set and the independent validation set for the asthma and COPD group. (**a**) Ever-smokers with asthma and COPD: training set (n = 469) 95% CI: 0.70–0.77 area under curve (AUC): 0.74; validation set (n = 124) 95% CI: 0.68–0.82 AUC: 0.75. (**b**) Time of last cigarette assumption in asthma and COPD patients (control: more than 24 h; case: less than 24 h): training set case = 127; control = 583; 95% CI: 0.54–0.65; AUC: 0.60; validation set case = 40; control = 84; 95% CI: 0.44–0.67; AUC: 0.55.

2.3. The Ability of the eNose to Discriminate Recent Cigarette Consumption in Asthma and COPD Patients

The eNose could less accurately identify patients with recent last cigarette consumption (<24 h) compared to smoking patients with a cigarette consumption >24 h in both the training and validation sets (ROC–AUC = 0.60; 95% CI = 0.54–0.65; ROC–AUC= 0.55; 95% CI = 0.44–0.67 respectively). eNose was not able to discriminate who smoked their last cigarette before and after 24 h before the visit (Figure 2b) with an accuracy of 51% after LOO-CV in both sets.

2.4. Does Smoking Influence eNose Results?

The same analysis was repeated with the healthy control group. Three PCs (out of 13 sensors) were selected that captured 61% of the variance within the dataset (PC1 34%,

PC2 14%, PC3 13%). The eNose was not able to distinguish among ever- and never-smokers in either the training (ROC–AUC: 0.54; 95% CI: 0.49–0.60) or the validation sets (ROC–AUC: 0.56; 95% CI: 0.50–0.69) with an accuracy of 53% after LOO-CV in both groups (Figure 3a). This was confirmed according to the number of pack/years among ever-smokers (see Supplementary Material, Figure S2b). Moreover, the eNose was not able to discriminate between subjects who smoked their last cigarette shorter or longer than 24 h prior to the exhaled breath measurement in both the training and the validation sets (training: area under curve (AUC): 0.60; 95% CI: 0.50–0.69; validation: AUC: 0.60; 95% CI: 0.47–0.70) (Figure 3b).

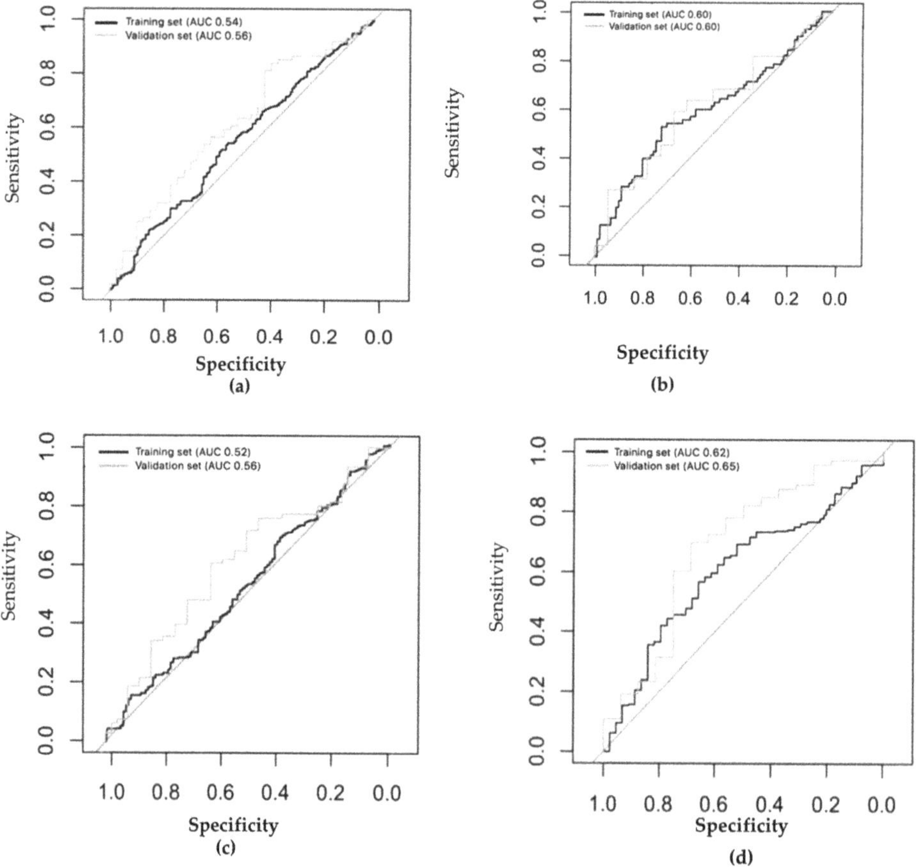

Figure 3. ROC analyses showing the accuracy of the linear discriminant model based on principal component reduction in the training set and the independent validation set for the healthy control group. (**a**) Ever-smokers in healthy subjects: training set (n = 452) AUC: 0.54 (95% CI: 0.49–0.60); validation set (n = 113) AUC: 0.60 (95% CI: 0.50–0.69). (**b**) Time of last cigarette consumption in healthy subjects (control: more than 24h; case: less than 24h) training set (n = 160) AUC: 0.60 (95% CI: 0.50–0.69); validation set (n = 39) AUC: 0.60 (95% CI: 0.42–0.70). (**c**) Ex- (n = 139) and never-smokers (n = 366): training set (ex-smokers = 116; never-smokers = 292), AUC: 0.52 (95% CI: 0.46–0.58); validation set (ex-smokers = 23; never-smokers = 74) AUC: 0.56 (95% CI: 0.46–0.60). (**d**) Active (n = 60) and never smokers (n = 366): training set (active smokers = 44; never-smokers = 292) AUC: 0.62 (95% CI: 0.51–0.67); validation set (active smokers = 16; never smokers = 74) AUC: 0.65 (95% CI: 0.53–0.71).

These results were further confirmed when repeating the analysis sub setting among ever-smoker healthy subjects, current smokers, and ex-smokers. The eNose was not accurate enough to distinguish between ex- (n = 139) and never-smokers (n = 366) in the training or the validation sets (training set AUC: 0.52; 95% CI: 0.46–0.58; validation set AUC: 0.56; 95% CI: 0.46–0.60) (Figure 3c). The eNose did not distinguish between ever-, current- (n = 60), and never-smokers (n = 366) in the training set (AUC: 0.62; 95% CI: 0.51–0.67) or the validation set (AUC: 0.65; 95% CI: 0.53–0.71) (Figure 3d).

Interestingly, the eNose was not able to distinguish between ever- and never-smokers in the asthma group (see Supplementary Material, Figure S3), but it could discriminate the diagnosis of asthma and COPD patients among the ever-smoker population (see Supplementary Material, Figure S4).

3. Discussion

In this study, we demonstrated that the use of the eNose to analyze exhaled breath can discriminate patients with a chronic respiratory disease (asthma or COPD) with and without a smoking history, but we could not make a distinction between smokers that did or did not smoke a cigarette in the last 24 h. We also demonstrated that the eNose is not influenced by smoking history in healthy volunteers; therefore, we can assume smoking may not be considered as a confounder that interferes with eNose measurements. These results were internally validated and were confirmed in an independent validation set.

Exhaled breath measurement by an eNose device has been used as a non-invasive tool for detecting several diseases with screening and diagnostic implications [10,14–16]. To our knowledge, this is the first study that evaluates the ability of the eNose to discriminate patients with chronic airway diseases according to their smoking status and the possible detection of the influence of smoking in both pathologies.

We can assume that the association of exhaled breath and eNose signals most likely reflects pathophysiological modifications related to the underlying chronic airway disease and combined airway alterations due to chronic smoking exposure. These results are in line with other studies. In a dual-center study [17] that recruited 222 smokers and non-smokers, with or without COPD, the eNose was able to classify COPD never- and ex-smokers and COPD active-smokers. Interestingly, a proportion of current smokers (9.3%) was misclassified non-smokers according to the analysis of their CO levels, which seems to confirm that, in line with our results, the eNose can distinguish among patients with chronic smoking habits, whereas it is not able to detect smoking in patients depending on the time of last cigarette consumption. Also, Papaefstathiou et al. [18] recently demonstrated that exhaled breath can be used to discriminate between smokers, non-smokers, and e-cigarette users in a population of healthy subjects and, in particular, that relevant VOCs can be identified among these three groups. Moreover, the diagnostic accuracy of exhaled breath analysis, linked to routine spirometry for chronic airway diseases, was previously assessed by De Vries et al. [10]. eNose patterns were found to be predictive for the differential diagnosis of asthma and COPD (AUC 0.88) and, in line with our results, eNose breath profiles did not show any ability to discriminate between current and ex-smokers (AUC 0.52) among patients with COPD, even though ROC analyses showed high accuracy in detecting exacerbations with the population stratified for pack-years [19]. Compared to our results, we are able to moderately distinguish between ever- and never-smokers (AUC 0.74), and we conclude that smoking could play a role in the VOCs contributing to the ability of the eNose to distinguish between asthma and COPD patients, but the eNose likely also detects other pathological factors that may be characteristic for these chronic respiratory diseases.

Furthermore, VOCs have been previously used to discriminate different inflammatory patterns in several chronic respiratory diseases, enabling researchers to obtain subgroups based on molecular characteristics [20–22]. Recently, Caruso et al. [23] used metabolomic analysis of VOCs in exhaled breath to identify a "severe asthma smoking phenotype," showing that, in line with our results, the analysis of VOCs identified differences among severe asthmatic smokers and ex-smokers, compared to never smokers. However, the severe asthmatics with smoking history were almost all ex-smokers, meaning that the differences may not have been related to active smoking, but potentially to the eNose results reflecting damage in the airways caused by smoking in the past [24].

We also considered whether smoking is a confounder of eNose results with respect to chronic respiratory disease phenotypes. We therefore also included data on the healthy controls. Smoking could influence the levels of individual compounds present in the exhaled breath and could therefore hamper the implementation of exhaled breath analysis as a diagnostic tool [25]. To our knowledge, this is the first study that demonstrates that recent or past smoking does not influence breath patterns in healthy subjects, in line with the hypothesis that breath patterns most likely reflect airway alterations caused by past smoking.

The strengths of our study are the relatively large sample size, the BreathCloud cohort, which recruited patients from different centers, obtaining a mixed population resembling the general COPD and asthma population, and the use of standardized methods for the analysis, including the internal and external validations that were performed to support the obtained results. The limitations of our study were that there was no information collected related to passive smoking that may have indirectly influenced never smokers and we had no information related to urinary nicotine concentration, which could be a more accurate measure, but more burdensome for patients [26]. A further limitation of our study is that the eNose can identify patterns of VOC mixture rather than the individual compounds that are driving the signal, even though this characteristic makes eNose breath profiles as suitable as composite multidimensional biomarkers in providing numerical probabilities for the presence or absence of a particular clinical condition [27,28]. On the other hand, the advantages are that this technology is noninvasive, easy to use, and results can be promptly available and interpretable for clinicians.

4. Materials and Methods

4.1. Study Design

We conducted a cross-sectional analysis using exhaled breath and clinical information obtained from the multicenter BreathCloud [9] database. BreathCloud enrolled patients with asthma, COPD, lung cancer, cystic fibrosis, and healthy volunteers from ten different centers in the Netherlands during routine outpatient visits. The following data from medical records collected general characteristics (age, BMI, gender, allergy history) symptoms assessment (asthma control questionnaire (ACQ), clinical COPD questionnaire (CCQ), oral corticosteroid assumption), functional tests (e.g., spirometry pre- and post- bronchodilator) and, among the ever-smokers, whether they smoked before and after 24 h. The exhaled breath measurements were collected in routine clinical practice and were subsequently handled in compliance with the Dutch Personal Data Protection Act (WPB).

4.2. Subject Selection

Patients and healthy controls were enrolled by six centers of primary, secondary and tertiary care in the Netherlands. Patients were included in this analysis if they were $\geq$18 years old, had a physician-reported diagnosis of asthma or COPD, and had answered the questions about smoking history. Healthy subjects were those who did not report any history of asthma or COPD, and who did not use any respiratory medications. Patients were stratified according to their smoking habits. Patients with a recent history of acute upper or lower respiratory tract infections were excluded because a history of upper [29–31] or lower [32–34] respiratory infections may influence the quality of breath samplings, and we did not know how much this could interfere with the resulting breath pattern profiles in patients with a diagnosis of asthma and COPD. The purpose of adding the SpiroNose to routine diagnostics was explained to the patients, who all gave their oral consent before enrollment. Due to the non-invasive nature of the BreathCloud study, the Amsterdam UMC medical ethical review board provided a waiver for ethical approval of the protocol (reference: W14_112#14.17.0147). All six sites made use of the same sampling protocol, which was part of the AMC MRB approval no: 14.17.0147.

4.3. Smoking Definitions

Patient-reported smoking history was chosen as an outcome, according to previous studies [35,36] that demonstrated that self-reported smoking is accurate. Smoking status was further divided into ever- and never-smokers; ever-smokers were considered active smokers who currently smoke cigarettes (number of pack/year) and former smokers who had smoked at least 100 cigarettes in their lifetime, but have quit smoking. Never-smokers were patients without any history of smoking habits, or who had smoked fewer than 100 cigarettes in their lifetime. The number of pack-years was calculated as (number of cigarettes smoked per day/20) × number of years smoked.

The second smoking definition was patient-reported recent cigarette smoking, which was defined as having smoked a cigarette in the 24 h prior to the exhaled breath measurement.

4.4. Exhaled Breath Measurements

Exhaled breath samples were collected using the SpiroNose [10]; an eNose composed of seven separate cross-reactive metal oxide semiconductor (MOS) sensors used to detect exhaled breath VOCs while monitoring for ambient VOCs. The SpiroNose comprises 7 different MOS sensors, each present in duplicate in both the reference and the exhaled breath sensor arrays. The MOS sensors (Figaro Engineering Inc., Osaka, Japan) were chosen based on their good stability and long-term performance [37]. MOS sensors operate with temperatures ranging between $-40\ °C$ and $+70\ °C$. Using thick film techniques, the sensor material was printed onto electrodes on an alumina substrate. Tin dioxide (SnO_2) was the main sensing material of the sensor element [38]. From each sensor signal, two variables were derived. The SpiroNose provided a spectrum of signals representing 13 data points originated by 6 sensor peaks normalized to sensor 2, the most stable sensor, and 7 peak/breath hold ratios. Each SpiroNose sensor signal had a high sensitivity to different mixtures of volatile organic compounds (VOCs)/gases in the exhaled breath and the reference (ambient air) sensor arrays [39].

Before breath measurement, patients had to rinse their mouth thoroughly with water three times, and then perform five tidal breaths, after which they maximally inhaled and held their breath for 5 s before slowly exhaling. The measurement was performed two times, with an interval of two minutes between maneuvers. Data were sent in real-time and stored on the online BreathCloud server.

4.5. Statistical Analysis

A descriptive analysis was performed to generate tables with the general characteristics of the population (Tables 1–3, Tables S1 and S2). A chi-squared test was used for categorical variables and a one-way ANOVA test was used for continuous variables. A principal component (PC) analysis was performed to summarize the eNose breath signals. According to the Kaiser criterion [40], all PCs with an eigenvalue >1 were considered for the analysis. PCs were constructed for the overall number of subjects (including both the training and validation sets). Furthermore, a linear discriminant analysis (LDA) was used to determine whether PCs could accurately classify patient-reported smoking history (ever-smokers vs. never-smokers) and, among ever-smokers, recent cigarette consumption (<24 h vs. >24 h). Internal validation was performed with leave-one-out cross-validation and by a split analysis in which the LDA model constructed with the training set was applied to the validation set. The area under the receiver–operator characteristic curve (ROC) was used to assess the accuracy of the models and it was obtained from the prediction made by the LDA model. The dataset was randomly divided into a training set containing 80% of the data (total asthma and COPD group = 716; asthma = 426; COPD = 288; healthy controls = 452) and a validation set including 20% of the data (total asthma and COPD group = 180; asthma = 106; COPD = 74; healthy controls = 113). The model acquired with the training set was used to retrieve similar variables in the validation set. A sensitivity analysis was also performed, using the number of pack-years of the ever-smokers (see Supplementary Materials, Figure S2a,b). Supplementary analyses concerning only the asthma group are reported in the Supplementary Materials (Figure S3); an additional analysis assessing the accuracy of the eNose in distinguishing asthma and COPD among a population of ever-smokers is also reported in Supplementary Materials (Figure S4). Data selected had no missing sensor values in BreathCloud.

The analysis was performed using R studio version 1.1.463 (R Studio Inc., Boston, MA, USA) and using R version 3.5.1 (The R Foundation for Statistical Computing, Vienna, Austria), with packages; dplyr, caret, pROC, and MASS [41,42].

5. Conclusions

We demonstrated that a smoking history might influence eNose breath profiles in patients with chronic airway diseases, while we cannot distinguish patients and healthy subjects according to recent cigarette consumption. This means that we can measure the influence of smoking on airways, but not the cigarette smoke itself. The present findings are in support of the usage of eNose technology as a quick and feasible technique for the diagnosis and phenotyping of chronic airway diseases in a clinical setting.

Supplementary Materials: The following are available online at https://www.mdpi.com/1420-304 9/26/5/1357/s1, Figure S1: scatter plot matrices displaying the correlation between each principal component (PC) in a population of ever- or non-smokers in the overall dataset (n = 896): dot plots represent never-smokers and triangles represent ever-smokers. Figure S2a and S2b: ROC analyses showing the accuracy of the linear discriminant model based on principal component reduction in the training set and the independent validation set according to pack-years (less than 5 pack-years, more than 5 pack-years) for (**a**) the asthma and COPD group (control: <5 pack/years; case: >5 pack/years). Training set: case = 397; control = 297; 95% CI: 0.72–0.78; AUC: 0.76. Validation set: case = 105; control = 72; 95% CI: 0.71–0.84; AUC: 0.76, and (**b**) for the healthy control group (control: <5 pack/years; case: >5 pack/years). Training set: AUC: 0.56 (95% CI: 0.50–0.61); validation set: AUC: 0.55 (95% CI: 0.58–0.74). Figure S3: A ROC curve showing the accuracy of the linear discriminant model based on principal component reduction in the asthmatic population (training and validation set), to distinguish ever- (n = 237) and never-smokers (n = 295). Training set case = 184; control = 241; AUC: 0.55; 95% CI: 0.49–0.60; validation set case = 53; control = 54; AUC 0.66; 95% CI: 0.56–0.77. Figure S4: ROC curve showing the accuracy of the linear discriminant model based on principal component reduction in the ever-smoker population (training and validation sets), to distinguish between asthma (n = 237) and COPD (n = 356) patients. Training set case = 182; control = 292; AUC: 0.88; 95% CI: 0.85–0.91; validation set case = 55; control = 64: AUC: 0.85; 95% CI: 0.78–0.92. Table S1: demographics of asthma patients stratified by smoking status. Data are expressed in number of patients, mean ± standard deviation or median and range for non-normal distributions. Table S2: demographics of COPD patients stratified by smoking status. Data are expressed in number of patients, mean ± standard deviation or median and range for non-normal distributions.

Author Contributions: Conceptualization, S.P., J.J.M.H.v.B., S.J.H.V. and A.H.M.-v.d.Z.; methodology, C.L, J.J.M.H.v.B., S.J.H.V.; software, S.P., J.J.M.H.v.B., C.L., S.J.H.V. and A.H.M.-v.d.Z.; validation, S.P., J.J.M.H.v.B., S.J.H.V. and A.H.M.-v.d.Z., P.J.S.; formal analysis, S.P., J.J.M.H.v.B., C.L.; investigation, J.J.M.H.v.B., R.d.V., P.J.S.; resources, R.d.V., P.J.S.; data curation S.P., J.J.M.H.v.B., S.J.H.V. and A.H.M.-v.d.Z.; writing—original draft preparation, S.P. and N.S.; writing—review and editing, S.P., J.J.M.H.v.B., C.L., R.d.V., P.J.S., N.S., S.J.H.V., A.H.M.-v.d.Z.; visualization, S.P., J.J.M.H.v.B., S.J.H.V. and A.H.M.-v.d.Z.; supervision, S.J.H.V., A.H.M.-v.d.Z. and N.S.; project administration, S.J.H.V., A.H.M.-v.d.Z. All authors have read and agreed to the published version of the manuscript.

Funding: The first author of the research leading to these results received funding from the SIP/IRS Fellowship 2020. Stefania Principe is the recipient of a SIP/IRS Fellowship 2020.

Institutional Review Board Statement: Due to the non-invasive nature of the BreathCloud study, the Amsterdam UMC medical ethical review board provided a waiver for ethical approval of the protocol (reference: W14_112#14.17.0147).

Informed Consent Statement: Informed consent was obtained from all subjects involved in the study.

Data Availability Statement: The data presented in this study are available on request from the corresponding author. The data are not publicly available due to privacy.

Acknowledgments: The authors wish to thank and acknowledge all BreathCloud partners, in particular the staff of Diagnostiek voor U (Eindoven, the Netherlands); Medisch Centrum Den Bosch Oost ('s-Hertogenbosch, the Netherlands); Medisch Spectrum Twente (Enschede, the Netherlands); Franciscus Gasthuis and Vlietland (Rotterdam, the Netherlands); and Amsterdam UMC, locations AMC and VUmc, (Amsterdam, the Netherlands) for their help in the recruitment of the subjects who have been enrolled in this study. Members of the Amsterdam UMC Breath Research Group: Anke H. Maitland-van der Zee, Peter J. Sterk, Paul Brinkman, Anne H. van Stuyvenberg-Neerincx, Cristina Longo, Lieuwe D. Bos, Anirban Sinha, Dominic W. Fenn, Marije Lammers, Levi B. Richards, Job J.M.H. van Bragt, Mahmoud I. Abdel-Aziz, Renate Kos, Yennece W.F. Dagelet, Marcus J. Schultz, Marry R. Smit and Laura A. Hagens (all Amsterdam UMC, University of Amsterdam, Depts of (Pediatric) Respiratory Medicine or Intensive Care Medicine).

Conflicts of Interest: S.P. reports an unrestricted research grant from SIP/IRS Fellowship 2020 during the conduct of the study. J.J.M.H.v.B., C.L., S.J.H.V. and A.H.M.-v.d.Z. have nothing to disclose. R.d.V.is has a considerable interest in the start-up company Breathomix BV. P.J.S. reports being scientific advisor to and having a formally inconsiderable interest in the start-up company Breathomix BV.

Sample Availability: Samples of the compounds are not available from the authors.

References

1. Han, M.L.K.; Agusti, A.; Calverley, P.M.; Celli, B.R.; Criner, G.; Curtis, J.L.; Fabbri, L.M.; Goldin, J.G.; Jones, P.W.; MacNee, W.; et al. Chronic Obstructive Pulmonary Disease Phenotypes: The Future of COPD. *Am. J. Respir. Crit. Care Med.* **2010**, *182*, 598–604. [CrossRef]
2. Wenzel, S.E. Asthma phenotypes: The evolution from clinical to molecular approaches. *Nat. Med.* **2012**, *18*, 716–725. [CrossRef]
3. Agusti, A.; Bel, E.; Thomas, M.; Vogelmeier, C.; Brusselle, G.; Holgate, S.; Humbert, M.; Jones, P.; Gibson, P.G.; Vestbo, J.; et al. Treatable traits: Toward precision medicine of chronic airway diseases. *Eur. Respir. J.* **2016**, *47*, 410–419. [CrossRef] [PubMed]
4. Lötvall, J.; Akdis, C.A.; Bacharier, L.B.; Bjermer, L.; Casale, T.B.; Custovic, A.; Lemanske, R.F., Jr.; Wardlaw, A.J.; Wenzel, S.E.; Greenberger, P.A. Asthma endotypes: A new approach to classification of disease entities within the asthma syndrome. *J. Allergy Clin. Immunol.* **2011**, *127*, 355–360. [CrossRef]
5. Bos, L.D.; Sterk, P.J.; Fowler, S.J. Breathomics in the setting of asthma and chronic obstructive pulmonary disease. *J. Allergy Clin. Immunol.* **2016**, *138*, 970–976. [CrossRef]
6. Buszewski, B.; Kęsy, M.; Ligor, T.; Amann, A. Human exhaled air analytics: Biomarkers of diseases. *Biomed. Chromatogr.* **2007**, *21*, 553–566. [CrossRef]
7. Van de Kant, K.D.G.; van der Sande, L.J.T.M.; Jöbsis, Q.; van Schayck, O.C.P.; Dompeling, E. Clinical Use of Exhaled Volatile Organic Compounds in Pulmonary Diseases: A Systematic Review. *Respir. Res.* **2012**, *13*, 1–23. [CrossRef]
8. Miekisch, W.; Schubert, J.K.; Noeldge-Schomburg, G.F. Diagnostic potential of breath analysis—focus on volatile organic compounds. *Clin. Chim. Acta* **2004**, *347*, 25–39. [CrossRef] [PubMed]
9. De Vries, R.; Dagelet, Y.W.; Spoor, P.; Snoey, E.; Jak, P.M.; Brinkman, P.; Dijkers, E.; Bootsma, S.K.; Elskamp, F.; De Jongh, F.H.; et al. Clinical and inflammatory phenotyping by breathomics in chronic airway diseases irrespective of the diagnostic label. *Eur. Respir. J.* **2018**, *51*, 1701817. [CrossRef]
10. De Vries, R.; Brinkman, P.; Van Der Schee, M.P.; Fens, N.; Dijkers, E.; Bootsma, S.; De Jongh, F.H.C.; Sterk, P.J. Integration of electronic nose technology with spirometry: Validation of a new approach for exhaled breath analysis. *J. Breath Res.* **2015**, *9*, 046001. [CrossRef]
11. Brinkman, P.; Der Zee, A.-H.M.-V.; Wagener, A.H. Breathomics and treatable traits for chronic airway diseases. *Curr. Opin. Pulm. Med.* **2019**, *25*, 94–100. [CrossRef] [PubMed]
12. Zarogoulidis, P.; Freitag, L.; Besa, V.; Teschler, H.; Kurth, I.; Khan, A.M.; Sommerwerck, U.; Baumbach, J.I.; Darwiche, K. Exhaled volatile organic compounds discriminate patients with chronic obstructive pulmonary disease from healthy subjects. *Int. J. Chronic Obstr. Pulm. Dis.* **2015**, *10*, 399–406. [CrossRef]
13. Tamimi, A.; Serdarevic, D.; Hanania, N.A. The effects of cigarette smoke on airway inflammation in asthma and COPD: Therapeutic implications. *Respir. Med.* **2012**, *106*, 319–328. [CrossRef]
14. Meerbeeck, J.V.; Lamote, K. Screening Tools for a High Risk Population-Can We Screen for Early Mesothelioma? *J. Thorac. Oncol.* **2013**, *8*, S107–S108.
15. Fens, N.; Gaarthuis, Y.; Bos, A.C.; Schlosser, N.J.J.; Sterk, P.J. Exclusion of Asthma for Screening Purposes Using Exhaled Air Molecular Profiling by Electronic Nose. *Eur. Respir. J.* **2011**, *38*, 4168.

16. Hubers, A.J.; Brinkman, P.; Boksem, R.J.; Rhodius, R.J.; Witte, B.I.; Zwinderman, A.H.; Heideman, D.A.M.; Duin, S.; Koning, R.; Steenbergen, R.D.M.; et al. Combined sputum hypermethylation and eNose analysis for lung cancer diagnosis. *J. Clin. Pathol.* **2014**, *67*, 707–711. [CrossRef] [PubMed]

17. Gaida, A.; Holz, O.; Nell, C.; Schuchardt, S.; Lavae-Mokhtari, B.; Kruse, L.; Boas, U.; Langejuergen, J.; Allers, M.; Zimmermann, S.; et al. A dual center study to compare breath volatile organic compounds from smokers and non-smokers with and without COPD. *J. Breath Res.* **2016**, *10*, 026006. [CrossRef] [PubMed]

18. Papaefstathiou, E.; Stylianou, M.; Andreou, C.; Agapiou, A. Breath analysis of smokers, non-smokers, and e-cigarette users. *J. Chromatogr. B* **2020**, 122349. [CrossRef]

19. Van Bragt, J.J.; Brinkman, P.; De Vries, R.; Vijverberg, S.J.; Weersink, E.J.; Haarman, E.G.; De Jongh, F.H.; Kester, S.; Lucas, A.; in't Veen, J.C.C.M.; et al. Identification of recent exacerbations in COPD patients by electronic nose. *ERJ Open Res.* **2020**, *6*. [CrossRef]

20. Brinkman, P.; Wagener, A.H.; Bansal, A.T.; Knobel, H.H.; Vink, T.J.; Rattray, N.; Santonico, M.; Pennazza, G.; Montuschi, P.; Fowler, S.J.; et al. Electronic Noses Capture Severe Asthma Phenotypes by Unbiased Cluster Analysis. *Am. J. Respir. Crit. Care Med.* **2014**, *189*, A2171.

21. De Groot, J.C.; Amelink, M.; Storm, H.; Reitsma, B.H.; Bel, E.; Ten Brinke, A. Identification of Three Subtypes of Non-Atopic Asthma Using Exhaled Breath Analysis by Electronic Nose. *Am. Thorac. Soc.* **2014**, *189*, A2170.

22. Fens, N.; De Nijs, S.B.; Peters, S.; Dekker, T.; Knobel, H.H.; Vink, T.J.; Willard, N.P.; Zwinderman, A.H.; Krouwels, F.H.; Janssen, H.-G.; et al. Exhaled air molecular profiling in relation to inflammatory subtype and activity in COPD. *Eur. Respir. J.* **2011**, *38*, 1301–1309. [CrossRef] [PubMed]

23. Caruso, M.; Emma, R.; Brinkman, P.; Sterk, P.J.; Bansal, A.T.; De Meulder, B.; Lefaudeux, D.; Auffray, C.; Fowler, S.J.; Rattray, N.; et al. Volatile Organic Compounds Breathprinting of U-BIOPRED Severe Asthma smokers/ex-smokers cohort. *Airw. Cell Biol. Immunopathol.* **2017**, *50*, PA2018. [CrossRef]

24. Thomson, N.C. Asthma and smoking-induced airway disease without spirometric COPD. *Eur. Respir. J.* **2017**, *49*, 1602061. [CrossRef] [PubMed]

25. Bosch, S.; Lemmen, J.P.M.; Menezes, R.; Van Der Hulst, R.; Kuijvenhoven, J.; Stokkers, P.C.F.; De Meij, T.G.J.; De Boer, N.K. The influence of lifestyle factors on fecal volatile organic compound composition as measured by an electronic nose. *J. Breath Res.* **2019**, *13*, 046001. [CrossRef] [PubMed]

26. Pinheiro, G.P.; De Souza-Machado, C.; Fernandes, A.G.O.; Mota, R.C.L.; Lima, L.L.; Vasconcellos, D.D.S.; Júnior, I.P.D.L.; Silva, Y.R.D.S.; Lima, V.B.; De Oliva, S.T.; et al. Self-reported smoking status and urinary cotinine levels in patients with asthma. *J. Bras. Pneumol.* **2018**, *44*, 477–485. [CrossRef] [PubMed]

27. Farraia, M.V.; Rufo, J.C.; Paciência, I.; Mendes, F.; Delgado, L.; Moreira, A. The electronic nose technology in clinical diagnosis: A systematic review. *Porto Biomed. J.* **2019**, *4*, e42. [CrossRef] [PubMed]

28. De Vries, R.; Sterk, P.J. ENose Breathprints as Composite Biomarker for Real-Time Phenotyping of Complex Respiratory Diseases. *J. Allergy Clin. Immunol.* **2020**, *146*, 995–996. [CrossRef]

29. Preti, G.; Thaler, E.; Hanson, C.W.; Troy, M.; Eades, J.; Gelperin, A. Volatile compounds characteristic of sinus-related bacteria and infected sinus mucus: Analysis by solid-phase microextraction and gas chromatography–mass spectrometry. *J. Chromatogr. B* **2009**, *877*, 2011–2018. [CrossRef]

30. Thaler, E.R.; Hanson, C.W. Use of an Electronic Nose to Diagnose Bacterial Sinusitis. *Am. J. Rhinol.* **2006**, *20*, 170–172. [CrossRef] [PubMed]

31. Dutta, R.; Dutta, R. Intelligent Bayes Classifier (IBC) for ENT infection classification in hospital environment. *Biomed. Eng. Online* **2006**, *5*, 65. [CrossRef]

32. Hanson, C.W.; Thaler, E.R. Electronic Nose Prediction of a Clinical Pneumonia Score: Biosensors and Microbes. *Anesthesiologists* **2005**, *102*, 63–68. [CrossRef]

33. Hockstein, N.G.; Thaler, E.R.; Lin, Y.; Lee, D.D.; Hanson, C.W. Correlation of Pneumonia Score with Electronic Nose Signature: A Prospective Study. *Ann. Otol. Rhinol. Laryngol.* **2005**, *114*, 504–508. [CrossRef] [PubMed]

34. Hockstein, N.G.; Thaler, E.R.; Torigian, D.; Miller, W.T.; Deffenderfer, O.; Hanson, C.W. Diagnosis of Pneumonia With an Electronic Nose: Correlation of Vapor Signature With Chest Computed Tomography Scan Findings. *Laryngoscope* **2004**, *114*, 1701–1705. [CrossRef] [PubMed]

35. Hilberink, S.R.; E Jacobs, J.; Van Opstal, S.; Van Der Weijden, T.; Keegstra, J.; Kempers, P.L.; Muris, J.W.; Grol, R.P.; De Vries, H. Validation of smoking cessation self-reported by patients with chronic obstructive pulmonary disease. *Int. J. Gen. Med.* **2011**, *4*, 85–90. [CrossRef]

36. Hirvonen, E.; Stepanov, M.; Kilpeläinen, M.; Lindqvist, A.; Laitinen, T.; Stepanov, M. Consistency and reliability of smoking-related variables: Longitudinal study design in asthma and COPD. *Eur. Clin. Respir. J.* **2019**, *6*, 1591842. [CrossRef] [PubMed]

37. Romain, A.; Nicolas, J. Long term stability of metal oxide-based gas sensors for e-nose environmental applications: An overview. *Sensors Actuators B: Chem.* **2010**, *146*, 502–506. [CrossRef]

38. Ibrahim, M.I.A.; Brinkman, P.; Vijverberg, S.J.H.; Neerincx, A.H.; Hashimoto, S.; De Vries, R.; Dagelet, Y.W.; Knipping, K.; Sterk, P.J.; Kraneveld, A.D.; et al. eNose breathprints as surrogate biomarkers for classifying asthma patients by atopy. *Allergy Immunol.* **2019**, *54*.

39. Kaiser, H.F. The Application of Electronic Computers to Factor Analysis. *Educ. Psychol. Meas.* **1960**, *20*, 141–151. [CrossRef]

40. Robin, X.A.; Turck, N.; Hainard, A.; Tiberti, N.; Lisacek, F.; Sanchez, J.-C.; Muller, M.J. pROC: An open-source package for R and S+ to analyze and compare ROC curves. *BMC Bioinform.* **2011**, *12*, 77. [CrossRef] [PubMed]
41. Kuhn, M. Building Predictive Models inRUsing thecaretPackage. *J. Stat. Softw.* **2008**, *28*, 1–26. [CrossRef]
42. Venables, W.N.; Ripley, B.D. *Modern Applied Statistics with S*, 4 ed.; Springer: New York, NY, USA, 2002.

molecules

Article

Detection of Paratuberculosis in Dairy Herds by Analyzing the Scent of Feces, Alveolar Gas, and Stable Air

Michael Weber [1], Peter Gierschner [2,3], Anne Klassen [1,4], Elisa Kasbohm [5], Jochen K. Schubert [2], Wolfram Miekisch [2], Petra Reinhold [1] and Heike Köhler [1,6,*]

[1] Institute of Molecular Pathogenesis at 'Friedrich-Loeffler-Institut' (Federal Research Institute for Animal Health), Naumburgerstr. 96a, 07743 Jena, Germany; michael.weber@fli.de (M.W.); annekuentzel@hotmail.de (A.K.); petra.reinhold@fli.de (P.R.)

[2] Rostock Medical Breath Research Analytics and Technologies (RoMBAT), Department of Anesthesia and Intensive Care, Rostock University Medical Center, Schillingallee 35, 18057 Rostock, Germany; peter.gierschner@gmx.net (P.G.); jochen.schubert@uni-rostock.de (J.K.S.); wolfram.miekisch@uni-rostock.de (W.M.)

[3] Albutec GmbH, Schillingallee 68, 18057 Rostock, Germany

[4] Thüringer Tierseuchenkasse, Rindergesundheitsdienst (Thuringian Animal Health Fund, Cattle Health Service), Victor-Goerttler-Straße 4, 07745 Jena, Germany

[5] Department of Mathematics and Computer Science, University of Greifswald, Walther-Rathenau-Straße 47, 17489 Greifswald, Germany; elisa.kasbohm@uni-greifswald.de

[6] National Reference Laboratory for Paratuberculosis, Naumburger Straße 96a, 07743 Jena, Germany

* Correspondence: heike.koehler@fli.de; Tel.: +49-3641-804-2240

Citation: Weber, M.; Gierschner, P.; Klassen, A.; Kasbohm, E.; Schubert, J.K.; Miekisch, W.; Reinhold, P.; Köhler, H. Detection of Paratuberculosis in Dairy Herds by Analyzing the Scent of Feces, Alveolar Gas, and Stable Air. *Molecules* **2021**, *26*, 2854. https://doi.org/10.3390/molecules26102854

Academic Editors: Ben de Lacy Costello and Natalia Drabińska

Received: 13 April 2021
Accepted: 6 May 2021
Published: 11 May 2021

Abstract: Paratuberculosis is an important disease of ruminants caused by *Mycobacterium avium* ssp. *paratuberculosis* (MAP). Early detection is crucial for successful infection control, but available diagnostic tests are still dissatisfying. Methods allowing a rapid, economic, and reliable identification of animals or herds affected by MAP are urgently required. This explorative study evaluated the potential of volatile organic compounds (VOCs) to discriminate between cattle with and without MAP infections. Headspaces above fecal samples and alveolar fractions of exhaled breath of 77 cows from eight farms with defined MAP status were analyzed in addition to stable air samples. VOCs were identified by GC–MS and quantified against reference substances. To discriminate MAP-positive from MAP-negative samples, VOC feature selection and random forest classification were performed. Classification models, generated for each biological specimen, were evaluated using repeated cross-validation. The robustness of the results was tested by predicting samples of two different sampling days. For MAP classification, the different biological matrices emitted diagnostically relevant VOCs of a unique but partly overlapping pattern (fecal headspace: 19, alveolar gas: 11, stable air: 4–5). Chemically, relevant compounds belonged to hydrocarbons, ketones, alcohols, furans, and aldehydes. Comparing the different biological specimens, VOC analysis in fecal headspace proved to be most reproducible, discriminatory, and highly predictive.

Keywords: classification models; dairy cows; exhaled breath; fecal headspace; *Mycobacterium avium* ssp. *paratuberculosis* (MAP); paratuberculosis; random forest; stable air; volatile organic compound (VOC)

1. Introduction

Paratuberculosis (paraTb) is one of the four economically most important infectious diseases of dairy cattle [1] in developed countries. Caused by *Mycobacterium avium* subsp. *paratuberculosis* (MAP), it is a chronic progressive granulomatous enteritis resulting in malnutrition, reduction in milk yield, weight loss, and eventually, death. Infection with MAP occurs in young animals and may remain clinically nonapparent for several years until clinical signs are observed [2]. Identification of affected herds is crucial for successful control of the disease; however, the existing diagnostic methods have limitations, either regarding their sensitivity (antibody detection) or due to high expenditure of time and labor (bacterial culture, molecular biological detection via PCR).

During recent years, the development of diagnostic tests that allow a rapid, economic, and reliable identification of animals or herds affected by infectious diseases on the spot, so-called pen-side tests, received growing attention. Chromatography-based lateral flow tests, for example, were developed for the detection of viruses [3–5] or antigen-specific antibodies [6,7] in serum samples. However, the application of these tests demands invasive blood sampling. Analysis of volatile organic compounds (VOCs) present in exhaled breath and headspace air of fecal samples was suggested as a novel, alternative, noninvasive approach to the diagnosis of infectious diseases, in particular *Mycobacteria* infections [8–11]. The relevant VOCs originate from different sources within the host-pathogen interaction, such as the bacterial metabolism and the inflammatory and immunological host response to the pathogen [12]. They belong to all classes of organic substances and appear in very low concentrations (ppbV to pptV). Due to their physicochemical properties, they transform into a gaseous state already at low temperatures [13]. Discrimination between infected and noninfected animals is not based on individual compounds, but on relative concentration changes of different informative VOCs, thus demanding multivariate data analysis [14].

The contribution of VOCs of variable origin to indoor air quality and their impact on human health has been studied extensively [15]. However, the diagnostic potential of VOC analysis in stable air, the equivalent to indoor air in animal husbandry, has not been assessed so far. This would be another option for pen-side diagnosis.

This proof-of-principle study performed in eight dairy herds of different farms was based on the hypothesis that it should be possible to discriminate between cattle with and without MAP infections by using VOCs as biomarkers of infection. Which biological specimen turns out as the most suitable one to detect MAP infection under farming conditions remained to be elucidated. Thus, (i) headspaces above fecal samples, (ii) the alveolar fractions of exhaled breath, and (iii) stable air samples were analyzed in parallel and were compared with respect to their diagnostically relevant VOC patterns.

2. Results

2.1. Visualization of VOC Datasets from Feces, Alveolar Gas, and Stable Air

The first step of the data analysis involved the explorative visualization of the three multidimensional datasets: fecal headspace data (F), alveolar gas data (A), and stable air data (S). Each is represented by a matrix, composed of a number of measured VOCs (columns) and samples (rows), which are either cattle associated (A,F) or stable associated (S). Additionally, stable-associated samples were subgrouped into S1, S2, S3, and S4, according to the location of sampling within the stable (see Methods). We aimed to visualize these VOC concentration matrices in a compact and illustrative way by using annotated heatmaps, which are shown in Figure 1. Each heatmap displays the VOC concentration levels in combination with annotation columns, which provide information about the MAP status and the corresponding farm of the sample.

In comparison, the heatmap plots reveal differences in the distribution of VOC concentration levels in the three datasets. Alveolar gas and stable air data have a large proportion of columns with values at the lowest level of the plotted color range (blue), while the majority of VOCs appeared to be present in fecal data. In total, 76 VOCs were found at detectable levels in feces, compared to 30 in alveolar gas. In stable air, there were 24 VOCs detectable in S1 and S2 samples (both collected at head-level of the animals), compared to 25 VOCs for S3 (collected close to the floor) and 23 VOCs for S4 (collected distant from animals). The varying number of detectable VOCs for each dataset resulted in a distinct number of potential features for the subsequent classification approach. Additionally, the farm annotation column does not indicate a distinct clustering of samples according to their farm but rather an overall mixed grouping of samples.

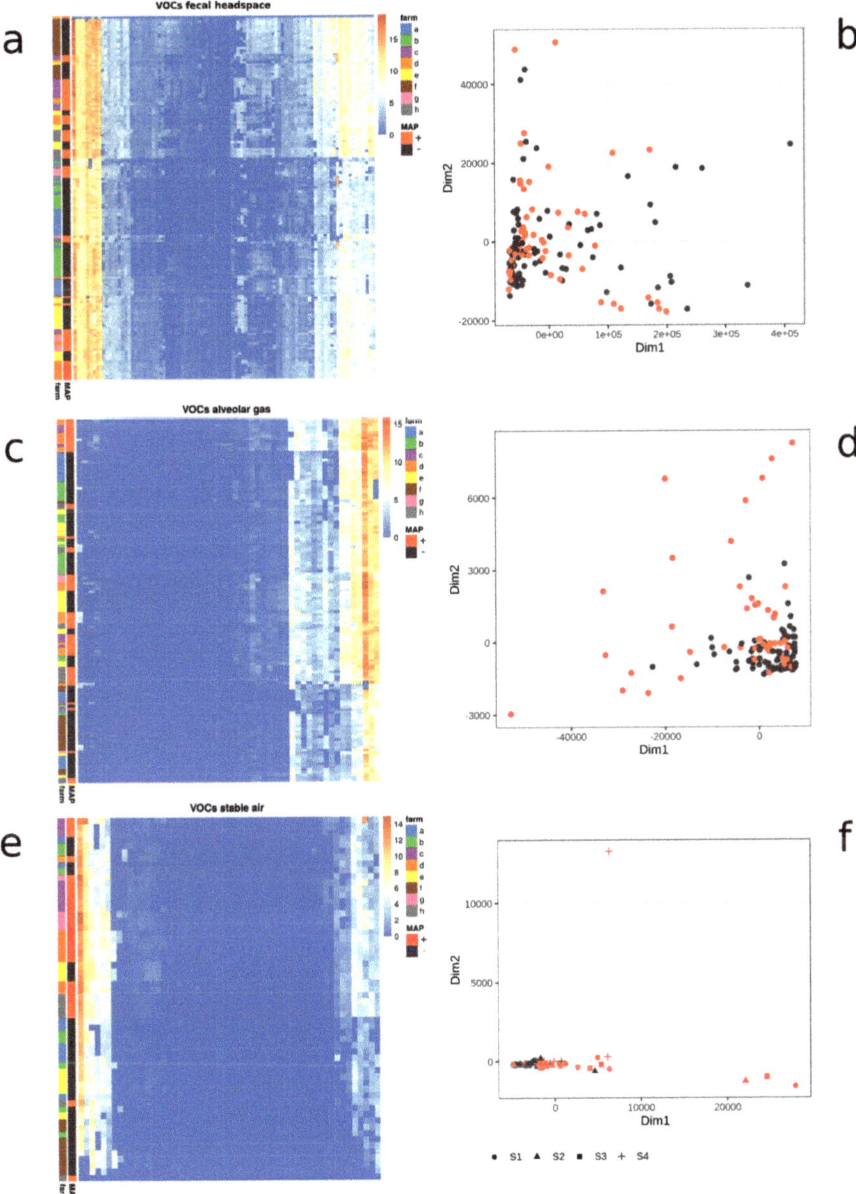

Figure 1. Heatmaps and two-dimensional scatterplots for each VOC dataset: fecal headspace (**a**,**b**), alveolar gas (**c**,**d**), and stable air (**e**,**f**). Heatmaps are composed of samples (rows), VOCs (columns), and two annotation columns. Heat colors indicate the log2-transformed substance concentration, which is explained by the colored range legend. Annotation columns indicate the associated farm and MAP status. Scatterplots display multidimensionally scaled (MDS) data points, each representing a VOC sample colored according to MAP status (red: MAP-positive, black: MAP-negative).

We also examined the clustering of the samples in each dataset in a two-dimensional scatter plot by using multidimensional scaling (MDS). These plots enable the analysis of relative dissimilarities between the VOC samples and the inspection of the presence of cluster structures. Fecal headspace samples partially form a cluster (Figure 1b), while the remaining samples are distributed in multiple different directions. MAP-positive and MAP-negative samples cannot be assigned to separate locations in the plot. Similarly, most alveolar gas samples group within a large cluster (Figure 1d), however, the outer samples are mainly assigned to MAP-positive animals. In the stable air plot (Figure 1f), there is generally less variability compared to the former two plots. Here, we observed a subset of MAP-positive samples, which is scattered along the x-axis. In summary, the MDS plots revealed some trends and suggested that a simple unsupervised separation of MAP samples is not feasible. Therefore, we decided to apply a multivariate machine learning approach to generate a classification model that is based on the combination of multiple VOC profiles. We analyzed each dataset individually to evaluate its performance in the classification of MAP.

2.2. Identification and Reproducibility of Significant VOCs Present in Fecal Headspaces and Alveolar Gas

Prior to the classification of the samples into MAP-positive and MAP-negative classes, we performed feature selection to identify VOCs that exhibit discriminatory and robust concentration levels. Therefore, we aimed to select VOCs that were measured across all farms and turned out as robust against the influence of the sampling time point. To investigate and quantify the latter temporal dependency, we performed the feature selection method Boruta on the two datasets representing headspace above fecal (F) and alveolar gas (A) samples. To compare the resulting importance scores, we generated dot charts, which are shown in Figure 2.

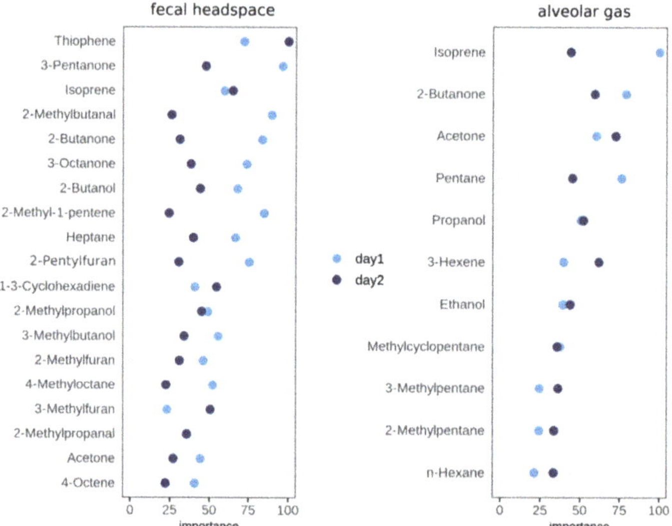

Figure 2. Variable importance plots displaying the importance values for all VOCs, which were selected with the Boruta method for each sampling day. The two panels show the resulting variables of the fecal headspace dataset and the alveolar gas dataset. The x-axis indicates the importance value, which is scaled to the maximum importance. Dot colors highlight the two sampling days.

From the 76 detectable VOCs above fecal samples, 19 fulfilled the inclusion criteria, i.e., succeeded in the Boruta feature selection. In contrast, 11 out of 30 VOCs, were selected in the alveolar gas dataset. Generally, the selected subset of features from the first sampling day was confirmed by significant importance scores from the second sampling day. However, particularly in the fecal headspace dataset, some VOCs showed high variance in their importance rank, e.g., 2-methylbutanal and 2-methyl-1-pentene. This provided evidence that day-dependent deviations in the VOC levels exist, which needed to be taken into account by the classification approach. Therefore, the following classification model validation included separate model testing for both sampling days.

2.3. Classification of MAP Status from VOCs Present above Feces or in Alveolar Gas

Using the VOC concentration data from the subset of previously selected VOCs, we aimed to build one classification model for each dataset, which correctly predicts the MAP disease status of the samples. We employed random forest models to perform the classification as described in Methods. To estimate the accuracy of the resulting models, we generated receiver operating characteristic (ROC) curves, which are displayed in Figure 3.

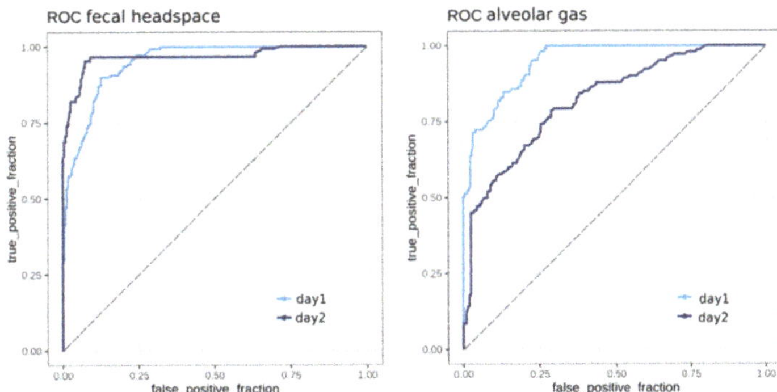

Figure 3. Receiver operating characteristic (ROC) curves for random forest classification models from subsets of fecal headspace data (model F, 19 variables) and alveolar gas data (model A, 11 variables). Each curve represents the outcome of repeated model cross-validation on unseen test data from either day 1 or day 2. Area under the curves (AUC) values for model F are 0.94 (day 1) and 0.96 (day 2) and for model A are 0.95 (day 1) and 0.82 (day 2).

Each curve represents the predictions of repeated 10-fold cross-validation for the respective dataset. To evaluate the model performance for both sampling days, we generated an individual ROC curve using the test predictions for each day. The area under the curves (AUC)–ROC values were calculated for all the curves.

The F-model (from fecal headspace samples) achieved comparably high levels (AUC–ROC day 1 = 0.94, AUC–ROC day 2 = 0.96), which indicated less model dependence on the sampling day and good reproducibility of the prediction accuracies. In contrast, the A-model (from samples of alveolar gas) showed promising results (AUC–ROC = 0.95) on day 1 but performed slightly worse on the second day (AUC–ROC = 0.82). Although the model was trained on data from both sampling days, it tended to predict day 1 samples more accurately, which in turn indicated that the VOC levels showed higher discriminatory power on the first day.

2.4. Identification of Significant VOCs Classifying for Paratuberculosis in Stable Air

In the next step, we analyzed the VOC composition in the stable air samples (collected at different locations) in order to investigate the variability and spatial dependencies of the VOC profiles. Hereby, we distinguished between four types of samples (S1, S2, S3, S4) as described in Methods. Initially, we analyzed if the presence of a face mask influenced the VOC profiles measured in front of the cow's head. Therefore, we statistically compared the mean concentration between S1 and S2 for each VOC using Wilcoxon-Mann–Whitney tests. Since no significant differences were found ($p > 0.05$), we merged groups S1 and S2 and treated them as a single head-level group (S1S2).

Next, we conducted a feature importance analysis as in Section 2.1 for the three groups of stable air (S1S2, S3, S4). As a result, 5 out of 25 detectable VOCs were selected from the

S1S2 dataset, compared to 4 out of 25 from the S3 group (close to the floor). The resulting VOCs are shown in Figure 4.

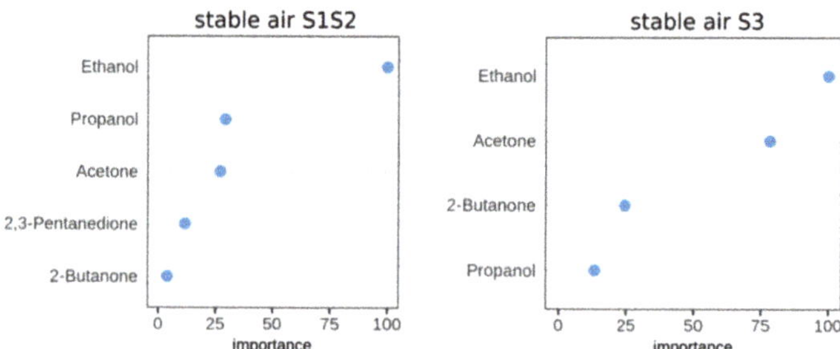

Figure 4. Variable importance plots displaying the importance values for all VOCs, which were selected with the Boruta method. The two panels show the resulting variables of the merged dataset of stable air collected at head level (S1S2), and the dataset of stable air collected close to the floor (S3). The x-axis indicates the importance value, which is scaled to the maximum importance.

Interestingly, no feature was confirmed relevant in the S4 group (stable air sampled distant from animals). Thus, two stable air groups remained (S1S2, S3), which represented VOC levels at head level and close to the floor, respectively. Both datasets were used to train a predictive classification model in the next step.

2.5. Classification of MAP Status from VOCs Present in Stable Air

In the second classification run, we aimed to build classification models from samples of two groups of stable air (S1S2, S3). Compared to the fecal headspace and alveolar gas datasets, the stable air dataset contains fewer samples (S1S2: n = 29, S3: n = 15); thus, we decided to perform cross-validation on all samples without creating separate datasets for each sampling day. Again, to evaluate the prediction performance of the model in cross-validation, we generated ROC curves for both models and calculated the associated AUC values. Stable air S1S2 achieved an averaged AUC value of 0.87, compared to an AUC value of 0.91 for S3 (Figure 5).

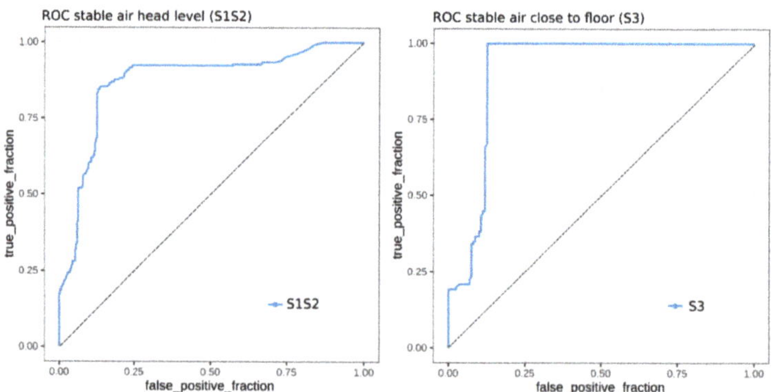

Figure 5. Receiver operating characteristic (ROC) curves for random forest classification models from subsets of stable air head-level data S1S2 (five variables) and stable air close to floor data S3 (four variables). Each curve represents the outcome of repeated model cross-validation. The area under the curve (AUC) for model S1S2 is 0.87 and for model S3 is 0.91.

2.6. Comparison of Resulting VOC Sets

In total, we identified 28 VOCs that were found to be relevant for the classification of MAP samples across all investigated resources (Supplementary Materials, Table S1). Two compounds, i.e., acetone and 2-butanone, were detected under all four investigated conditions. Ethanol and propanol were detected in the groups of A, S1S2, and S3. Isoprene was present in fecal headspaces and alveolar gas samples. However, the largest number of VOCs was only present in a single set: 16 VOCs in fecal headspaces, 6 VOCs in alveolar gas, and 1 VOC in stable air collected close to the animals (S1S2). Since the Venn diagram (Figure 6) is based on the presence or absence of VOC compounds, we also evaluated the regulation of the VOCs and found that 26 VOCs were upregulated in MAP-positive samples, compared to MAP-negative samples under the tested conditions. Interestingly, 2-butanone was upregulated in groups A, S1S2, and S3 but downregulated in F. Additionally, 4-octene was downregulated in group F (Supplementary Materials, Table S1).

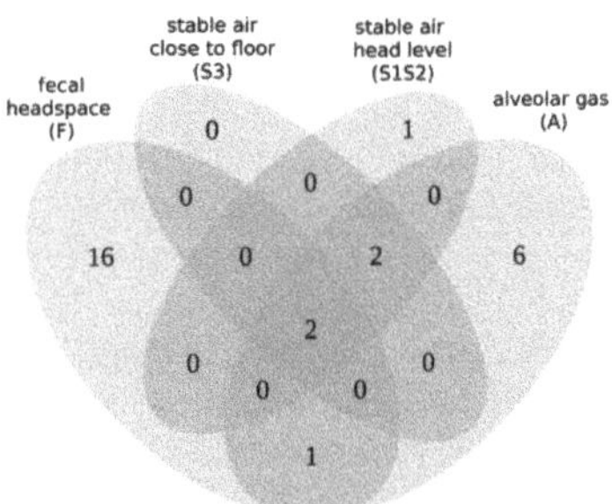

Figure 6. Venn diagram of four sets of selected VOCs (fecal headspace (F) n = 19, stable air close to floor (S3) n = 4, stable air head level (S1S2) n = 5, alveolar gas (A) n = 11), which represents the comparison of the selected VOCs. Values indicate the number of common VOCs among the sets of overlapping areas. The core area indicates the number of common VOCs, which are shared by all four sets (n = 2).

3. Discussion

Confirming the hypothesis that MAP infections in animals or herds, respectively, are detectable by means of volatile biomarkers, this study presents the first consideration of tracing VOC profiles as potential diagnostic markers even under field conditions of livestock farming. With respect to the suitability of different matrices, headspace above feces, the alveolar fractions of exhaled breath, and stable air samples revealed diagnostically relevant VOCs of unique but partly overlapping patterns. Biological and methodological aspects need to be taken into account when interpreting the results of this study.

3.1. Biological Aspects

MAP infection is characterized by a chronic local inflammation that mainly affects the gut-associated lymphoid tissue of the small intestine and proximal colon, the intestinal mucosa of jejunum and ileum, and the mesenteric and ileocolic lymph nodes. A wide variation of severity and distribution of lesions (focal to multifocal to diffuse), inflammatory cell infiltrates (lymphocytes, multinucleated giant cells, epitheloid macrophages), and mycobacteria within lesions (paucibacillary, multibacillary) of animals can be observed [16]. MAP is transferred into the intestinal content and is eventually shed within feces in varying concentrations [17]. Intestinal inflammation is accompanied by malabsorption and cachexia [18].

Based on these disease characteristics, we anticipated MAP-related VOCs predominantly in samples originating from feces acknowledging that (i) the pathogen itself and (ii) markers of intestinal inflammation and/or local immune response might contribute to the pattern of volatile compounds. Furthermore, the smell of feces will always be influenced by emissions related to the intestinal content, i.e., more or less digested feed components, cell debris of the mucosal epithelium, and constituents of the gut microbiota. Feed composition may vary between herds, but paraTb may also alter the intestinal interior due to inflammatory processes in the intestinal wall.

Exploiting the alveolar fraction of exhaled breath is based on the rationale that there is a huge contact surface between lung and blood allowing the transfer of blood-borne VOCs into the lung. Thus, alveolar gas is most likely representative of blood-borne volatiles. If all fractions of exhaled breath would be collected, a mixture between gas columns in the larger airways (airway dead space gas, composition roughly equivalent to ambient air) and alveolar gas would be analyzed. Compared to the human lung, the proportion of airway dead space volume per breath is much greater in large animals due to the anatomy of extra-thoracic airways [12]. Related to paraTb, blood-born VOCs may origin from (i) catabolic processes that are a consequence of malabsorption and inflammation and (ii) from systemic markers of inflammation and host response. In addition, processes localized in the gut or in draining lymph nodes as well as bacterial metabolism within the lesions might contribute to blood markers due to large blood–intestinal exchange areas within the body.

Both, VOC emissions from feces and from exhaled breath contribute to the VOC composition of stable air. However, the composition of stable air will be significantly dependent on the collection site within the stable.

3.1.1. Sources of Variability in VOC Profiles

From the 76 detectable VOCs above fecal samples, 19 fulfilled the inclusion criteria, and 11 out of 30 VOCs were selected in the alveolar gas dataset. VOCs selected for the first sampling day were confirmed for the second day, although a certain degree of variation was observed. Based on these sets of VOCs, we were able to classify MAP-negative and MAP-positive samples with high accuracy. This confirms the findings of previous experimental studies suggesting that discrimination between MAP-negative and MAP-positive animals is possible by the analysis of the VOC profiles in the headspace of fecal samples and in exhaled breath, i.e., alveolar gas, of goats [8,14]. Despite standardized feeding, housing, and management conditions, physiological variability of discriminatory VOC profiles was noted [14]. The patterns of the most prominent substances changed in the course of infection; however, differences between inoculated and noninoculated animals remained detectable at any time in fecal samples and breath [8]. The results of two consecutive experimental studies in goats revealed that certain VOCs contributed reproducibly to the discriminatory VOC profile, although the effect size of the most important substances varied [14]. In contrast to standardized conditions in animal experiments, there are additional sources of variability of VOC emissions in the field, such as other diseases, differences in feed composition, bedding material, floor conditions, air exchange rates in the stable, numbers of animals per square meter. Such factors can influence the VOC measurements and therefore require the application of suitable statistical methods. Despite all these potential confounding factors, we successfully extracted VOC sets indicative of MAP infection from different sources.

3.1.2. VOCs Indicative of MAP Infection

VOCs indicative of MAP infection differed between the three matrices explored in this study. Our results indicated that the number of VOCs contributing to discrimination was higher in fecal headspace (19 VOCs), compared to alveolar gas (11 VOCs), which is in line with previous reports from goat experiments [8,14]. The lowest number of discriminatory VOCs was observed in stable air (4–5 VOCs). Interestingly, some of the VOCs with high effect sizes in the goat experiments [14] were also considered important for the identification of MAP infection in cattle under field conditions, such as 2-butanone, acetone, and propanol in alveolar gas, and isoprene, 2-butanone, 3-octanone, heptane, 2-pentylfuran, 2-methylfuran, 3-methylfuran and acetone in fecal headspace. It seems that, in both animal species, these VOCs originate from similar metabolic and pathophysiologic processes.

There was only a partial overlap of indicative VOCs between the three matrices. Only two substances, acetone and 2-butanone, were generally considered important. Ethanol and propanol were discriminatory in alveolar gas and stable air, and isoprene in samples of alveolar gas and fecal headspace. As already discussed previously, VOCs indicative of MAP infection in fecal headspace and alveolar gas originate from different sources, which may explain these observations. Relevant VOCs in fecal headspace seemed predominantly related to MAP metabolism, in particular fatty acid turnover and carbon metabolism. This is supported by the fact that eight compounds, namely, heptane, 3-methylbutanol, acetone, 2-butanone, 3-pentanone, 3-octanone, 3-methylfuran, and 2-pentylfuran, belonged to the recently published VOC core profile of cultivated MAP strains [19]. Furthermore, five compounds, namely 2-methylpropanol, 3-methylbutanol, 3-pentanone, isoprene, and 2-methylfuran, were indicative for MAP growth during cultural isolation from clinical samples of cattle and goats [20].

In alveolar gas, however, compounds originating from inflammatory processes seemed to predominate. VOCs indicative of inflammatory bowel disease (IBD), such as 1-propanol, ethanol, and pentane [21], were also indicative of MAP infection. The local inflammatory processes in the intestine share features in both diseases. Inflammation may lead to oxidative stress and increased production of reactive oxygen species (ROS). ROS oxidize biologically important molecules and cause lipid peroxidation of polyunsaturated fatty acids, generating alkanes and methylated alkanes [22], such as pentane, hexane, 2- and 3-methylpentane, and methylcyclopentane, which belonged to the most important discriminatory substances in alveolar gas of MAP-infected cows in our study.

Some of the indicative VOCs, in particular acetone and isoprene, are likely to originate from both, mycobacterial metabolism and host response. Mycobacteria are able to produce methyl ketones, such as acetone [23], but acetone has also been linked to fat catabolism in cattle and humans [24,25]. One major source of isoprene is the bacterial metyl-erythritol phosphate pathway [26,27]; on the other hand, isoprene formation in humans was shown to correlate with cholesterol biosynthesis [28] and, more importantly, with IBD [29].

Stable air samples only turned out meaningful when the collection was performed close to the animals, i.e., either in front of the animal's head (breathing area) or near to the stable's floor (coated with feces or manure). Despite these different sites of collection, the interesting VOCs were nearly identical (Figure 4). However, compared to fecal headspaces or alveolar gas samples, the number of indicative VOCs found in relevant stable air samples was significantly lower. Multiple sources contribute to the VOC composition of stable air, such as emissions from feces, urine, and skin of the animals, gases eructated from the forestomaches of ruminants, components of breath, feedstuffs, bedding material, fuel, building materials, and VOCs contained in outdoor air. This may impair the identification of relevant VOCs. In addition, stable air is continuously exchanged by fresh outdoor air, which may result in a decrease of VOC concentrations below the limit of detection even in the vicinity of the animals.

3.2. Methodological Aspects

Sampling was conducted by preconcentration of VOCs either directly in the stable (alveolar gas and stable air) or from fecal headspace after filling feces into sealed headspace vials. Volatiles were identified later offline by GC–MS. This enabled the detection of VOCs in very low concentrations in the range of ppbV to pptV. Utilization of VOC analysis for practical pen-side diagnosis would demand a different approach. VOC emission has to be measured directly on the spot. Analytical platforms that allow an online analysis of VOC emissions, such as ion mobility spectrometry (IMS), ion flow tube–mass spectrometry (SIFT–MS), or proton transfer reaction–mass spectrometry (PTR–MS), respectively, are available and could be adapted for this purpose. Finally, the discriminatory performance of the adapted analysis systems compared to established diagnostic methods has to be evaluated.

Demands on test quality largely depend on the final purpose of testing, either identification of MAP-positive herds or individual MAP-positive animals. Analysis of fecal headspace or alveolar gas would enable identification of both, positive individuals as well as positive herds. Analysis of stable air would only allow identification of positive herds.

From the diagnostic perspective, test accuracy and validity of the classification model are of particular importance. Here, we applied an adapted multivariate classification analysis based on the random forest method, which has been used in various cases in the field of metabolomics [20,30,31]. In this context, some considerations were taken into account. First, overfitting of the predictive models, i.e., the loss of general prediction validity can occur if the model fitted too tightly to the training data. Random forests employ several techniques to avoid or minimize overfitting, for instance, the random selection of variables at each node, the error estimation on the unseen out-of-bag data, and the large number of trees, each of which is only based on a subset of the input data. Additionally, we applied repeated cross-validation to improve the error estimation of the model, due to the varying composition of the training dataset. This subsetting approach is essential to account for the variability of different animals, farms, and sampling days on the classification results. In this context, we also evaluated the reproducibility between consecutive measurement days. For validation, we generated model test sets for both days separately in order to evaluate one model on test data from two time points. Nevertheless, the validation of the model on an independently sampled population remains to be shown.

With AUC values of 0.94 (day 1) and 0.96 (day 2), VOC analysis in fecal headspace proved to be more reproducible and discriminatory than alveolar gas analysis for the identification of MAP-positive animals. Furthermore, analysis of fecal headspace was highly predictive, because the ROC curves closely approach the upper left corner of optimal prediction. VOC analysis in alveolar gas was less reproducible, because of marked differences in the AUC values between the two analysis days (0.95 on day 1 vs. 0.82 on day 2), and less predictive, too.

The AUC values calculated for VOC analysis in fecal headspace of individual cattle indicate that this approach has higher discriminatory power than established indirect antibody ELISAs for the identification of MAP-infected animals, where AUC values of 0.55–0.57 [32] or 0.77–0.91 [33] were calculated, depending on the reference method used. It is also superior to recently published novel biomarker-based diagnostic tests for paratuberculosis (ABCA13-based ELISA, SPARC ELISA, MMP8 ELISA) with AUC values of 0.79–0.85 [34]. However, these are only preliminary data, which have to be confirmed in future field studies covering larger numbers of animals and herds. Likewise, the number of MAP-positive and MAP-nonsuspect herds included in this study is too low to evaluate the diagnostic performance of the approach on the herd level.

This limitation also applies to VOC analysis in stable air. However, our preliminary data indicate that, despite relatively high AUC values (0.87 and 0.91), VOC analysis in stable air is not predictive enough for the identification of MAP-positive herds. This is mainly due to the fact that sensitivity is increased only to the cost of specificity, and a high rate of false-positive results has to be anticipated.

4. Animals, Materials, and Methods

4.1. Legislation and Ethical Approval

This study was carried out in strict accordance with European and National Law for the Care and Use of Animals. The protocol was approved by the Animal Health and Welfare Unit of the Thüringer Landesamt für Verbraucherschutz (Permit Number: 04-102/16; date of permission: 20.04.2016). The experiments were conducted under the supervision of the authorized institutional Animal Protection Officer. Every effort was made to minimize discomfort and suffering throughout the duration of the study. No sedation or anesthesia was applied to the animals.

4.2. Characteristics of Herds and Animals Used for Sample Collection

In total, 8 dairy herds and 77 individual dairy cows were included in this explorative study. All herds were enrolled in the voluntary paratuberculosis control program of the

German federal state of Thuringia. The paratuberculosis status of the herds was defined depending on the results of annual whole herd tests for the presence of MAP in individual fecal samples by bacterial culture (fecal culture). Accordingly, four herds were classified as MAP-positive herds, because MAP was detected in the fecal samples of animals from these herds, and four herds were classified as MAP-nonsuspect herds, because MAP was not identified in any fecal sample of animals from these herds during at least three preceding annual whole herd tests.

The individual dairy cows included in this study were selected based on their fecal culture results during whole herd testing. Animals from MAP-nonsuspect herds were classified as MAP negative. Cows from MAP-positive herds with at least one positive test result of fecal culture were classified as MAP positive. During the study period, 30 MAP-positive cows were available, and 47 MAP-nonsuspect cows could be recruited. The animals were examined for fecal shedding of MAP at the time of VOC sampling by fecal culture [35]; and for the presence of antibodies against MAP in blood serum by a commercially available enzyme-linked immunosorbent assay (IDEXX Paratuberculosis Screening ELISA, IDEXX, Montpellier, France).

4.3. Sampling and Preconcentration of Headspace above Feces, Alveolar Gas, and Stable Air

Each animal and each herd were sampled twice in a one-week interval. The numbers of samples included in this study are given in Tables 1 and 2.

Table 1. Numbers of alveolar gas and fecal samples obtained from dairy herds and included in VOC analyses.

Biological Specimen	Herd Status: MAP-Negative Herds/Animals/Samples	Herd Status: MAP-Positive Herds/Animals/Samples
alveolar gas	4/46/85	4/30/49
headspace above feces	4/47/93	4/30/58

Table 2. Numbers of stable air samples obtained from dairy herds and included in VOC analyses.

Locations of Stable Air Sampling	Herd Status: MAP-Negative Herds/Samples	Herd Status: MAP-Positive Herds/Samples
S1: head level, without face mask	4/8	4/7
S2: head level, through face mask	4/8	4/6
S3: close to floor contaminated with manure or feces	4/8	4/8
S4: distant from animals, slurry, or wastes, partially floated by fresh air	4/8	4/7

Fecal samples (F) were collected on an individual basis directly in a clean sampling container, either through rectal manipulation or during spontaneous defecation. From this, smaller portions of about 3 g of fresh feces per cow and time point were filled into a 20 mL headspace vial sealed with Teflon-coated rubber septa in combination with magnetic crimp caps, as described elsewhere [36]. The vials were stored at 4 °C and processed within 24–36 h after sampling. For preconcentration with needle trap microextraction (NTME), the vials were heated up to 37 °C. A needle trap device (NTD, packed with 1 cm divinylbenzene, Carbopack X and Carboxen) was connected to a 1 mL syringe and inserted through the Teflon-coated rubber. One mL of headspace gas was manually moved through the NTD into the syringe and back through the NTD into the headspace volume 20 times.

Collecting alveolar gas (A) was based on CO_2-controlled sampling of exhaled breath using a tightly fitting face mask and combining mainstream capnometry with needle-trap microextraction (NTME). The technical setup designed particularly for large animals has been described in detail elsewhere [12,36,37]. Shortly, a fast responding capnometer ensured continuous measurement of CO_2 in exhaled breath. Above a defined threshold

test. Introducing VOC analysis as a practical pen-side diagnostic test would ideally require online analysis of VOC emissions. The performance of such analytic systems, compared to established diagnostic methods, has yet to be evaluated.

Supplementary Materials: The following are available online, Table S1: List of important VOCs.

Author Contributions: Conceptualization, P.R., H.K., W.M., and J.K.S.; methodology, A.K., E.K., P.G., and W.M.; software, E.K. and M.W.; validation, P.G. and M.W.; formal analysis, P.G. and M.W.; investigation, A.K. and P.G.; resources, A.K., H.K., P.G, P.R., W.M., and J.K.S.; data curation, P.G. and M.W.; writing—original draft preparation, M.W., P.R., and H.K.; writing—review and editing, P.R. and W.M.; visualization, M.W.; supervision, W.M. and P.R.; project administration, P.R. and W.M.; funding acquisition, P.R. and J.K.S. All authors have read and agreed to the published version of the manuscript.

Funding: This research was funded by Deutsche Forschungsgemeinschaft (DFG), Grant Numbers RE 1098/4-2 and SCHU 1960/4-2.

Institutional Review Board Statement: The protocol was approved by the Animal Health and Welfare Unit of the Thüringer Landesamt für Verbraucherschutz (Permit Number: 04-102/16; date of permission: 20.04.2016). The experiments were conducted under the supervision of the authorized institutional Animal Welfare Officer.

Informed Consent Statement: Not applicable.

Data Availability Statement: Data are contained within the article or supplementary material.

Acknowledgments: The authors are very grateful for the excellent technical support of Dietmar Hein (PAS Technology Deutschland GmbH, Magdala, Germany). In addition, we thank the veterinarians of Thüringer Tiergesundheitsdienst who helped with establishing contact with all farmers. Furthermore, we wish to extend our gratitude to Ingolf Rücknagel and his crew of animal handlers (FLI Jena, Germany) for professional assistance when handling the animals.

Conflicts of Interest: The authors declare no conflict of interest. The funders had no role in the design of the study; in the collection, analyses, or interpretation of data; in the writing of the manuscript, or in the decision to publish the results.

References

1. Tiwari, A.; Vanleeuwen, J.A.; Dohoo, I.R.; Keefe, G.P.; Haddad, J.P.; Tremblay, R.; Scott, H.M.; Whiting, T. Production effects of pathogens causing bovine leukosis, bovine viral diarrhea, paratuberculosis, and neosporosis. *J. Dairy Sci.* **2007**, *90*, 659–669. [CrossRef]
2. Sweeney, R.W. Pathogenesis of Paratuberculosis. *Vet. Clin. N. Am. Food A* **2011**, *27*, 537–546. [CrossRef] [PubMed]
3. Brüning, A.; Bellamy, K.; Talbot, D.; Anderson, J. A rapid chromatographic strip test for the pen-side diagnosis of rinderpest virus. *J. Virol. Methods* **1999**, *81*, 143–154. [CrossRef]
4. Ferris, N.P.; Clavijo, A.; Yang, M.; Velazquez-Salinas, L.; Nordengrahn, A.; Hutchings, G.H.; Kristersson, T.; Merza, M. Development and laboratory evaluation of two lateral flow devices for the detection of vesicular stomatitis virus in clinical samples. *J. Virol. Methods* **2012**, *180*, 96–100. [CrossRef]
5. Reid, S.M.; Ferris, N.P.; Bruning, A.; Hutchings, G.H.; Kowalska, Z.; Akerblom, L. Development of a rapid chromatographic strip test for the pen-side detection of foot-and-mouth disease virus antigen. *J. Virol. Methods* **2001**, *96*, 189–202. [CrossRef]
6. Fleming, J.R.; Sastry, L.; Wall, S.J.; Sullivan, L.; Ferguson, M.A. Proteomic Identification of Immunodiagnostic Antigens for *Trypanosoma vivax* Infections in Cattle and Generation of a Proof-of-Concept Lateral Flow Test Diagnostic Device. *PLoS Negl. Trop. Dis.* **2016**, *10*, e0004977. [CrossRef]
7. Hanon, J.B.; Vandenberge, V.; Deruelle, M.; De Leeuw, I.; De Clercq, K.; Van Borm, S.; Koenen, F.; Liu, L.; Hoffmann, B.; Batten, C.A.; et al. Inter-laboratory evaluation of the performance parameters of a Lateral Flow Test device for the detection of Bluetongue virus-specific antibodies. *J. Virol. Methods* **2016**, *228*, 140–150. [CrossRef]
8. Bergmann, A.; Trefz, P.; Fischer, S.; Klepik, K.; Walter, G.; Steffens, M.; Ziller, M.; Schubert, J.K.; Reinhold, P.; Köhler, H.; et al. In Vivo Volatile Organic Compound Signatures of *Mycobacterium avium* subsp. *paratuberculosis*. *PLoS ONE* **2015**, *10*, e0123980. [CrossRef]
9. Ellis, C.K.; Rice, S.; Maurer, D.; Stahl, R.; Waters, W.R.; Palmer, M.V.; Nol, P.; Rhyan, J.C.; VerCauteren, K.C.; Koziel, J.A. Use of fecal volatile organic compound analysis to discriminate between nonvaccinated and BCG-Vaccinated cattle prior to and after *Mycobacterium bovis* challenge. *PLoS ONE* **2017**, *12*, e0179914. [CrossRef]

10. Ellis, C.K.; Stahl, R.S.; Nol, P.; Waters, W.R.; Palmer, M.V.; Rhyan, J.C.; VerCauteren, K.C.; McCollum, M.; Salman, M.D. A Pilot Study Exploring the Use of Breath Analysis to Differentiate Healthy Cattle from Cattle Experimentally Infected with *Mycobacterium bovis*. *PLoS ONE* **2014**, *9*, e89280.

11. Peled, N.; Ionescu, R.; Nol, P.; Barash, O.; McCollum, M.; VerCauteren, K.; Koslow, M.; Stahl, R.; Rhyan, J.; Haick, H. Detection of volatile organic compounds in cattle naturally infected with *Mycobacterium bovis*. *Sensors Actuators B Chem.* **2012**, *171*, 588–594. [CrossRef]

12. Reinhold, P.E.; Gierschner, P.; Küntzel, A.; Schubert, J.K.; Miekisch, W.; Köhler, H.U. Chapter 27: Ruminants. In *Breathborne Biomarkers and the Human Volatilome*, 2nd ed.; Beauchamp, J., Davis, C., Pleil, J., Eds.; Elsevier: Amsterdam, The Netherlands, 2020; pp. 441–460.

13. Bos, L.D.; Sterk, P.J.; Schultz, M.J. Volatile metabolites of pathogens: A systematic review. *PLoS Pathog.* **2013**, *9*, e1003311. [CrossRef]

14. Kasbohm, E.; Fischer, S.; Küntzel, A.; Oertel, P.; Bergmann, A.; Trefz, P.; Miekisch, W.; Schubert, J.K.; Reinhold, P.; Ziller, M.; et al. Strategies for the identification of disease-related patterns of volatile organic compounds: Prediction of paratuberculosis in an animal model using random forests. *J. Breath Res.* **2017**, *11*, 047105. [CrossRef]

15. Sundell, J. Reflections on the history of indoor air science, focusing on the last 50 years. *Indoor Air* **2017**, *27*, 708–724. [CrossRef]

16. Krüger, C.; Köhler, H.; Liebler-Tenorio, E.M. Sequential Development of Lesions 3, 6, 9, and 12 Months After Experimental Infection of Goat Kids with *Mycobacterium avium* subsp. *paratuberculosis*. *Vet. Pathol.* **2015**, *52*, 276–290. [CrossRef]

17. Crossley, B.M.; Zagmutt-Vergara, F.J.; Fyock, T.L.; Whitlock, R.H.; Gardner, I.A. Fecal shedding of *Mycobacterium avium* subsp *paratuberculosis* by dairy cows. *Vet. Microbiol.* **2005**, *107*, 257–263. [CrossRef]

18. Manning, E.J.B.; Collins, M.T. *Mycobacterium avium* subsp *paratuberculosis*: Pathogen, pathogenesis and diagnosis. *Rev. Sci. Tech. OIE* **2001**, *20*, 133–150. [CrossRef]

19. Küntzel, A.; Weber, M.; Gierschner, P.; Trefz, P.; Miekisch, W.; Schubert, J.K.; Reinhold, P.; Köhler, H. Core profile of volatile organic compounds related to growth of *Mycobacterium avium* subspecies *paratuberculosis*—A comparative extract of three independent studies. *PLoS ONE* **2019**, *14*, e0221031. [CrossRef]

20. Vitense, P.; Kasbohm, E.; Klassen, A.; Gierschner, P.; Trefz, P.; Weber, M.; Miekisch, W.; Schubert, J.K.; Möbius, P.; Reinhold, P.; et al. Detection of *Mycobacterium avium* ssp. *paratuberculosis* in Cultures from Fecal and Tissue Samples Using VOC Analysis and Machine Learning Tools. *Front. Vet. Sci.* **2021**, *8*, 620327. [CrossRef]

21. Van Malderen, K.; De Man, J.; De Winter, B.Y.; De Schepper, H.U.; Lamote, K. Volatomics in Inflammatory Bowel Disease and Irritable Bowel Syndrome: Present and Future. *Gastroenterology* **2020**, *158*, S884. [CrossRef]

22. Ahmed, I.; Fayyaz, F.; Nasir, M.; Niaz, Z.; Furnari, M.; Perry, L. Extending landscape of volatile metabolites as novel diagnostic biomarkers of inflammatory bowel disease—A review. *Scand. J. Gastroenterol.* **2016**, *51*, 385–392.

23. Lukins, H.B.; Foster, J.W. Methyl Ketone Metabolism in Hydrocarbon-Utilizing Mycobacteria. *J. Bacteriol.* **1963**, *85*, 1074–1087 [CrossRef]

24. Anderson, J.C. Measuring Breath Acetone for Monitoring Fat Loss: Review. *Obesity* **2015**, *23*, 2327–2334. [CrossRef]

25. Gross, J.J.; Bruckmaier, R.M. Review: Metabolic challenges in lactating dairy cows and their assessment via established and novel indicators in milk. *Animal* **2019**, *13*, S75–S81. [CrossRef]

26. Eisenreich, W.; Bacher, A.; Arigoni, D.; Rohdich, F. Biosynthesis of isoprenoids via the non-mevalonate pathway. *Cell. Mol. Life Sci.* **2004**, *61*, 1401–1426. [CrossRef]

27. Eisenreich, W.; Schwarz, M.; Cartayrade, A.; Arigoni, D.; Zenk, M.H.; Bacher, A. The deoxyxylulose phosphate pathway of terpenoid biosynthesis in plants and microorganisms. *Chem. Biol.* **1998**, *5*, R221–R233. [CrossRef]

28. Salerno-Kennedy, R.; Cashman, K.D. Potential applications of breath isoprene as a biomarker in modern medicine: A concise overview. *Wien. Klin. Wochenschr.* **2005**, *117*, 180–186. [CrossRef]

29. Kurada, S.; Alkhouri, N.; Fiocchi, C.; Dweik, R.; Rieder, F. Review article: Breath analysis in inflammatory bowel diseases. *Aliment. Pharmacol. Ther.* **2015**, *41*, 329–341. [CrossRef] [PubMed]

30. Itoh, T.; Koyama, Y.; Shin, W.; Akamatsu, T.; Tsuruta, A.; Masuda, Y.; Uchiyama, K. Selective Detection of Target Volatile Organic Compounds in Contaminated Air Using Sensor Array with Machine Learning: Aging Notes and Mold Smells in Simulated Automobile Interior Contaminant Gases. *Sensors* **2020**, *20*, 2687. [CrossRef] [PubMed]

31. Jaeger, D.M.; Runyon, J.B.; Richardson, B.A. Signals of speciation: Volatile organic compounds resolve closely related sagebrush taxa, suggesting their importance in evolution. *New Phytol.* **2016**, *211*, 1393–1401. [CrossRef] [PubMed]

32. McKenna, S.L.B.; Keefe, G.P.; Barkema, H.W.; Sockett, D.C. Evaluation of three ELISAs for *Mycobacterium avium* subsp *paratuberculosis* using tissue and fecal culture as comparison standards. *Vet. Microbiol.* **2005**, *110*, 105–111. [CrossRef]

33. Köhler, H.; Burkert, B.; Pavlik, I.; Diller, R.; Geue, L.; Conraths, F.J.; Martin, G. Evaluation of five ELISA test kits for the measurement of antibodies against *Mycobacterium avium* subspecies *paratuberculosis* in bovine serum. *Berl. Münch. Tierärztl. Wochenschr.* **2008**, *121*, 203–210.

34. Blanco Vazquez, C.; Alonso-Hearn, M.; Juste, R.A.; Canive, M.; Iglesias, T.; Iglesias, N.; Amado, J.; Vicente, F.; Balseiro, A.; Casais, R. Detection of latent forms of *Mycobacterium avium* subsp. *paratuberculosis* infection using host biomarker-based ELISAs greatly improves paratuberculosis diagnostic sensitivity. *PLoS ONE* **2020**, *15*, e0236336. [CrossRef]

35. Paratuberkulose. Amtliche Methode und Falldefinition. 2016. Available online: https://www.openagrar.de/receive/openagrar_mods_00058039 (accessed on 15 April 2016).

36. Fischer, S.; Trefz, P.; Bergmann, A.; Steffens, M.; Ziller, M.; Miekisch, W.; Schubert, J.S.; Köhler, H.; Reinhold, P. Physiological variability in volatile organic compounds (VOCs) in exhaled breath and released from faeces due to nutrition and somatic growth in a standardized caprine animal model. *J. Breath Res.* **2015**, *9*, 027108. [CrossRef]
37. Küntzel, A.; Oertel, P.; Trefz, P.; Miekisch, W.; Schubert, J.K.; Köhler, H.; Reinhold, P. Animal science meets agricultural practice: Preliminary results of an innovative technical approach for exhaled breath analysis in cattle under field conditions. *Berl. Münch. Tierärztl. Wochenschr.* **2018**, *131*, 444–452.
38. R Core Team. R: A Language and Environment for Statistical Computing. 2019. Available online: https://www.r-project.org/ (accessed on 31 March 2021).
39. Kolde, R. Pheatmap: Pretty Heatmaps. R Package Version 1.0.8. Available online: https://CRAN.R-project.org/package=pheatmap (accessed on 28 March 2021).
40. Kursa, M.B.; Rudnicki, W.R. Feature Selection with the Boruta Package. *J. Stat. Softw.* **2010**, *36*, 1–13. [CrossRef]
41. Oertel, P.; Küntzel, A.; Reinhold, P.; Köhler, H.; Schubert, J.K.; Kolb, J.; Miekisch, W. Continuous real-time breath analysis in ruminants: Effect of eructation on exhaled VOC profiles. *J. Breath Res.* **2018**, *12*, 036014. [CrossRef]
42. Breiman, L. Random forests. *Mach. Learn.* **2001**, *45*, 5–32. [CrossRef]
43. Kuhn, M. Building Predictive Models in R Using the caret Package. *J. Stat Softw.* **2008**, *28*, 1–26. [CrossRef]
44. Carter, J.V.; Pan, J.; Rai, S.N.; Galandiuk, S. ROC-ing along: Evaluation and interpretation of receiver operating characteristic curves. *Surgery* **2016**, *159*, 1638–1645. [CrossRef]

Article

Exploratory Study Using Urinary Volatile Organic Compounds for the Detection of Hepatocellular Carcinoma

Ayman S. Bannaga [1,2], Heena Tyagi [3], Emma Daulton [3], James A. Covington [3] and Ramesh P. Arasaradnam [1,2,4,5,*]

1 Department of Gastroenterology and Hepatology, University Hospital, Coventry CV2 2DX, UK; ayman.bannaga@warwick.ac.uk
2 Warwick Medical School, University of Warwick, Coventry CV4 7HL, UK
3 School of Engineering, University of Warwick, Coventry CV4 7AL, UK; heena.tyagi@warwick.ac.uk (H.T.); e.daulton@warwick.ac.uk (E.D.); j.a.covington@warwick.ac.uk (J.A.C.)
4 Faculty of Health & Life Sciences, Coventry University, Coventry CV1 5FB, UK
5 Leicester Cancer Research Centre, University of Leicester, Leicester LE1 7RH, UK
* Correspondence: r.arasaradnam@warwick.ac.uk; Tel.: +44-2476-966087

Abstract: Hepatocellular carcinoma (HCC) biomarkers are lacking in clinical practice. We therefore explored the pattern and composition of urinary volatile organic compounds (VOCs) in HCC patients. This was done in order to assess the feasibility of a potential non-invasive test for HCC, and to enhance our understanding of the disease. This pilot study recruited 58 participants, of whom 20 were HCC cases and 38 were non-HCC cases. The non-HCC cases included healthy individuals and patients with various stages of non-alcoholic fatty liver disease (NAFLD), including those with and without fibrosis. Urine was analysed using gas chromatography–ion mobility spectrometry (GC–IMS) and gas chromatography–time-of-flight mass spectrometry (GC–TOF-MS). GC–IMS was able to separate HCC from fibrotic cases with an area under the curve (AUC) of 0.97 (0.91–1.00), and from non-fibrotic cases with an AUC of 0.62 (0.48–0.76). For GC-TOF-MS, a subset of samples was analysed in which seven chemicals were identified and tentatively linked with HCC. These include 4-methyl-2,4-*bis*(*p*-hydroxyphenyl)pent-1-ene (2TMS derivative), 2-butanone, 2-hexanone, benzene, 1-ethyl-2-methyl-, 3-butene-1,2-diol, 1-(2-furanyl)-, bicyclo(4.1.0)heptane, 3,7,7-trimethyl-, [1S-(1a,3β,6a)]-, and sulpiride. Urinary VOC analysis using both GC–IMS and GC-TOF-MS proved to be a feasible method of identifying HCC cases, and was also able to enhance our understanding of HCC pathogenesis.

Keywords: urinary biomarkers; hepatocellular carcinoma; diagnosis; volatile organic compounds; headspace analysis

Citation: Bannaga, A.S.; Tyagi, H.; Daulton, E.; Covington, J.A.; Arasaradnam, R.P. Exploratory Study Using Urinary Volatile Organic Compounds for the Detection of Hepatocellular Carcinoma. *Molecules* **2021**, *26*, 2447. https://doi.org/10.3390/molecules26092447

Academic Editors: Natalia Drabińska and Ben de Lacy Costello

Received: 25 March 2021
Accepted: 20 April 2021
Published: 22 April 2021

Publisher's Note: MDPI stays neutral with regard to jurisdictional claims in published maps and institutional affiliations.

1. Introduction

Hepatocellular carcinoma (HCC) is the third most common cause of cancer-related death worldwide [1]. In most cases, HCC is considered a consequence of liver fibrosis/cirrhosis, with chronic viral hepatitis, alcoholic liver disease, and non-alcoholic fatty liver disease (NAFLD) being the most common underlying causes [2]. Early detection of HCC is usually reliant on ultrasound scan (USS) surveillance of cirrhotic patients. In these patients, the USS detection of HCC lesions varies according to the experience of the USS operator. Detection sensitivity can range from 40% to 80%. Another test that can be used for cirrhotic patients is the serum marker alpha-fetoprotein (AFP). AFP has poor sensitivity and relies on the cut-off being applied. Due to this, the clinical guidelines in 2018 recommended that AFP should no longer be used in routine clinical practice [3,4].

HCC diagnosis relies on advanced contrast-enhanced scans, which are either computed tomography (CT) or magnetic resonance (MR). HCC tissue biopsy is reserved for the confirmation of inconclusive HCC lesions found on a scan, or for determining the choice of palliative chemotherapy in case there is need to differentiate between HCC and other hepatobiliary malignancies [3–5]. HCC is often diagnosed late due to inaccessibility to CT and/or MR scans, especially in low-resource settings. Another factor involved in delayed diagnosis is the absence of symptoms until late in the disease. In addition, HCC has no approved screening programme for the general population—unlike colorectal, breast, or cervical cancers [1–5].

Given these factors there is still a need for ways to diagnose and understand the pathogenesis of HCC. One of the described mechanisms in HCC pathogenesis involves the impairment of hepatic metabolic pathways. The literature suggests that HCC development could be related to the malfunction of the cytochrome polysubstrate 450 (CYP450). These are heme-containing monooxygenases located in the endoplasmic reticula of the hepatic cells. The main function of cytochromes is to detoxify chemicals that could be harmful to tissues. However, this detoxification may produce harmful metabolites that could disrupt the hepatic cellular DNA division mechanisms required to maintain hepatic cellular proliferation, with subsequent cancer formation [6–11]. Because HCC is a vascularized tumour, we hypothesized that the byproducts of CYP450, including different volatile organic compounds (VOCs), would be found in the urine following the homeostatic HCC cells' secretion of these compounds into systemic circulation, and subsequent kidney filtration. We therefore designed a pilot study with the aim of assessing this hypothesis.

2. Results

Figure 1a,b shows the outputs from GC-IMS and GC-TOF-MS, respectively. For the GC–IMS output, the background is defined in blue, with the red peaks showing areas of high intensity. The long red line is the output of the instrument to the carrier gas (in this case, nitrogen). The results show that we were able to separate different chemicals within the urine sample without saturating the machine and without chemical overlap. For the GC-TOF-MS output, we see a broad range of chemical peaks throughout the spectra, with good separation. On average, the total number of peaks detected using GC-TOF-MS, after analysing HCC and fibrosis samples, was 112, and the total number of peaks detected among HCC and non-fibrosis samples was 74. Similarly, for fibrosis and non-fibrosis samples, 79 peaks were detected on average.

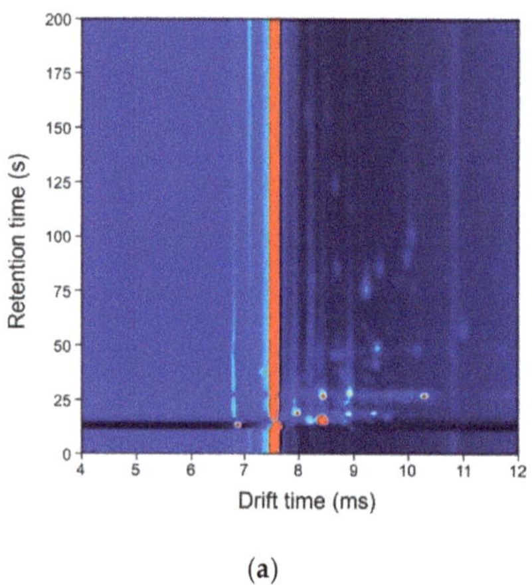

(a)

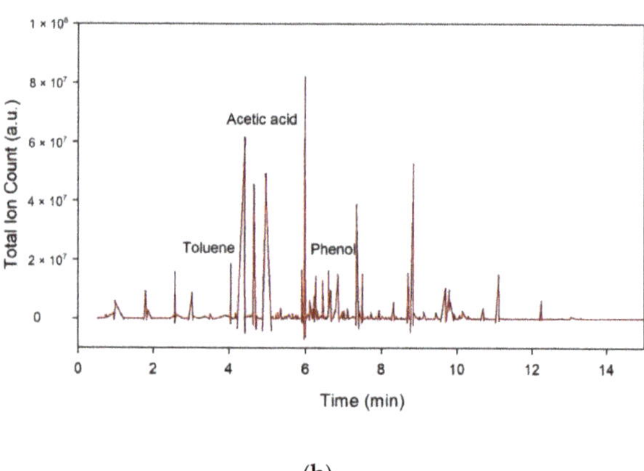

(b)

Figure 1. Example outputs of the instruments to a urine sample: (**a**) gas chromatography-ion mobility spectrometry (GC-IMS); (**b**) gas chromatography-time-of-flight mass spectrometry (GC–TOF-MS).

2.1. Results from GC-IMS

Table 1 shows the results of the separation of those with HCC from non-HCC patients with liver fibrosis. The area under the curve (AUC), sensitivity, and specificity were 0.97, 0.43, and 0.95, respectively. Conversely, the separation of those with HCC compared to non-HCC patients without liver fibrosis shows modest separation, with an AUC, sensitivity, and specificity of 0.62, 0.60, and 0.74, respectively. Comparison of both fibrosis and non-fibrosis patients revealed an AUC, sensitivity, and specificity of 0.63, 0.29, and 0.90, respectively. The receiver operating characteristic (ROC) curves for the different liver groups, using GC-IMS, are presented in Figure 2. The optimal threshold values were applied for the comparison of HCC and fibrosis samples, HCC and non-fibrosis samples, and fibrosis and non-fibrosis samples, and were 0.39, 0.35, and 0.52, respectively.

Table 1. Statistical results from the GC–IMS analysis (95% confidence intervals are in brackets). Positive predictive value (PPV); negative predictive value (NPV).

Comparison	Classifier	AUC	Sensitivity	Specificity	PPV	NPV
HCC vs. Fibrosis	Random Forest	0.97 (0.91–1.00)	0.43 (0.13–0.75)	0.95 (0.86–1.00)	0.75 (0.33–1.00)	0.83 (0.68–0.95)
HCC vs. Non-Fibrosis	Random Forest	0.62 (0.48–0.76)	0.60 (0.41–0.78)	0.74 (0.61–0.87)	0.60 (0.42–0.78)	0.74 (0.61–0.88)
Fibrosis vs. Non-Fibrosis	Linear Regression	0.63 (0.36–0.89)	0.29 (0.00–0.60)	0.90 (0.81–0.97)	0.40 (0.00–0.83)	0.85 (0.74–0.94)

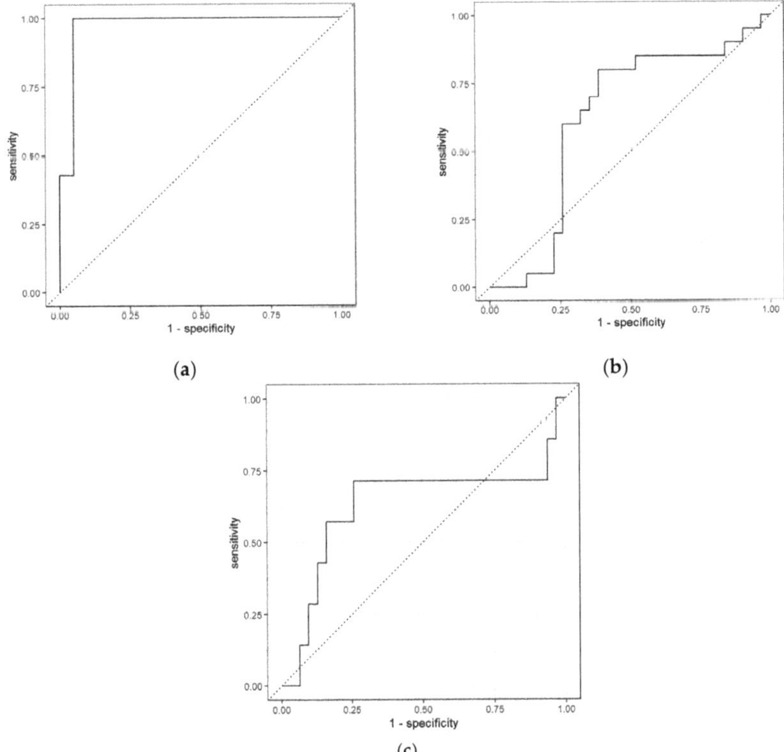

Figure 2. Receiver operating characteristic (ROC) curves for the GC–IMS analysis: (**a**) HCC vs. fibrosis; (**b**) HCC vs. non-fibrosis; (**c**) fibrosis vs. non-fibrosis.

The results showed that the diagnostic tests gave four false positives for comparison between HCC and fibrosis samples, eight false positive tests for HCC and non-fibrosis samples, and only three false positive tests for fibrosis and non-fibrosis samples. Moreover, the number of false negative tests for HCC and fibrosis samples was only 1, whereas the number of false negative tests for HCC and non-fibrosis samples, and for fibrosis and non-fibrosis samples, was 12 and 5, respectively.

2.2. Results from GC-TOF-MS Chemical Identification

Test accuracy for HCC and non-HCC cases using GC–TOF-MS is provided in Supplementary Materials Table S2 and Figure S1. This includes ROC curves for the different liver disease groups. From the total list of more than 200 chemicals identified using the National Institute of Standards and Technology (NIST) software, 5 were found to be statistically significant between the groups, with p-values of <0.05. Further analysis was undertaken comparing HCC with fibrosis and with non-fibrosis, and an additional two chemicals were identified from HCC versus fibrosis in the same way. No additional chemicals were identified when comparing HCC with non-fibrosis. These chemicals are listed in Table 2, with numbers 1–5 for HCC vs. non-HCC, and the remaining two associated with HCC vs. fibrosis. This table also includes the chemical retention time, the p-value between the groups, and whether the abundance of a chemical increased or decreased with HCC. We have not attempted to quantify these changes here due to the small sample size.

Table 2. List of the relevant chemicals identified using GC-TOF-MS for HCC vs. non-HCC.

No.	Retention Time (min)	Chemical	p-value	Abundance Change
1	15.25	4-Methyl-2,4-*bis*(*p*-hydroxyphenyl)pent-1-ene, 2TMS derivative	<0.01	Lower for HCC
2	2.5998	2-Butanone	0.03637	Higher for HCC
3	4.5684	2-Hexanone	0.04309	Lower for HCC
4	6.3215	Benzene, 1-ethyl-2-methyl-	0.04183	Lower for HCC
5	12.1318	3-Butene-1,2-diol, 1-(2-furanyl)-	0.03247	Lower for HCC
6	8.2054	Bicyclo[4.1.0]heptane, 3,7,7-trimethyl-, [1S-(1a,3ß,6a)]-	0.03553	Lower for HCC
7	13.861	Sulpiride	0.04369	Lower for HCC

In addition, fibrosis and non-fibrosis samples were analysed in the same way. Table S3 in the Supplementary Materials provides a list of the relevant chemicals found in this analysis.

3. Discussion

In this study we investigated the use of VOCs as a means of providing biomarkers for the diagnosis of HCC. Here, VOCs were analysed using GC-IMS and GC-TOF-MS, which we have previously used in other clinical studies [12–14]. Importantly, this study further consolidates existing published studies utilizing urinary VOCs for cancer detection. The non-HCC group included both those with and without liver fibrosis, to reflect clinical HCC screening scenarios. The high specificity of 0.95 (0.86–1.00) in separating HCC from those with liver fibrosis offers important insights into the role of urinary VOCs as a screening modality. The hypothesis that the hepatic CYP450 byproducts (VOCs) related to HCC could be detected in different biological samples has been previously described. Two studies have shown that VOCs can be detected in the headspace of incubated in vitro HCC cells, supporting the use of VOC analysis for the assessment of hepatic enzyme function, as well as for the prediction of HCC progression and metastasis [15,16]. Qin et al. [17] utilized VOCs in the breath to identify HCC, independent of AFP levels or the disease's clinical stage. A recent study by Miller-Atkins et al. [18] showed that the use of 22 VOCs in the breath could detect HCC with 0.73 sensitivity, compared with 0.53 for AFP in the same cohort.

Urine is a stable sample medium, and easier to collect for VOC analysis [19]. We have previously reported that urinary VOC analysis using solid-phase microextraction (SPME) was able to differentiate HCC and non-liver disease cases. The SPME AUC for HCC with negative alpha fetoprotein (AFP) was 0.68, and it rose to 0.83 when combined with raised AFP [20]. This was comparable to current findings reported here, where the HCC AUC was 0.62 using GC-IMS, and 0.79 using GC-TOF-MS. The study reported here also demonstrated the feasibility of urinary VOCs for differentiating between non-fibrotic, fibrotic, and HCC cases, as demonstrated in Table 1 and Supplementary Tables S1 and S2.

Using GC-TOF-MS, we tentatively identified seven VOCs related to HCC, as shown in Table 2. Though we did not perform verification and quantification of these chemicals, we did undertake a search of these VOCs in relation to the development of HCC as per the current literature. We found out that the most described VOC in HCC was 2-butanone. In experimental models, exposure to 2-butanone led to hepatotoxicity by potentiating dihydronicotinamide adenine dinucleotide phosphate (NADPH) cytochrome c reductase activity, along with the concentration of cytochrome P450 enzymes. In addition, 2-butanone exposure in these models, concomitantly with the known hepatocarcinogenic agent carbon tetrachloride (CCl4), accelerated the formation of hepatotoxic metabolites and HCC. 2-Butanone was also found to inhibit the activity of membrane-bound monoamine oxidase. This is important because monoamine oxidase was found to suppress HCC metastasis and progression by inhibiting the adrenergic system and its transactivation of epidermal growth factor receptor (EGFR) signalling [21–30]. In human studies, 2-butanone was found in the breath of HCC patients, and was found to have the best diagnostic value among other organic compounds [17]. In NAFLD paediatric patients, 2-butanone appeared at significantly higher levels in the faeces and was related to faecal *Lachnospiraceae*—a family of anaerobic, spore-forming bacteria. Additionally, the study found that *Oscillospirae* decrease relative to 2-butanone upregulation [31]. 2-Butanone was found to be elevated in cirrhotic patients who underwent liver transplantation [32]. 2-Butanone levels in the blood were found to be significantly discriminant in liver cancer patients, in comparison to healthy individuals [33]. In breath studies looking into cirrhotic and non-cirrhotic liver patients, serum bilirubin showed a positive correlation with 2-butanone. The 2-butanone in the breath also distinguished different classes of liver cirrhosis, demonstrated by Child-Turcotte-Pugh (CTP) scores of A, B and C [34,35].

We also tentatively identified 4-methyl-2,4-*bis*(*p*-hydroxyphenyl)pent-1-ene (MBP), which is a derivative of bisphenol A (BPA), a major pollutant. In the liver, MBP metabolic activation from BPA occurs via the cytochrome P450 system [36]. MBP can induce the function of oestrogen in experimental models via activation of the oestrogen receptor (ER) [37]. In patients with HCC, ERs are present and functional in around 50% of cases, but their role in promoting carcinogenesis is still not fully clear [38]. The presence of urinary MBP in HCC patients in this study suggests that MBP plays a role in HCC, perhaps via the activation of ERs, but this requires further research.

Another VOC possibly found in this study related to HCC is 2-hexanone, which was found to have a potentiating effect on the hepatotoxic agent chloroform, and subsequent liver injury, in experimental animal models [39,40]. The mechanism for this was found to be due to the induction of the CYP450 system [41–43]. Chronic inhalation of an isomer of 2-hexanone (methyl isobutyl ketone, MIBK) was found to cause hepatocellular adenomas and HCC in mice [44–46]. This was shown to be in part due to the activation of the pregnane X and constitutive androstane nuclear receptors; these receptors are responsible for the regulation of CYP450 activity [44].

Benzene, 1-ethyl-2-methyl- has been identified as a blood biomarker of HCC in a study using SPME-GC-MS [47]. Sulpiride is another chemical found in our study that is closely related to many chronic liver diseases. In particular, sulpiride was found to be related to biliary liver cirrhosis [48], NAFLD [49], and cholestatic hepatitis [50]. Though it has not been identified as a biomarker for HCC, the presence of sulpiride indicates that it may be a significant chemical for HCC. A study has suggested 3-butene-1,2-diol,

1-(2-furanyl)- as an important VOC for lung cancer [51], but it has not been verified as an HCC biomarker. Similarly, bicyclo[4.1.0]heptane, 3,7,7-trimethyl-, [1S-(1a,3β,6a)]-, found in our study, has not been identified as a biomarker. Further investigation is needed to confirm these chemicals in a larger cohort.

Our study was limited in not accounting for other factors that can be involved in the production of VOCs, such as occupational environmental factors, diet, smoking, and drug use. Another limitation was the small number of study participants. Nevertheless, this study has answered the question of whether VOCs related to the function of CYP450 in HCC can be detected in the urine. In particular, as discussed earlier, the tentative identification of urinary VOCs in this study has been seen previously in various experimental and clinical studies. The strong literature around 2-butanone encourages further study to identify the exact biochemical pathways of this compound during HCC pathogenesis. However, we did not validate these chemicals, nor did we quantify them; this effort will be undertaken in a larger study. In addition, the data from the GC-IMS system were analysed using a pattern recognition approach, and we did not attempt to identify chemical components. Again, we propose to look further into this in the next study.

4. Materials and Methods

This pilot study was approved by the Coventry and Warwickshire and Northeast Yorkshire NHS Ethics Committees (Ref 18717 and Ref 260179). The study conformed to the ethical principles of the Declaration of Helsinki. Study participants were recruited from University Hospital Coventry and the Warwickshire NHS Trust, UK. All participants provided written informed consent. Five-millilitre urine samples were collected into universal bottles from each study participant. These samples were then immediately frozen at $-80\,^{\circ}$C within 1 to 2 h. The samples were then stored until further sample analysis at the end of the recruitment process. We have previously tested the stability of urine samples in storage, and all methods were in line with these findings [52,53].

4.1. Study Characteristics

There were a total of 58 participants. These included 20 HCC cases and 38 non-HCC cases. The non-HCC cases were recruited from two sources in order to decrease bias: The first source consisted of healthy individuals without liver disease. The second source consisted of patients with different stages of NAFLD. The advantage here is that these patients represent those at risk of becoming HCC cases in the future. The non-HCC cases were then further divided into 31 non-fibrotic and 7 fibrotic/cirrhotic cases. The exclusion criteria were pregnancy and age <18 years. All of the participants were recruited prior to any anticancer treatment.

HCC diagnosis was made according to the current international guidelines, with all inconclusive cases being confirmed by a liver biopsy. Liver fibrosis/cirrhosis was confirmed by clinical examination and different radiological tests. In case of ambiguity about the clinical diagnosis, a liver biopsy was performed so as to ascertain the cause of the liver disease, and to look for the presence or absence of liver fibrosis/cirrhosis. We further collected other clinical covariates of interest, including gender, age at the time of urine sampling, history of absence or presence of diabetes, and the extent of HCC spread. We also collected liver function tests at the time of urine sampling, including AFP, alanine aminotransferase (ALT), alkaline phosphatase (ALP), albumin, and bilirubin. The study participants' characteristics are further detailed in Table 3.

Table 3. Clinical and biochemical characteristics of the recruited study participants at the time of obtaining their urine samples.

Covariate	HCC Cases	Non-HCC Cases
No. of Patients	20	38
Age: Mean (Range)	73 (53–84)	58.08 (29–89)
Gender: Female/Male	2/18	11/27
Cause of Liver Disease	3 Alcohol 1 HBV 1 HCV 13 NASH 2 Primary/Idiopathic	1 HBV Cirrhosis 9 NAFLD 10 NASH 6 NASH Cirrhosis 12 without Liver Disease
Histological/Radiological Features of Liver Cirrhosis: Present/Absent	16/4	7/31
Diabetes: Present/Absent	11/9	7/31
AFP: Mean (Range), KU/L	1380.60 (1–9400)	-
ALT: Mean (Range), U/L	44.60 (13–149)	50.74 (5–304)
ALP: Mean (Range), U/L	150.90 (83–326)	89.76 (53–279)
Albumin: Mean (Range), g/L	39 (24–44)	43.87 (28–50)
Bilirubin: Mean (Range), μmol/L	24.30 (5–84)	7.97 (5–21)
Stage of the HCC: Hepatic/Extra-Hepatic	13/7	-

Characteristics of the HCC and non-HCC groups. HCC diagnosis was made in line with international guidelines. Liver disease was established using a combination of radiological scans, FibroScan, laboratory markers, and histology. All covariates were collected at the time of urine collection. Abbreviations: AFP, alpha-fetoprotein; ALT, alanine aminotransferase; ALP, alkaline phosphatase; HBV, hepatitis B virus; HCV, hepatitis C virus; NAFLD, non-alcoholic fatty liver disease; NASH, non-alcoholic steatohepatitis.

4.2. GC-IMS Methodology

Samples were shipped from University Hospital Coventry and from Warwickshire in universal sample containers, on dry ice, to the School of Engineering, University of Warwick, where they were stored at −20 °C until use. Prior to testing, the samples were thawed overnight in a laboratory fridge at 4 °C. Once thawed, 5 mL of each urine sample was aliquoted into 20 mL glass vials (Thames Restek, UK), and sealed with a PTFE crimp cap (Thames Restek, UK). Samples were then analysed using a FlavourSpec GC-IMS (G.A.S, Dortmund, Germany). The FlavourSpec was fitted with a CombiPAL autosampler, allowing for high-throughput automatic analysis of the samples. The samples were loaded into a cooled autosampler tray, keeping the samples at 4 °C. Each sample was heated to 40 °C and then agitated for 10 min prior to analysis. A 0.5 mL sample of the headspace was then taken using the autosampler syringe and injected directly into the GC-IMS for sampling. The GC–IMS settings were as follows: drift gas flow of 150 mL/m, and a carrier gas flow rate of 20 mL/min. The drift gas used was 99.99% nitrogen. The IMS was heated to 45 °C (T1), the GC to 40 °C (T2), the injector to 80 °C (T3), the T4 transfer line to 80 °C, and the T5 transfer line to 45 °C. Sample analysis took 10 min. Once completed, the data acquired were viewed using LAV software (G.A.S, Dortmund, Germany) and then exported for further analysis. This method has been developed over several urinary VOC studies, and is designed to maximize information content and chemical separation [12,54]. This includes the volume of urine, agitation period, and temperature. For quality control, blank samples were added at the beginning and end of each run, with the instrument having regular calibration checks run. Furthermore, the information content of each sample was checked, which included a visual inspection of each sample file.

4.3. GC-TOF-MS Methodology

A subset of samples was also analysed using GC-TOF-MS (Markes International, UK), with a UNITY-xr thermal desorber and ULTRA-xr autosampler (Markes International, UK).

Urine samples for GC-TOF-MS were aliquoted as outlined, with about 5 mL of each sample in a 20 mL vial, which was sealed with a crimp camp. The headspace of each urine sample was then adsorbed onto a Markes bio-monitoring tube (C2-AAXX-5149). The septum of the vial was pierced, and the sorbent tube pushed through into the headspace in the vial. The samples were then heated to 40 °C for 20 min, before a pump was attached to the sorbent tube and the sample was pulled through onto the sorbent bed of the tube for 20 min whilst still being heated to 40 °C. Once complete, the tube was removed from the vial and placed into the Markes ULTRA-xr autosampler. The ULTRA-xr autosampler was set to run with a standby split of 150 °C, and a GC temperature ramp of 20 °C per minute, heating from 40 °C to 280 °C with a GC run time of 25 min. The samples were each pre-purged for 1 min, following which the sorbent tube was desorbed onto the trap for 10 min at 250 °C. Once complete, the trap was purged for a further minute and then cooled to 30 °C, before being heated to 300 °C for 3 min. Post-analysis, a dynamic baseline correction (DBS) was applied using the native TOF-DS software, and the chromatogram was integrated and deconvoluted with the following settings: global height reject of 10,000, global width reject of 0.01, baseline threshold of 3, and global area reject of 10,000. The peaks identified were then compared with the NIST list, with a match (forward and reverse) factor of 450, to identify the compounds present. As with GC–IMS, this method has been used in a number of VOC studies, including those associated with cancer, and has been previously reported on [52].

4.4. Statistical Analysis

The analysis of the data was undertaken using our previously reported data analysis pipeline for GC-IMS and GC-TOF-MS data, using "R" (version 3.6.3) [12–14]. In brief, for GC–IMS data, we applied a two-stage pre-processing step. This was undertaken because the dataset has high dimensionality (typically 11 million data points), but low chemical information. The first step was to crop the central section of the output data, where all of the chemical information is located. This was followed by the application of a threshold, below which all values were given a value of zero. This was undertaken to remove the background, leaving just the chemical information. The crop parameters were manually selected, and the same values were applied to all of the data. The threshold was defined by the value of the background noise. The data were then processed using a 10-fold cross validation. Here, the data were split into a 90% training set and a 10% test set. Within each fold, a Wilcoxon rank sum test was undertaken, and the 100 features with the lowest p-value were extracted. Classification models were constructed using two classifiers (eXtreme Gradient Boosting (XGBoost), and logistic regression). This process was repeated until all of the samples had been in the test group. The results were then collated, and from the resultant probabilities, statistical parameters, including sensitivity and specificity, were calculated.

For GC-TOF-MS, a similar process was undertaken. However, in this case, we used chemical identification to create features and, due to the much lower dimensionality, these were used directly by the classifier with no additional feature reduction. A further step used here was to undertake the statistical analysis of each chemical. A non-parametric t-test was undertaken in order to calculate the p-value of each chemical, comparing the samples in the two groups. Those chemicals found to have a p-value of <0.05 were considered statistically important.

5. Conclusions

Urinary VOCs can identify HCC cases non-invasively. The putative VOCs are likely related to CYP450 function in HCC. Our study further highlights how urine can provide a good medium for the investigation of metabolic function in HCC for further work on the cellular level.

Supplementary Materials: The following are available online: Table S1 compares HCC with non-HCC cases using GC–IMS analysis, providing AUC, sensitivity, specificity, thresholds, negative predictive value, and positive predictive value. Table S2 compares HCC with non-HCC cases using GC–TOF-MS analysis, providing AUC, sensitivity, specificity, thresholds, negative predictive value, and positive predictive value. Table S3 shows the identified chemicals for Fibrosis vs Non-Fibrosis that were statistically relevant using GC-TOF-MS. Figure S1 provides ROCs for HCC and Fibrosis samples, HCC and Non-Fibrosis and Fibrosis and Non-Fibrosis using GC-TOF-MS.

Author Contributions: Conceptualization, A.S.B., H.T., J.A.C. and R.P.A.; methodology, H.T., E.D. and J.A.C.; formal analysis, E.D. and J.A.C.; investigation, H.T. and E.D.; resources, A.S.B., J.A.C. and R.P.A.; data curation, A.S.B., H.T. and J.A.C.; writing—original draft preparation, A.S.B.; writing—review and editing, A.S.B., E.D., H.T., J.A.C. and R.P.A.; visualization; A.S.B., H.T., J.A.C.; supervision, J.A.C. and R.P.A. All authors have read and agreed to the published version of the manuscript.

Funding: This research received no external funding.

Institutional Review Board Statement: This pilot study was approved by the Coventry and Warwickshire and Northeast Yorkshire NHS Ethics Committees (Ref 18717 and Ref 260179). The study conformed to the ethical principles of the Declaration of Helsinki.

Informed Consent Statement: Written informed consent was obtained from all subjects involved in the study.

Data Availability Statement: All data are available in this manuscript and its supplementary files.

Acknowledgments: We acknowledge Sean James (head of Arden tissue bank) and Parmjit Dahaley (tissue bank biomedical assistants) for their help in the storage and transfer of samples. We acknowledge the UHCW R&D for their help in implementing the study. We also acknowledge all of the patients who kindly provided their samples for use in this study.

Conflicts of Interest: The authors declare no conflict of interest.

References

1. International Agency for Research on Cancer; Liver; World Health Organization. 2020. Available online: http://gco.iarc.fr/today/data/factsheets/cancers/11-Liver-fact-sheet.pdf (accessed on 30 January 2021).
2. Forner, A.; Reig, M.; Bruix, J. Hepatocellular carcinoma. *Lancet* **2018**, *391*, 1301–1314. [CrossRef]
3. Galle, P.R.; Forner, A.; Llovet, J.M.; Mazzaferro, V.; Piscaglia, F.; Raoul, J.-L.; Schirmacher, P.; Vilgrain, V. EASL Clinical Practice Guidelines: Management of hepatocellular carcinoma. *J. Hepatol.* **2018**, *69*, 182–236. [CrossRef]
4. Heimbach, J.K.; Kulik, L.M.; Finn, R.S.; Sirlin, C.B.; Abecassis, M.M.; Roberts, L.R.; Zhu, A.X.; Murad, M.H.; Marrero, J.A. AASLD guidelines for the treatment of hepatocellular carcinoma. *Hepatol.* **2018**, *67*, 358–380. [CrossRef]
5. Villanueva, A. Hepatocellular Carcinoma. *New Engl. J. Med.* **2019**, *380*, 1450–1462. [CrossRef]
6. Huang, Q.; Tan, Y.; Yin, P.; Ye, G.; Gao, P.; Lu, X.; Wang, H.; Xu, G. Metabolic Characterization of Hepatocellular Carcinoma Using Nontargeted Tissue Metabolomics. *Cancer Res.* **2013**, *73*, 4992–5002. [CrossRef] [PubMed]
7. Thomas, M.; Bayha, C.; Vetter, S.; Hofmann, U.; Schwarz, M.; Zanger, U.M.; Braeuning, A. Activating and Inhibitory Functions of WNT/β-Catenin in the Induction of Cytochromes P450 by Nuclear Receptors in HepaRG Cells. *Mol. Pharmacol.* **2015**, *87*, 1013–1020. [CrossRef]
8. Hamamoto, I.; Tanaka, S.; Maeba, T.; Chikaishi, K.; Ichikawa, Y. Microsomal cytochrome P-450-linked monooxygenase systems and lipid composition of human hepatocellular carcinoma. *Br. J. Cancer* **1989**, *59*, 6–11. [CrossRef]
9. Gao, P.; Liu, Z.-Z.; Yan, L.-N.; Dong, C.-N.; Ma, N.; Yuan, M.-N.; Zhou, J. Cytochrome P450 family members are associated with fast-growing hepatocellular carcinoma and patient survival: An integrated analysis of gene expression profiles. *Saudi J. Gastroenterol.* **2019**, *25*, 167–175. [CrossRef]
10. Tsutsumi, M.; Matsuda, Y.; Takada, A. Role of ethanol-inducible cytochrome P-450 2E1 in the development of hepatocellular carcinoma by the chemical carcinogen, N-nitrosodimethylamine. *Hepatology* **1993**, *18*, 1483–1489. [CrossRef]
11. Eun, H.S.; Cho, S.Y.; Lee, B.S.; Seong, I.O.; Kim, K. HProfiling cytochrome P450 family 4 gene expression in human hepatocellular carcinoma. *Mol. Med. Rep.* **2018**, *18*, 4865–4876. [CrossRef]
12. Daulton, E.; Wicaksono, A.N.; Tiele, A.; Kocher, H.M.; Debernardi, S.; Crnogorac-Jurcevic, T.; Covington, J.A. Volatile organic compounds (VOCs) for the non-invasive detection of pancreatic cancer from urine. *Talanta* **2021**, *221*, 121604. [CrossRef]
13. Tiele, A.; Wicaksono, A.; Daulton, E.; Ifeachor, E.; Eyre, V.; Clarke, S.; Timings, L.; Pearson, S.; A Covington, J.; Li, X. Breath-based non-invasive diagnosis of Alzheimer's disease: A pilot study. *J. Breath Res.* **2019**, *14*, 026003. [CrossRef]
14. Daulton, E.; Wicaksono, A.; Bechar, J.; Covington, J.A.; Hardwicke, J. The Detection of Wound Infection by Ion Mobility Chemical Analysis. *Biosensors* **2020**, *10*, 19. [CrossRef]
15. Mochalski, P.; Sponring, A.; King, J.; Unterkofler, K.; Troppmair, J.; Amann, A. Release and uptake of volatile organic compounds by human hepatocellular carcinoma cells (HepG2) in vitro. *Cancer Cell Int.* **2013**, *13*, 72. [CrossRef]

16. Haick, H.; Amal, H.; Ding, L.; Liu, B.; Tisch, U.; Xu, Z.-Q.; Shi, D.-Y.; Zhao, Y.; Chen, J.; Sun, R.-X.; et al. The scent fingerprint of hepatocarcinoma: In-vitro metastasis prediction with volatile organic compounds (VOCs). *Int. J. Nanomed.* **2012**, *7*, 4135–4146. [CrossRef]

17. Qin, T.; Liu, H.; Song, Q.; Song, G.; Wang, H.-Z.; Pan, Y.-Y.; Xiong, F.-X.; Gu, K.-S.; Sun, G.-P.; Chen, Z.-D. The Screening of Volatile Markers for Hepatocellular Carcinoma. *Cancer Epidemiology Biomarkers Prev.* **2010**, *19*, 2247–2253. [CrossRef]

18. Miller-Atkins, G.; Acevedo-Moreno, L.; Grove, D.; Dweik, R.A.; Tonelli, A.R.; Brown, J.M.; Allende, D.S.; Aucejo, F.; Rotroff, D.M. Breath Metabolomics Provides an Accurate and Noninvasive Approach for Screening Cirrhosis, Primary, and Secondary Liver Tumors. *Hepatol. Commun.* **2020**, *4*, 1041–1055. [CrossRef]

19. Becker, R. Non-invasive cancer detection using volatile biomarkers: Is urine superior to breath? *Med. Hypotheses* **2020**, *143*, 110060. [CrossRef]

20. Bannaga, A.S.I.; Kvasnik, F.; Persaud, K.C.; Arasaradnam, R.P. Differentiating cancer types using a urine test for volatile organic compounds. *J. Breath Res.* **2020**, *15*, 017102. [CrossRef]

21. Traiger, G.J.; Bruckner, J.V.; Jiang, W.; Dietz, F.K.; Cooke, P.H. Effect of 2-butanol and 2-butanone on rat hepatic ultrastructure and drug metabolizing enzyme activity. *J. Toxicol. Environ. Heal. Part A* **1989**, *28*, 235–248. [CrossRef] [PubMed]

22. Toftgard, R.; Nilsen, O.G.; Gustafsson, J.Å. Changes in rat liver microsomal cytochrome P-450 and enzymatic activities after the inhalation of n-hexane, xylene, methyl ethyl ketone and methylchloroform for four weeks. *Scand. J. Work. Environ. Heal.* **1981**, *7*, 31–37. [CrossRef]

23. Wlodzimirow, K.; Abu-Hanna, A.; Schultz, M.; Maas, M.; Bos, L.; Sterk, P.; Knobel, H.; Soers, R.; Chamuleau, R.A. Exhaled breath analysis with electronic nose technology for detection of acute liver failure in rats. *Biosens. Bioelectron.* **2014**, *53*, 129–134. [CrossRef] [PubMed]

24. Raunio, H.; Liira, J.; Elovaara, E.; Riihimäki, V.; Pelkonen, O. Cytochrome P450 isozyme induction by methyl ethyl ketone and m-xylene in rat liver. *Toxicol. Appl. Pharmacol.* **1990**, *103*, 175–179. [CrossRef]

25. Peng, H.; Raner, G.; Vaz, A.; Coon, M. Oxidative Cleavage of Esters and Amides to Carbonyl Products by Cytochrome P450. *Arch. Biochem. Biophys.* **1995**, *318*, 333–339. [CrossRef] [PubMed]

26. Brown, E.M.; Hewitt, W.R. Dose-response relationships in ketone-induced potentiation of chloroform hepato- and nephrotoxicity. *Toxicol. Appl. Pharmacol.* **1984**, *76*, 437–453. [CrossRef]

27. Raymond, P.; Plaa, G.L. Ketone potentiation of haloalkane-induced hepato- and nephrotoxicity. I. dose-response relationships. *J. Toxicol. Environ. Heal. Part A* **1995**, *45*, 465–480. [CrossRef]

28. Fowler, C.J.; Oreland, L. The effect of lipid-depletion on the kinetic properties of rat liver monoamine oxidase-B. *J. Pharm. Pharmacol.* **1980**, *32*, 681–688. [CrossRef]

29. Kinemuchi, H.; Sunami, Y.; Sudo, M.; Suh, Y.H.; Arai, Y.; Kamijo, K. Membrane lipid environment of carp brain and liver mitochondrial monoamine oxidase. *Comp. Biochem. Physiol. Part C Comp. Pharmacol.* **1985**, *80*, 245–252. [CrossRef]

30. Li, J.; Yang, X.-M.; Wang, Y.-H.; Feng, M.-X.; Liu, X.-J.; Zhang, Y.-L.; Huang, S.; Wu, Z.; Xue, F.; Qin, W.-X.; et al. Monoamine oxidase A suppresses hepatocellular carcinoma metastasis by inhibiting the adrenergic system and its transactivation of EGFR signaling. *J. Hepatol.* **2014**, *60*, 1225–1234. [CrossRef]

31. Del Chierico, F.; Nobili, V.; Vernocchi, P.; Russo, A.; De Stefanis, C.; Gnani, D.; Furlanello, C.; Zandonà, A.; Paci, P.; Capuani, G.; et al. Gut microbiota profiling of pediatric nonalcoholic fatty liver disease and obese patients unveiled by an integrated meta-omics-based approach. *Hepatology* **2017**, *65*, 451–464. [CrossRef]

32. Del Río, R.F.; O'Hara, M.; Holt, A.; Pemberton, P.; Shah, T.; Whitehouse, T.; Mayhew, C. Volatile Biomarkers in Breath Associated With Liver Cirrhosis—Comparisons of Pre- and Post-liver Transplant Breath Samples. *EBioMedicine* **2015**, *2*, 1243–1250. [CrossRef] [PubMed]

33. Wu, S.; Cai, C.; Cheng, J.; Cheng, M.; Zhou, H.; Deng, J. Polydopamine/dialdehyde starch/chitosan composite coating for in-tube solid-phase microextraction and in-situ derivation to analysis of two liver cancer biomarkers in human blood. *Anal. Chim. Acta* **2016**, *935*, 113–120. [CrossRef] [PubMed]

34. Morisco, F.; Aprea, E.; Lembo, V.; Fogliano, V.; Vitaglione, P.; Mazzone, G.; Cappellin, L.; Gasperi, F.; Masone, S.; De Palma, G.D.; et al. Rapid "Breath-Print" of Liver Cirrhosis by Proton Transfer Reaction Time-of-Flight Mass Spectrometry. A Pilot Study. *PLoS ONE* **2013**, *8*, e59658. [CrossRef] [PubMed]

35. Velde, S.V.D.; Nevens, F.; Van Hee, P.; Van Steenberghe, D.; Quirynen, M. GC–MS analysis of breath odor compounds in liver patients. *J. Chromatogr. B* **2008**, *875*, 344–348. [CrossRef]

36. Liu, S.-H.; Su, C.-C.; Lee, K.-I.; Chen, Y.-W. Effects of Bisphenol A Metabolite 4-Methyl-2,4-bis(4-hydroxyphenyl)pent-1-ene on Lung Function and Type 2 Pulmonary Alveolar Epithelial Cell Growth. *Sci. Rep.* **2016**, *6*, 39254. [CrossRef]

37. Hirao-Suzuki, M.; Takeda, S.; Okuda, K.; Takiguchi, M.; Yoshihara, S. Repeated Exposure to 4-Methyl-2,4-*bis*(4-hydroxyphenyl)pent-1-ene (MBP), an Active Metabolite of Bisphenol A, Aggressively Stimulates Breast Cancer Cell Growth in an Estrogen Receptor β (ERβ)–Dependent Manner. *Mol. Pharmacol.* **2018**, *95*, 260–268. [CrossRef]

38. Villa, E. Role of Estrogen in Liver Cancer. *Women's Heal.* **2008**, *4*, 41–50. [CrossRef]

39. Cowlen, M.S.; Hewitt, W.R.; Schroeder, F. 2-Hexanone potentiation of [14C]chloroform hepatotoxicity: Covalent interaction of a reactive intermediate with rat liver phospholipid. *Toxicol. Appl. Pharmacol.* **1984**, *73*, 478–491. [CrossRef]

40. Hewitt, L.A.; Valiquette, C.; Plaa, G.L. The role of biotransformation–detoxication in acetone-, 2-butanone-, and 2-hexanone-potentiated chloroform-induced hepatotoxicity. *Can. J. Physiol. Pharmacol.* **1987**, *65*, 2313–2318. [CrossRef]

41. Nakajima, T.; Elovaara, E.; Park, S.S.; Gelboin, H.V.; Vainio, H. Immunochemical detection of cytochrome P450 isozymes induced in rat liver byn-hexane, 2-hexanone and acetonyl acetone. *Arch. Toxicol.* **1991**, *65*, 542–547. [CrossRef]

42. Nakajima, T.; Elovaara, E.; Okino, T.; Gelboin, H.; Klockars, M.; Riihimaki, V.; Aoyama, T.; Vainio, H. Different Contributions of Cytochrome P450 2E1 and P450 2B1/2 to Chloroform Hepatotoxicity in Rat. *Toxicol. Appl. Pharmacol.* **1995**, *133*, 215–222. [CrossRef] [PubMed]

43. Cowlen, M.S.; Hewitt, W.R.; Schroeder, F. Mechanisms in 2-hexanone potentiation of chloroform hepatotoxicity. *Toxicol. Lett.* **1984**, *22*, 293–299. [CrossRef]

44. Hughes, B.; Thomas, J.; Lynch, A.; Borghoff, S.; Green, S.; Mensing, T.; Sarang, S.; LeBaron, M. Methyl isobutyl ketone-induced hepatocellular carcinogenesis in B6C3F1 mice: A constitutive androstane receptor (CAR)-mediated mode of action. *Regul. Toxicol. Pharmacol.* **2016**, *81*, 421–429. [CrossRef] [PubMed]

45. Stout, M.D.; Herbert, R.A.; Kissling, G.E.; Suarez, F.; Roycroft, J.H.; Chhabra, R.S.; Bucher, J.R. Toxicity and carcinogenicity of methyl isobutyl ketone in F344N rats and B6C3F1 mice following 2-year inhalation exposure. *Toxicology* **2008**, *244*, 209–219. [CrossRef]

46. National Toxicology Program. Toxicology and carcinogenesis studies of methyl isobutyl ketone (Cas No. 108-10-1) in F344/N rats and B6C3F1 mice (inhalation studies). *Natl. Toxicol. Prog. Tech. Rep. Ser.* **2007**, *538*, 1–236.

47. Xue, R.; Dong, L.; Zhang, S.; Deng, C.; Liu, T.; Wang, J.; Shen, X. Investigation of volatile biomarkers in liver cancer blood using solid-phase microextraction and gas chromatography/mass spectrometry. *Rapid Commun. Mass Spectrom.* **2008**, *22*, 1181–1186. [CrossRef]

48. Ohmoto, K.; Yamamoto, S.; Hirokawa, M. Symptomatic Primary Biliary Cirrhosis Triggered by Administration of Sulpiride. *Am. J. Gastroenterol.* **1999**, *94*. [CrossRef]

49. Zhou, X.; Ren, L.; Yu, Z.; Huang, X.; Li, Y.; Wang, C. The antipsychotics sulpiride induces fatty liver in rats via phosphorylation of insulin receptor substrate-1 at Serine 307-mediated adipose tissue insulin resistance. *Toxicol. Appl. Pharmacol.* **2018**, *345*, 66–74. [CrossRef]

50. Gustafsson, F.; Foster, A.J.; Sarda, S.; Bridgland-Taylor, M.H.; Kenna, J.G. A Correlation Between the In Vitro Drug Toxicity of Drugs to Cell Lines That Express Human P450s and Their Propensity to Cause Liver Injury in Humans. *Toxicol. Sci.* **2013**, *137*, 189–211. [CrossRef] [PubMed]

51. Chen, K.-C.; Tsai, S.-W.; Zhang, X.; Zeng, C.; Yang, H.-Y. The Investigation of the Volatile Metabolites of Lung Cancer from the Microenvironment of Malignant Pleural Effusion. 2021. Available online: https://www.researchsquare.com/article/rs-144572/v1 (accessed on 25 March 2021).

52. Esfahani, S.; Sagar, N.M.; Kyrou, I.; Mozdiak, E.; O'Connell, N.; Nwokolo, C.; Bardhan, K.D.; Arasaradnam, R.P.; Covington, J.A. Variation in Gas and Volatile Compound Emissions from Human Urine as It Ages, Measured by an Electronic Nose. *Biosensors* **2016**, *6*, 4. [CrossRef]

53. McFarlane, M.; Mozdiak, E.; Daulton, E.; Arasaradnam, R.; Covington, J.; Nwokolo, C. Pre-analytical and analytical variables that influence urinary volatile organic compound measurements. *PLoS ONE* **2020**, *15*, e0236591. [CrossRef] [PubMed]

54. Mozdiak, E.; Wicaksono, A.N.; Covington, J.A.; Arasaradnam, R.P. Colorectal cancer and adenoma screening using urinary volatile organic compound (VOC) detection: Early results from a single-centre bowel screening population (UK BCSP). *Tech. Coloproctology* **2019**, *23*, 343–351. [CrossRef] [PubMed]

Article

Searching for Potential Markers of Glomerulopathy in Urine by HS-SPME-GC×GC TOFMS

Tomasz Ligor [1,2,*], **Joanna Zawadzka** [3], **Grzegorz Strączyński** [4], **Rosa M. González Paredes** [5], **Anna Wenda-Piesik** [6], **Ileana Andreea Ratiu** [2,7] and **Marek Muszytowski** [3]

1   Department of Environmental Chemistry and Bioanalytics, Faculty of Chemistry, Nicolaus Copernicus University, 87-100 Toruń, Poland

2   Interdisciplinary Centre of Modern Technologies, Nicolaus Copernicus University, 87-100 Toruń, Poland; andreea_ratiu84@yahoo.com

3   Department of Nephrology, Diabetology and Internal Medicine, Nicolaus Copernicus University, Rydygier Hospital, 87-100 Toruń, Poland; as.zawadzka@gmail.com (J.Z.); marek.muszytowski@gmail.com (M.M.)

4   USL Ltd., 43-110 Tychy, Poland; grzegorz.straczynski@usl.com.pl

5   Department of Analytical Chemistry, Nutrition and Food Sciences, University of Salamanca, 37008 Salamanca, Spain; rosamgonzal@usal.es

6   Department of Plant Growth Principles and Experimental Methods, UTP University of Science and Technology, 85-796 Bydgoszcz, Poland; apiesik@utp.edu.pl

7   "Raluca Ripan" Institute for Research in Chemistry, Babes-Bolyai University, 30 Fantanele, RO-400239 Cluj Napoca, Romania

*   Correspondence: tligor@umk.pl

Citation: Ligor, T.; Zawadzka, J.; Strączyński, G.; González Paredes, R.M.; Wenda-Piesik, A.; Ratiu, I.A.; Muszytowski, M. Searching for Potential Markers of Glomerulopathy in Urine by HS-SPME-GC×GC TOFMS. *Molecules* **2021**, *26*, 1817. https://doi.org/10.3390/molecules 26071817

Academic Editors: Natalia Drabińska and Ben de Lacy Costello

Received: 8 March 2021
Accepted: 21 March 2021
Published: 24 March 2021

Abstract: Volatile organic compounds (VOCs) exiting in urine are potential biomarkers of chronic kidney diseases. Headspace solid phase microextraction (HS-SPME) was applied for extraction VOCs over the urine samples. Volatile metabolites were separated and identified by means of two-dimensional gas chromatography and time of flight mass spectrometry (GC × GC TOF MS). Patients with glomerular diseases ($n = 27$) and healthy controls ($n = 20$) were recruited in the study. Different VOCs profiles were obtained from patients and control. Developed methodology offers the opportunity to examine the metabolic profile associated with glomerulopathy. Four compounds found in elevated amounts in the patients group, i.e., methyl hexadecanoate; 9-hexadecen-1-ol; 6,10-dimethyl-5,9-undecadien-2-one and 2-pentanone were proposed as markers of glomerular diseases.

Keywords: volatile organic compounds; urine analysis; comprehensive two-dimensional gas chromatography; kidney diseases

1. Introduction

Urine contains a multitude of organic substances, mainly products of metabolism, the majority including nitrogenous compounds. Thus, urine is also a rich source of volatile organic metabolites. For centuries organoleptic analysis of urine facilitated diagnosing illnesses, the most characteristic examples being the specific odor of urine present in diabetes and urinary tract infections. At that time, however, there were no methods that made it possible to ascertain what was responsible for the particular smell of the diseases. Shirasu et al. reviewed odoriferous compounds which are identified in urine, breath, sweat and other human secretions of ill patients and they observed to which diseases they were related [1]. A detailed study on characterization of odor active compounds in urine was conducted by Wagenstaller and coworkers. They evaluated urine samples by means of two-dimensional gas chromatographic system combined with mass spectrometry and sniffing technique [2]. At present, urine analysis constitutes an important element of medical diagnosis. However, currently used diagnostic tests do not provide information on volatile organic compounds (VOCs) present in urine, except for ketone bodies. Developments in gas chromatography and mass spectrometry as well as in sample preparation techniques naturally led to growing interest in determining VOCs in urine. The importance of VOCs for clinical diagnosis was reviewed as early as in 1981 [3]. One of the earliest applications of GC-MS in urine analysis was detection and quantitation of trimethylamine in urine of patients with fish odor syndrome [4]. Mills et al. used GC-MS and SPME for profiling VOCs in urine collected from patients with ketoacidosis, homocystinuria, hepatitis as well as from healthy persons [5]. Smith et al. identified the total of 92 substances in samples coming from 24 healthy males [6]. More recently, de Lacy Costello et al. reviewed and classified 1840 VOCs secreted from a healthy human body. They reported 279 volatiles which are identified in human urine [7]. Silva et al. used GC MS and SPME to study volatile metabolites in urine which are potentially important as cancer biomarkers. Samples were collected from a group of 33 cancer patients and 21 healthy individuals. The authors found 82 different VOCs in the group of cancer patients and in the control group [8]. Santos and coworkers analyzed ketones in urine samples to discriminate between lung cancer patients and healthy controls [9,10]. Comprehensive two-dimensional gas chromatography-time of flight mass spectrometry (GC × GC TOFMS) is a powerful tool which has been successfully used in metabolomics and biomarker discovery. This technique offers high resolution, ordered structure of chromatograms and high peak capacity and is efficient in urine analysis. GC × GC quadrupole MS or TOFMS has been used to determine anabolic steroids and their metabolites in human urine samples [11–13], to quantify salvinorin A in urine [14] and to detect acidic compounds in children's urine [15]. Among the most detailed studies were those conducted by Rocha et al. exploring human urine metabolomics. They applied GC × GC-TOFMS and SPME to study VOCs in the urine headspace of healthy persons and detected ca. 700 compounds in each sample of which 294 were tentatively identified [16].

A variety of renal injuries may lead to chronic kidney diseases (CKD) [17], which is a group of pathologies, where renal excretion is chronically compromised. Most often, CKD is irreversible and progressive. Patients with end-stage kidney diseases need renal replacement therapy (RRT) such as kidney transplant or dialysis. Worldwide, about 2 million people are receiving RRT [18,19]. Numerous inflammatory and non-inflammatory diseases affect the renal glomerulus and lead to glomerular kidney diseases [20,21]. However, searching for specific VOCs in urine can be promising way to find biomarkers of glomerular diseases. The aim of the study was method development based on SPME extraction and GC × GC TOFMS analysis of VOCs from human urine. Consequently, our research work focused on identification of VOCs in urine of patients with glomerular diseases and healthy controls. We developed SPME and GC × GC TOF MS method for extraction and analysis of volatiles. An automatic method of chromatographic data processing was applied. The ability of the developed method to differentiate between the two investigated groups of subjects was proved and with it the usefulness of GC × GC TOFMS in search for glomerulopathy-specific substances in urine was demonstrated as well.

2. Results

2.1. Fiber Selection

Three SPME fibers with different coatings-PDMS, CAR/PDMS and PDMS/DVB-were evaluated in terms of number of peaks obtained and identified as compounds, peak area and reproducibility. All fibers were exposed to the sample headspace for 30 min of incubation and 30 min of extraction, while temperature was 45 °C. As presented in Figure 1A, the PDMS/DVB fiber provided the highest extraction efficiency, since the total peak area was at the level of 4.2×10^6 and RSD 10.1% was obtained with this coating. CAR/PDMS provided total peak area at the level of 3.0×10^6 and RSD 12.2%. The lowest extraction efficiency was observed with the PDMS coating, with total peak area 1.4×10^6 and RSD 8.4%. Thus, the PDMS/DVB fiber was selected as the SPME fiber for the analysis of the volatile compounds of urine. In terms of number of the peaks extracted, PDMS/DVB extracted more peaks (324 ± 11) than the other 2 fibers (288 ± 7 in case of CAR/PDMS and

173 ± 6 in case of PDMS), as presented in Figure 1A1. Figure 1 was created using IBM SPSS Statistics 21. The center lines of the boxes represent the mean; boxes represent mean ± SD, while whiskers represent min–max values.

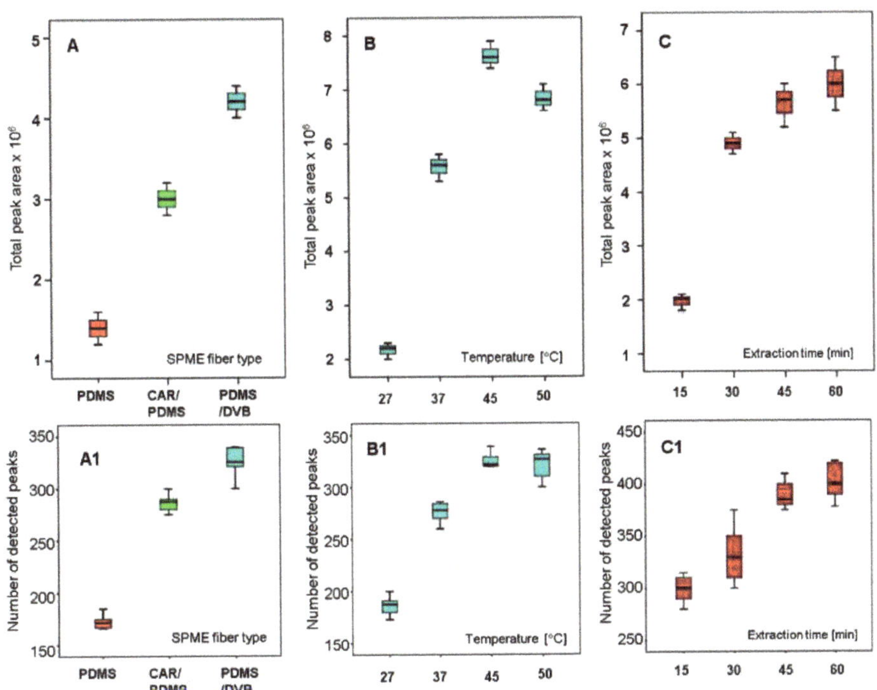

Figure 1. Optimization of extraction parameters, including comparison of PDMS/DVB, Carboxen/PDMS and PDMS coatings for extraction of VOC (part (**A,A1**)), different extraction temperatures (part (**B,B1**)) and extraction time (part (**C,C1**)). The boxplots in the upper were drawn according with total peak area of extracted VOCs (subfigures **A–C**), while the bottom box plots represent the number of extracted VOCs (subfigures **A1–C1**). Optimization of extraction parameters, including comparison of Polydimethylsiloxane/Divinylbenzene (PDMS/DVB), Carboxen/PDMS (CAR/PDMS), Polydimethylsiloxane (PDMS) for extraction of VOC (part **A**), different extraction temperatures (part **B**) and extraction time (part **C**). The boxplots in the upper were drawn according with total peak area of extracted VOCs (subfigures **A–C**), while the bottom box plots represent the number of extracted VOCs (subfigures **A1–C1**).

2.2. Extraction Temperature

With the use of the PDMS/DVB fiber, 30 min of incubation time, 30 min of extraction time and with 10 mL aliquots of the same urine from a healthy volunteer, the effect of the extraction temperature was studied at 27, 37, 45 and 50 °C (Figure 1B). It was observed that both signal areas and detected number of the peaks increased gradually with increasing the temperature from 27 °C up to 45 °C. At 50 °C, the signal areas of most of the peaks decreased, while the number of the peaks remained constant with the number detected at 45 °C (Figure 1B1). According to these results and with the aim of preventing degradation of the sample at high temperature, 45 °C was chosen as the optimum value of extraction temperature. Our results regarding temperature optimization are in agreement with the results obtained by other authors. For example, Monteiro et al. applied SPME and GC-MS to the analysis of renal carcinoma patients' urine. The authors carried out a detailed optimization of the SPME extraction process, considering SPME sorbents, urine sample pH, extraction time and temperature, etc. It was temperature that had the greatest influence on the obtained results, followed by extraction time and salt addition [10].

2.3. Extraction Time

The influence of the extraction time was studied as the last step of optimization. Incubation time was 30 min, followed by PDMS/DVB fiber exposure to the urine headspace at 45 °C for 15, 30, 45 and 60 min. We observed an increasing trend in the signal areas and peaks number that occurred from minute 15 to 45 (Figure 1C). In case of signal area, the very noticeable increasing trend was from 15 min to 30 min, while in case of peaks number the highest increase was from 30 min (325 peaks) to 45 min (390 peaks). After 60 min of extraction the increasing trend was insignificant in case of both signal areas and peaks number. These results showed that 45 min of the extraction time was not enough to reach the equilibrium. However, the extension of the extraction time to 60 min caused an increase in the extraction efficiency by only 5%. Similar situation was observed in case of the number of the peaks. Thus, 45 min was selected as an adequate value of extraction time.

Random variability of these signals was evaluated as well. For the whole experiment the extractions of VOCs from human urine samples were performed in triplicate, except three cases when patients did not provide a sufficient amount of samples. Repeatability was satisfactory, with RSD values lower than 7.1% for each urine sample. For variability investigation, 6 urine samples were kept in the freezer. The samples were defrosted and analyzed 45 days after sample taking at hospital sampling. Samples were analyzed by means of SPME and GC × GC TOF MS. The RSD of the measurements ($n = 18$) performed after 45 days of sample storage was in the range 6.1−9.2%.

2.4. Identification

The identification of volatile metabolites was performed on the basis of similarity of measured mass spectra to MS libraries (match factor > 900) and signal-to-noise (S/N) ratio (>10). For this purpose, the mass spectrum of each compound was automatically matched to those in MS libraries Wiley 9-th Ed./NIST 2011.The unique mass for each peak was chosen by the software algorithm and was used for peak area calculations. The automatically identified substances were manually verified in order to remove artefacts (mainly contaminants, silicones, column bleed, plasticizers, etc.) and compared with literature data describing the substances detected in urine headspace. For urine normalization, we selected patients and healthy persons for experiments, if the specific gravity of urine ranged from 1010 to 1030.

2.5. Statistical Analyses

The peak areas of identified substances were used to build the data matrix for chemometric analyses. The dataset representing distributions of 282 investigated compounds in urine was used to build the network analysis model (Figure 2). These were all the components that appear in more than 5 samples for a given group. Consequently, based on the obtained profiles and using the incidence of the peaks, a network analysis was created in order to obtain a preliminary exploration of the data. R studio with console (version 3.6.3, Boston, MA, USA) was used for network analyses. The applied model successfully separated the two investigated groups, by leading into the formation of two cluster groups (group of diseased patients including subjects P1 to P27, clustered into the left part and group of healthy controls, represented by numbers H1 to H20, in the right down part), as presented in Figure 2. In addition, VOCs detected just in diseased patients have been dispersed around the patients group, common components were located mostly between the two groups, while VOCs detected just in healthy subjects were scattered into the right-up part (green diamonds). Regarding the number of VOCs used in network analysis, 11 were specific for healthy group, 90 for diseased group, while 181 VOCs were common between the groups. We assume that most of compounds presented in Figure 2 are endogenous generated by the organism, as a normal process of metabolism or as a response to the pathology. However, parts of VOCs are exogenous absorbed by the organism from the environment and eliminated through urine. High variability in detected VOCs coming from different subjects was found. Such phenomenon is related to several factors, among which diet, living style, personal habits, etc.

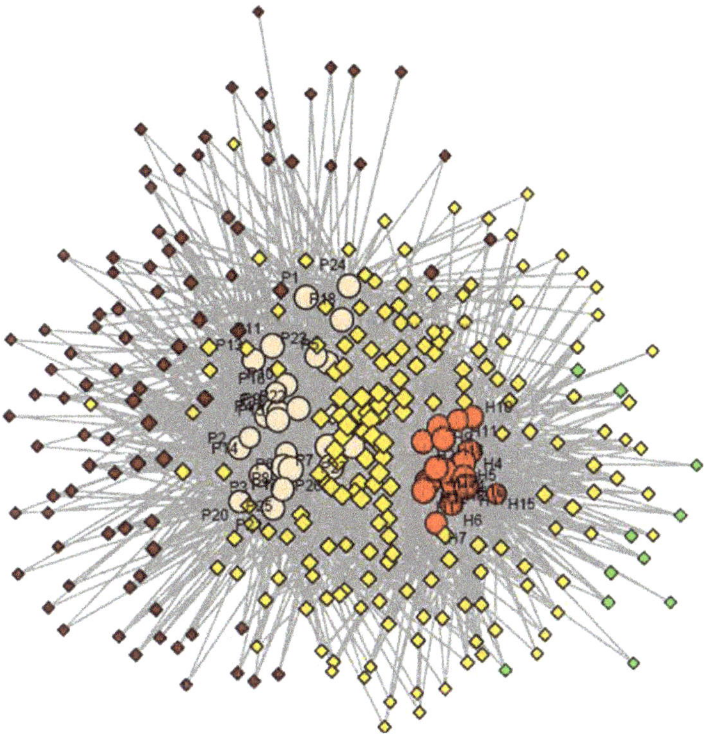

Figure 2. Network analyses presenting separation between the two investigated groups based on the VOCs specific for patients with glomerular diseases (brown diamonds), VOCs common between groups (yellow diamonds) and VOCs characteristic for healthy group (green diamonds), where H 1 to 20 represents the number of healthy control and P 1 to 27 the number of patients.

In the next step VOCs areas were used and hierarchical clusters analyses (nearest neighbor method) based on Squared Euclidean distance were created. The heat map color code, grey-red-yellow-green, is according with increasing value of peaks area, from absence (o value, highlighted in grey) to the highest area (green). Cluster segregation according with patients and control groups was obtained, as presented in Figure 3. Nevertheless, healthy group clustered in two groups that uncompressed between the diseases group. The patients P 26 and 27, expressed different characteristics and fused together in one cluster with similar distance level at the end of the dendrogram.

The peak areas of identified substances were finally used to build the data matrix for chemometric analyses, in attempt to obtain data set reduction. The aim was to search for some VOCs with discriminative features, able to be used as biomarkers of glomerular diseases. From the whole dataset representing distributions of the 146 VOCs found in urine that could be analyzed by ANOVA, thirteen compounds varied quantitatively between subjects with glomerular diseases and healthy controls ($p < 0.05$) and two at the level $0.05 < p < 0.1$ that can be also acceptable for screening study in life science. Based on the grouping of volatile compounds using the k-mean method, all subjects with glomerular diseases had elevated level of 40 compounds (Table 1).

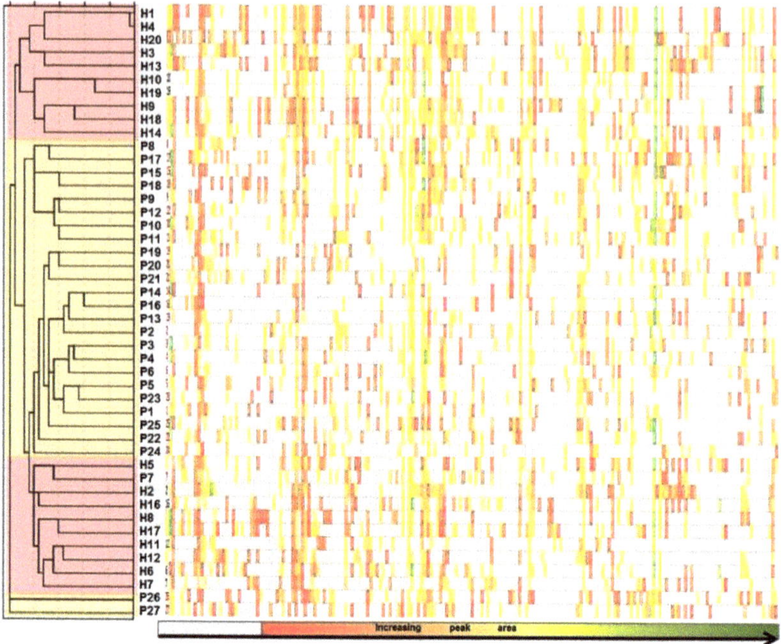

Figure 3. Heat map combined with dendrogram presenting hierarchical clusters segregation according with patient and control groups, where H 1to 20 represents the number of healthy control and P 1 to 27 the number of patients.

Generally, the significant effect was obtained for a number of 15 compounds, namely: cyclohexanone; 3-ethylcyclopentanone; 3-hexanone; 3-heptanone; methyl hexadecanoate; 9-hexadecen-1-ol; 3-methyl-2-pentanone; 6,10-dimethyl-5,9-undecadien-2-one; 2-pentanone; acetophenone; 2-methoxy-4-vinylphenol; 1-decanol; N-acetylpyrrole; 6-methylhept-5-en-2-one; dimethyl sulfone, presented in Table 1.

They play opposite role then biomarkers, while the compounds: 1, 4, 5, 6, 7, 8, 9, 10, 14, 16, 19, 21, 22, 24, 26, 27, 30, 31, 34, 35, 36, 37, 38 and 39 represent the moderate, (in ANOVA p-value is below 0.05). However, all classified compounds are presented in Table 1 with the information resulting on the F statistic.

Principal component analyses (PCA) were used to display total variation in the meaning of two main components results presented in Tables 1 and 2. Contribution of individual compounds to the C1 explained 30.5% of variance. As can be observed the VOCs that loaded most positive strength on 1st components are: 5, 6, 8, 9 and they had the significant status as biomarkers, displaying higher and significant concentration in urine coming from diseased persons (they are marked in the circle). The compounds 31, 32, 33 and 35 clustered positively in C2 gathering 13.6% of total variance (Figure 4) and are both speared by the others with higher concentration in the healthy than in diseased group.

Table 1. ANOVA grouping results of VOC in k-mean analysis (disease group vs. control). where: SS Group–squared sum, df-degree of freedom, SS Error-error sum of square, F-F test, p-probability value.

ID	Compound	SS_{Group}	df	SS_{Error}	df	F	p
1	Cyclohexanone	160.14	1	99.80	39	54.56	0.000
2	3-Ethylcyclopentanone	60.33	1	121.36	39	16.90	0.000
3	3-Hexanone	87.83	1	166.31	39	17.96	0.000
4	3-Heptanone	102.78	1	145.92	39	23.95	0.000
5	Methyl hexadecanoate	68.88	1	134.76	39	17.38	0.000
6	9-Hexadecen-1-ol	75.72	1	142.22	39	18.10	0.000
7	3-Methyl-2-pentanone	73.83	1	209.25	39	12.00	0.001
8	6,10-Dimethyl-5,9-undecadien-2-one	49.61	1	117.49	39	14.36	0.001
9	2-Pentanone	44.38	1	144.91	39	10.41	0.003
10	Acetophenone	25.60	1	117.55	39	7.40	0.010
11	2-Methoxy-4-vinylphenol	19.72	1	119.53	39	5.61	0.024
12	1-Decanol	28.56	1	184.65	39	5.26	0.028
13	N-Acetylpyrrole	28.44	1	204.56	39	4.73	0.037
14	6-Methylhept-5-en-2-one	22.96	1	198.97	39	3.92	0.056
15	Dimethyl sulfone	3.24	1	37.79	39	2.91	0.097
16	1-Tetradecanol	14.96	1	190.06	39	2.68	0.111
17	4-Heptanone	9.43	1	128.86	39	2.49	0.124
18	Benzaldehyde	1.66	1	23.90	39	2.36	0.134
19	2-Nonanone	14.35	1	213.06	39	2.29	0.139
20	5-Methyl-3-hexanone	13.07	1	210.41	39	2.11	0.155
21	Dimethyl trisulfide	10.23	1	190.57	39	1.83	0.186
22	2-Aminobenzaldehyde	9.40	1	222.11	39	1.44	0.239
23	3-Methylcyclopentanone	8.12	1	230.47	39	1.20	0.281
24	Hexanal	3.85	1	217.07	39	0.60	0.443
25	1-Octanol	3.59	1	245.37	39	0.50	0.485
26	Benzeneacetaldehyde	3.43	1	234.68	39	0.50	0.486
27	2,5-Dimethylpyrazine	2.85	1	198.49	39	0.49	0.490
28	Nonanal	1.50	1	155.73	39	0.33	0.571
29	9-Octadecen-1-ol	1.47	1	183.26	39	0.27	0.605
30	Indole	0.35	1	49.34	39	0.24	0.626
31	Theaspirane	1.37	1	218.11	39	0.21	0.647
32	Benzonitrile	0.52	1	185.47	39	0.09	0.760
33	2-Heptanone	0.27	1	157.55	39	0.06	0.811
34	4-Methylphenol	0.27	1	170.99	39	0.05	0.818
35	Phenol	0.24	1	166.54	39	0.05	0.828
36	Decanal	0.09	1	73.76	39	0.04	0.837
37	1-Methyl-4-(1-methylethenyl)-benzene	0.08	1	192.35	39	0.01	0.905
38	N-Phenylformamide	0.00	1	227.31	39	0.00	0.979
39	Ethyl acetate	0.00	1	328.20	39	0.00	0.986
40	Octanal	0.00	1	199.51	39	0.00	0.986

Table 2. Descriptive statistics of the VOCs defined in higher concentration in urine by patients suffering from glomerular diseases, where: min–minimum, max–maximum.

ID	Compound	Diseased Group			Healthy Group		
		Mean	Min	Max	Mean	Min	Max
1	Cyclohexanone	3.28×10^5	1.33×10^4	9.43×10^5	3.54×10^4	2.23×10^4	4.64×10^4
2	3-Ethylcyclopentanone	1.10×10^5	5.14×10^3	4.53×10^5	3.41×10^4	9.57×10^3	6.58×10^4
3	3-Hexanone	9.41×10^5	1.65×10^4	1.35×10^6	3.44×10^4	2.92×10^4	1.06×10^6
4	3-Heptanone	2.84×10^6	6.06×10^4	2.57×10^7	1.75×10^5	7.46×10^4	3.90×10^6
5	Methyl hexadecanoate	7.63×10^4	7.64×10^3	3.64×10^5	2.19×10^4	6.99×10^3	8.38×10^4
6	9-Hexadecen-1-ol	7.39×10^4	5.28×10^3	4.34×10^5	9.39×10^3	2.78×10^3	1.93×10^5
7	3-Methyl-2-pentanone	2.69×10^6	2.90×10^4	2.81×10^7	1.81×10^5	8.95×10^3	1.19×10^7
8	6,10-Dimethyl-5,9-undecadien-2-one	1.41×10^5	2.47×10^4	3.91×10^5	7.41×10^4	1.93×10^4	1.73×10^5
9	2-Pentanone	1.94×10^7	6.39×10^4	7.62×10^7	3.18×10^6	1.46×10^6	1.05×10^8
10	Acetophenone	1.08×10^5	4.37×10^4	1.40×10^5	1.23×10^4	8.19×10^4	1.79×10^5
11	2-Methoxy-4-vinylphenol	1.95×10^5	1.76×10^4	9.97×10^5	1.07×10^5	1.19×10^4	4.58×10^5
12	1-Decanol	1.62×10^5	2.46×10^4	5.20×10^5	2.82×10^5	1.99×10^4	7.00×10^5
13	N-Acetylpyrrole	3.12×10^5	1.11×10^4	1.06×10^6	6.87×10^4	3.17×10^4	7.43×10^5
14	6-Methylhept-5-en-2-one	1.84×10^5	9.19×10^3	9.31×10^5	2.00×10^5	2.30×10^4	7.26×10^5
15	Dimethyl sulfone	3.24×10^4	1.10×10^4	1.97×10^5	2.97×10^4	4.16×10^3	9.29×10^4
16	1-Tetradecanol	2.93×10^5	1.33×10^4	1.23×10^6	2.11×10^5	3.10×10^4	4.69×10^5
17	4-Heptanone	6.50×10^6	2.24×10^4	3.82×10^7	5.73×10^6	1.77×10^4	2.25×10^7
18	Benzaldehyde	6.23×10^4	6.18×10^3	1.89×10^5	1.01×10^5	1.26×10^4	3.58×10^5
19	2-Nonanone	1.56×10^5	1.31×10^4	7.72×10^5	1.54×10^5	1.07×10^4	5.24×10^5
20	5-Methyl-3-hexanone	1.89×10^6	5.78×10^4	1.48×10^7	3.22×10^6	1.92×10^4	1.90×10^7
21	Dimethyl trisulfide	6.17×10^4	5.14×10^3	1.88×10^5	4.51×10^4	5.76×10^3	7.46×10^4
22	2-Aminobenzaldehyde	1.93×10^5	6.12×10^3	6.80×10^5	2.65×10^5	8.50×10^4	5.20×10^5
23	3-Methylcyclopentanone	2.76×10^5	4.29×10^4	8.61×10^5	2.05×10^5	3.19×10^4	7.21×10^5
24	Hexanal	1.80×10^6	1.16×10^5	6.48×10^6	9.01×10^5	8.86×10^4	1.86×10^6
25	1-Octanol	5.13×10^5	1.01×10^4	3.61×10^6	6.37×10^5	7.98×10^4	3.42×10^6
26	Benzeneacetaldehyde	1.43×10^5	2.67×10^4	1.25×10^6	1.47×10^5	5.34×10^4	5.60×10^5
27	2,5-Dimethylpyrazine	1.19×10^5	1.06×10^4	5.67×10^5	7.50×10^4	1.82×10^4	2.36×10^5
28	Nonanal	2.26×10^5	2.04×10^4	1.09×10^6	1.28×10^5	7.82×10^4	1.77×10^5
29	9-Octadecen-1-ol	4.98×10^5	3.25×10^4	1.79×10^6	4.60×10^5	3.20×10^4	9.43×10^5
30	Indole	1.20×10^5	2.19×10^4	3.47×10^5	9.08×10^4	1.45×10^4	2.00×10^5
31	Theaspirane	1.09×10^5	7.87×10^3	2.62×10^5	1.29×10^5	4.64×10^4	3.72×10^5
32	Benzonitrile	4.53×10^4	6.57×10^3	6.30×10^4	1.67×10^4	8.76×10^3	3.10×10^4
33	2-Heptanone	1.92×10^5	1.13×10^4	1.41×10^6	1.75×10^5	7.46×10^4	5.69×10^5
34	4-Methylphenol	1.22×10^5	1.71×10^4	5.13×10^5	1.71×10^5	2.71×10^4	1.18×10^6
35	Phenol	3.52×10^4	9.95×10^3	4.67×10^5	2.90×10^4	2.08×10^4	1.11×10^5
36	Decanal	1.39×10^5	1.62×10^4	7.38×10^5	1.56×10^4	2.06×10^4	2.06×10^4
37	1-Methyl-4-(1-methylethenyl)-benzene	1.13×10^5	1.12×10^4	1.41×10^5	1.41×10^5	1.90×10^4	6.09×10^5
38	N-Phenylformamide	1.17×10^5	1.59×10^4	3.32×10^5	1.54×10^5	3.02×10^4	3.35×10^5
39	Ethyl acetate	1.56×10^7	2.02×10^5	7.11×10^7	1.41×10^7	1.21×10^1	6.41×10^{57}
40	Octanal	2.17×10^5	5.62×10^4	6.97×10^5	1.62×10^5	1.49×10^5	4.34×10^5

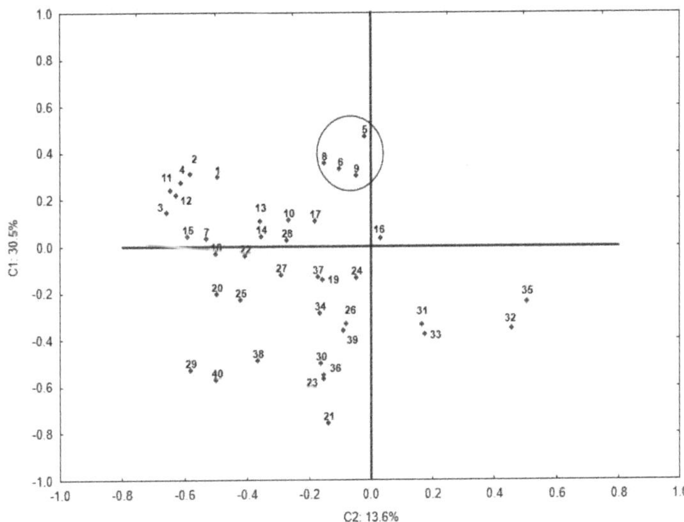

Figure 4. Principal component analyses (PCA) used to display variation between statistically significant determined VOCs, where: 1. cyclohexanone; 2. 3-ethylcyclopentanone; 3. 3-hexanone; 4. 3-heptanone; 5. methyl hexadecanoate; 6. 9-hexadecen-1-ol; 7. 3-methyl-2-pentanone; 8. 6,10-dimethyl-5,9-undecadien-2-one; 9. 2-pentanone; 10. acetophenone; 11. 2-methoxy-4-vinylphenol; 12. 1-decanol; 13. N-acetylpyrrole; 14. 6-methylhept-5-en-2-one; 15. dimethyl sulfone; 15. dimethyl sulfone; 16. 1-tetradecanol, 17. 4-heptanone; 18. benzaldehyde; 19. 2-nonanone; 20. 5-methyl-3-hexanone; 21. dimethyl trisulfide; 22. 2-aminobenzaldehyde; 23. 3-methylcyclopentanone; 24. hexanal; 25. 1-octanol; 26. benzeneacetaldehyde; 27. 2,5-dimethylpyrazine; 28. nonanal; 29. 9-octadecen-1-ol; 30. indole; 31. theaspirane; 32. benzonitrile; 33. 2-heptanone; 34. 4-methylphenol; 35. phenol; 36. decanal; 37. 1-methyl-4-(1-methylethenyl)-benzene; 38. N-phenylformamide, 39. ethyl acetate; 40. octanal.

In Table 2, the VOCs are listed in the order from the highest to the lowest level of significance at which they were present in urine samples. They were obtained by the standardization of the total matrix, in order to divide the VOCs into significantly distinct groups. All the components in Table 2 were classified based on the grouping of volatile compounds using the k-mean method. The method shows that all persons with glomerular diseases had elevated level of all 40 compounds presented in Table 2.

Additionally, the compounds that present a decreasing trend can be treated as secondary chemo indicators of this category of diseases. Discrepancies between the results of statistical analyses in this respect (ANOVA and grouping of k-means) result from the fact that the tested group was a statistically small sample.

3. Discussion

On urine sample chromatograms, varying numbers of particular chromatographic signals were observed from 100 to 250 peaks coming from different substances present in the samples. Moreover, we observed variations in the number of samples in which certain compounds were detected. This phenomenon occurred regardless of which group the samples were collected, from the patients or healthy control. Such a large number of varied compounds posed a significant difficulty in classifying the substances and selecting potential biomarkers. This may result from a large number of factors influencing biosynthesis of volatile metabolites (metabolic pathways, genetic differences, consumed food, age, sex, physiological state, addictions, etc.). Substances present in urine can be divided into characteristic chemical groups such as ketones, aldehydes, hydrocarbons, volatile sulfur compounds, heterocycles, alcohols, phenols and terpenes. The substances most numerously represented were ketones. They amounted to almost 46% of all volatile substances identified in the samples from healthy volunteers and 49% of those found in samples from ill persons. Ketones are the products of metabolism and result from oxidation of secondary alcohols and fatty acids. Part of ketones can be also of dietary origin [6]. The most frequently observed ketones included acetone, acetophenone, 3-ethylcyclopentanone, 5-methylohexan-3-one, benzophenone, 2-pentanone, 2-heptanone, 2-butanone, 3-hexanone, 4-heptanone, 3-methylcyclopentanone, cyclohexanone and 1-octen-3-one. The next groups were aldehydes and alcohols, ca. 10 and 12% of all the chemical compounds respectively. We observed the presence of a series of aliphatic aldehydes from acetaldehyde to decanal, including also unsaturated and methylated compounds (i.e., 2-methylbutanal, 2-methyl-2-butenal). This is similar to the observations of Smith and Ratclife [6,7]. Nonanal was identified in the majority of samples. We also found aldehydes containing a benzene ring, i.e., benzaldehyde, 2,4-dimethylbenzaldehyde, 2-hydroxybenzaldehyde, 4-methylbenzaldehyde, 4-(1-methylethyl)-benzaldehyde and alpha-methylbenzeneacetaldehyde. As for the presence of alcohols in the samples, the most frequently observed were 1-dodecanol, 1-octanol, 1-tetradecanol, benzyl alcohol, 1-hexanol, 1-nonanol, 1-octen-3-ol and 1-butanol. Many terpenes were identified among the detected substances. The substances most frequently present in the samples included limonene, pinene, menthadienes, mentol, mentone, valencene, geraniol, linalool, thujene, myrcene, sabinene hydrate, β-caryophyllene, linalool oxides, etc. Considering that terpenes are produced by plant organisms and not the human body, we assumed that the source of terpenes and their metabolites in urine is food. This assumption is supported also by the works of other [6,7]. Due to this fact, we excluded this group of substances from the set of potential disease markers.

Another exogenous group were plasticizers, most frequently diethyl phthalate and antioxidants (BHA, BHT), which we excluded from the analysis as they constitute an addition to polymers. Similarly, we removed silicones and oximes as these substances mostly come from polymers in urine containers, plastic tips, septa for HS vials etc. This is supported by blank analyses done according to the same procedure as urine sample analysis but containing only distilled water and NaCl. Phenol and cresols were found in urine samples, as well as dimethylphenols, guaiacol, 2-methoxy-4-vinylphenol and eugenol. Phenol and cresols are typical metabolites present in urine, identified by many authors. There is a correlation between the content of phenols in urine and the amount of consumed proteins [7].

A numerous group of substances were also N-, O-, S-heterocycles. The most frequently identified of them was indole and less common were 3-methylindole (skatole) as well as substituted pyridines, pyrroles, pyrazines, furans and benzothiazoles. The source of indole and skatole may be bacterial metabolism of aromatic amino acids (tyrosine, phenylalanine and tryptophan) occurring in the intestines. These substances may subsequently be absorbed into the blood and excreted with urine. We also observed series of gamma and delta lactones, i.e., nonalactone, decalactone and undecalactone as well as coumarin. However, the origin of such substances in urine has not been explained. Another group of chemicals are benzene and its alkyl derivatives (toluene, dimethyl benzenes, ethyl benzene, propylbenzene, styrene, etc.). On the one hand, these substances are known to be typical environmental pollutants; on the other, many studies consider them to be probable disease markers. We decided that we can overlook benzene, toluene and xylene isomers as exogenous substances. Nevertheless, isomers of trimetylbenzenes, ethylmethylbenzenes, naphtalene and its derivatives are included in our statistical analysis. Figure S1 shows structure ordered GC×GC chromatogram of urine (Supplementary Material).

Little is known, at present, whether the substances identified in the urine samples from ill and healthy people are created as a result of metabolism in the cells of a human organism, whether they originate from the diet or from metabolic changes occurring under the influence of an illness in the body. We observed elevated levels of benzeneacetaldehyde; 1-octanol; 1-decanol, 6-methylhept-5-en-2-one in urine of patients. Regarding the origin, phenol is very common metabolite existing in urine. Increased level of phenol can be explained by extensive metabolism of proteins or increasing of bacterial activity in the colon. 9-Octadecene-1-ol may be oxidation products of certain hydrocarbons conducted by cytochrome P450 or there are products of oxidative stress. 6,10-Dimethyl-5,9-undecadiene-2-one can be formed by the decarboxylation of keto acids generated during fatty acids metabolism. Their origin can be associated to oxidation of fatty acids. Moreover, inflammation processes are important factor in glomerular diseases. In this case a key role is played by cellular and humoral responses involving creating immuno complexes (circulating and in situ-formed) and complement pathways [21]. Oxidative stress and inflammation are initiated by the reactive oxygen species (ROS). Hence, ROS induce formation of several by products such as fatty acids, hydrocarbons, aldehydes and alcohols. The origin of the proposed markers can be connected mainly with oxidative stress. The knowledge regarding the biosynthesis of VOCs in the organism is very limited and covers only a small number of volatiles identified in bodily fluids and tissue. Parts of the substances present in urine are exogenous substances which enter the organism as food or flavors (terpenes), as well as environmental pollutants (aromatic hydrocarbons). Another group are exogenous substances that are transdermally absorbed into the organism, where they can be metabolized or not and then excreted with urine. At the moment there are studies to define new specific biomarkers of kidney damage, detected in both serum and urine. These include cystatin C, neutrophil gelatinase-associated lipocalin, kidney injury molecule-1 and interleukin 18 [21]. Nevertheless, our study proved the discrimination between two investigated groups (the group with glomerular diseases and the control group) was possible based on the VOCs released from urine samples. Moreover, 4 VOCs that presented statistically significant differences between the groups can be assumed as markers of glomerular diseases. These are significantly increased peaks area in the patients group and they can be assumed as direct chemo indicators for glomerular diseases, while the other four were significantly lower and they may be considered as secondary chemo indicators. However, our research should be treated as a preliminary study, as the number of persons participating in the study was too small to draw more unequivocal conclusions, but using the developed methodology and involving higher number of patients, more deep investigations will be realized, with respect of non-proliferative or proliferative types. This will make the object of another study.

4. Materials and Methods

4.1. Materials

SPME device as well as Carboxene/PDMS, PDMS/Divinylbenzene and PDMS coated fibers were purchased from Supelco (Bellefonte, PA, USA). The screw top headspace glass vials with silicon/PTFE septa and caps were supplied by Supelco. Sodium chloride was purchased from Sigma-Aldrich (Steinheim, Germany) and Sil Tite micro union from Trajan (Trajan, Rigwood Victoria, Australia) Ultrahigh purity helium BIP 5.5 was purchased from Air Liquide, Poland.

4.2. Apparatus

The analysis was carried out using a Leco Pegasus 4D GCGC TOF-MS instrument (Leco Corp., St. Joseph, NH, USA) equipped with a dual stage jet cryogenic modulator. The mass spectrometer was hyphenated with an Agilent 6890 gas chromatograph (Agilent Technologies, Waldbronn, Germany). The gas chromatograph was equipped with a PTV injector (Gerstel, Mulhheim, Germany) with 0.75 mm ID liner and MPS-2 autosampler, (Gerstel, Mulhheim, Germany) for automatic SPME extraction and fiber desorption. The first-dimension column was RTX-Wax capillary (30 m × 0.25 mm × 0.20 μm) integrated with deactivated guard column (5 m × 0.25 mm) and the second-dimension column was a 1.5 m Rtx–1 capillary (1.5 m × 0.18 mm × 0.20 μm), both supplied by Restek (Restek, Bellefonte, PA, USA). The first column was connected to the second analytical column with an SGE micro union. The injector temperature was kept at 230 °C. SPME desorptions were made in the splitless mode within 1 min. Helium was used as the carrier gas in constant flow mode at 1 mL/min. The mass spectrometer was operating as follows both ion source temperature and transfer line 225 °C; ionization energy 70 eV (electron impact ionization); acquisition range 35–350 amu; acquisition range 180 Hz. The first-dimension column temperature was programmed as follows: the initial temperature of 40 °C was held for 3 min and then the temperature was increased by 10 °C/min to 235 °C and maintained for 5 min at this value. The second-dimension column temperature was maintained 5 °C higher than the corresponding first dimension column. The modulator temperature was maintained 15 °C higher than the corresponding one of the second-dimension column and the modulation period was 3 s, hot 0.9 s and cold 0.6 s. The programming rate and hold times were the same for the two columns and modulator.

4.3. Data Processing

Data processing parameters were as follow tune check on, baseline offset 1, peak width 2 s. for 2-nd dimension, match spectra require to combine was 700, minimum S/N 11 subpeak to be retained, maximum number of unknown peaks to find 10,000, automatic peak finding S/N = 5, library search mode-forward, library identify-normal, maximum mass 350, library mass threshold 5, minimum similarity match 900, peak mass to use for area calculation-unique mass. For mass spectrometry QC method, the instrument was adjusted to optimization mass 219 m/z, with minimum intensity 20,000 and maximum intensity 100,000. Chromatof (Leco) software version 4.50.8.0 was used for data acquisition and processing.

4.4. Sample Pre-Processing and Extraction

Obtained urine samples were immediately frozen and stored at −20 °C. In this study, frozen samples were not stored longer than 5 days before analysis. Prior to analyses, the samples were defrosted and then immediately pre-processed and analyzed as follows. NaCl (3.6 g) was weighed in a 20 mL HS glass vial and 10.0 mL of urine sample was added. Then, the vial was capped with a PTFE septum and a screw cap. The sample vial was incubated for 30 min at 45 °C. The fiber was exposed to the headspace at 45 °C for 45 min. After sampling, the SPME fiber was withdrawn into the needle, removed from the vial and inserted into the GC injector port for 1 min in the splitless mode, wherein the metabolites were thermally desorbed and transferred directly into the column.

4.5. Human Subjects

The patients were recruited among the patients of Nephrology, Diabetology and Internal Medicine Department (Collegium Medicum, Nicolaus Copernicus University, Rydygier Hospital, Torun, Poland). We selected a population of 27 adult patients (14 female, 13 male) with confirmed glomerular diseases and enrolled healthy control volunteers (20 persons). The mean age was 48.0 years. One male patient was a smoker and 3 patients (15%) had kidney biopsy in the past. The patients and the controls were not restricted to any particular diet. Serum creatinine concentration ranged from 0.5 to 3.87 mg/dl (mean 1.26). Estimated glomerular filtration rate (eGFR) ranged from 15 to 126 mL/min/1.73 m². Serum urea concentration ranged from 24 to 236 mg/dl (mean 58). C-reactive protein average was 9.1 mg/L (range 0.1–79). Early morning mid-stream urine samples were collected in 100 mL sterile plastic containers at hospital. Afterwards, samples were immediately frozen and stored at −20 °C. Prior to analyses, the samples were defrosted.

All subjects gave their informed consent for inclusion before they participated in the study. The study was conducted in accordance with the Declaration of Helsinki and the protocol was approved by the Ethics Committee of Collegium Medicum in Bydgoszcz (No. KB 621/2016-25.10.2016).

5. Conclusions

Our work focused on the development a methodology able to differentiate patients with glomerular diseases from healthy persons. Our study highlighted that volatile profiles coming from the two groups, diseased and controls can be easily discriminated by network analysis. Cluster analysis (based on Squared Euclidean distance) also segregated the two investigated groups. Dataset reduction highlighted that all persons with glomerular diseases had elevated level of several dozens of compounds with different origins; however, four compounds (methyl hexadecanoate; 9-hexadecen-1-ol; 6,10-dimethyl-5,9-undecadien-2-one, 2-pentanone were finally, proposed as markers of glomerular diseases. Identified compounds may be promising biomarker candidates for discrimination of patients with glomerular diseases and healthy volunteers. Moreover, deeper investigations of the biochemical pathways of the particular compounds as well selection of larger group of participants, with respect of non-proliferative or proliferative types, are essential. In the future perspective, investigations focused on person's diet will be particularly interesting.

Supplementary Materials: The following are available online at https://www.mdpi.com/1420-3049/26/7/1817/s1, Figure S1: structure ordered GCxGC chromatogram of urine sample from patient. Peak groups: 1—Ketones (2-pentanone, 3-methyl-2-pentanone, 3-hexanone, 3-heptanone, 5-methyl-3-hexanone, acetophenone), 2—Aldehydes (hexanal, heptanal, octanal, nonanal, decanal, benzaldehyde, benzeneacetaldehyde), 3—Alcohols (1-hexanol, 1-octanol, 1-nonanol, menthadienol, 1-decanol, verbenol, 1-dodecanol, 1-tetradecanol, 9-hexadecen-1-ol, 9-octadecen-1-ol), 4—Phenols (phenol, 4-methylphenol).

Author Contributions: Conceptualization, T.L. and J.Z.; methodology, T.L., G.S., I.A.R., R.M.G.P., M.M.; software, A.W.-P., I.A.R.; validation, T.L., G.S., I.A.R.; formal analysis, T.L., I.A.R.; investigation, T.L, R.M.G.P., G.S., J. Z.; data curation, A.W.-P., I.A.R.; writing—original draft preparation, T.L., I.A.R.; writing—review and editing, T.L., J.Z., G.S., R.M.G.P., A.W.-P., I.A.R., M.M.; supervision, T.L.; funding acquisition, T.L. All authors have read and agreed to the published version of the manuscript.

Funding: This work was supported by Toruń Center of Excellence "Towards Personalized Medicine" operating under Excellence Initiative-Research University (T. Ligor, B. Buszewski).

Institutional Review Board Statement: The study was conducted according to the guidelines of the Declaration of Helsinki and approved by the Ethics Committee of Collegium Medicum in Bydgoszcz (No. KB 621/2016-25.10.2016).

Informed Consent Statement: Informed consent was obtained from all subjects involved in the study.

Data Availability Statement: Data is contained within the article.

Acknowledgments: The authors would like to thanks to Aneta Cieślak for technical support.

Conflicts of Interest: The authors declare no conflict of interest. The funders had no role in the design of the study; in the collection, analyses, or interpretation of data; in the writing of the manuscript, or in the decision to publish the results.

Sample Availability: Samples of the compounds are not available from the authors.

References

1. Shirasu, M.; Touhara, K. The scent of disease: Volatile organic compounds of the human body related to disease and disorder. *J. Biochem.* **2011**, *150*, 257–266. [CrossRef]
2. Wagenstaller, M.; Buettner, A. Characterization of odorants in human urine using a combined chemo-analytical and human-sensory approach: A potential diagnostic strategy. *Metabolomics* **2013**, *9*, 9–20. [CrossRef]
3. Zlatkis, A.; Brazell, R.S.; Poole, C.F. The role of organic volatile profiles in clinical diagnosis. *Clin. Chem.* **1981**, *27/6*, 789–797. [CrossRef]
4. Mills, G.A.; Walker, V.; Mughal, H. Quantitative determination of trimethylamine in urine by solidphase microextraction and gas chromatography–mass spectrometry. *J. Chromatogr. B* **1999**, *723*, 281–285. [CrossRef]
5. Smith, S.; Burden, H.; Persad, R.; Whittington, K.; de Lacy Costello, B.; Ratcliffe, N.M.; Probert, C.S. A comparative study of the analysis of human urine headspace using gas chromatography–mass spectrometry. *J. Breath Res.* **2008**, *2*, 037022. [CrossRef]

6. de Lacy Costello, B.; Amann, A.; Al-Kateb, H.; Flynn, C.; Filipiak, W.; Khalid, T.; Osborne, D.; Ratcliffe, N.M. A review of the volatiles from the healthy human body. *J. Breath Res.* **2014**, *8*, 014001. [CrossRef]

7. Silva, C.L.; Passos, M.; Camara, J.S. Investigation of urinary volatile organic metabolites as potential cancer biomarkers by solid-phase microextraction in combination with gas chromatography-mass spectrometry. *Br. J. Cancer* **2011**, *105*, 1894–1904. [CrossRef] [PubMed]

8. Santos, P.M.; del Nogal Sánchez, M.; Pozas, Á.P.C.; Pavón, J.L.P.; Cordero, B.M. Determination of ketones and ethyl acetate—a preliminary study for the discrimination of patients with lung cancer. *Anal. Bioanal. Chem.* **2017**, *409*, 5689–5696. [CrossRef] [PubMed]

9. Monteiro, M.; Carvalho, M.; Henrique, R.; Jeronimo, C.; Moreira, N.; de Lourdes Bastos, M.; Guedes de Pinho, P. Analysis of volatile human urinary metabolome by solid-phase microextraction in combination with gas chromatography–mass spectrometry for biomarker discovery: Application in a pilot study to discriminate patients with renal cell carcinoma. *Eur. J. Cancer* **2014**, *50*, 1993–2002. [CrossRef]

10. Silva, A.I., Jr.; Pereira, H.M.G.; Casilli, A.; Conceic, F.C.; Aquino Neto, F.R. Analytical challenges in doping control: Comprehensive two-dimensional gas chromatography with time of flight mass spectrometry, a promising option. *J. Chromatogr. A* **2009**, *1216*, 2913–2922. [CrossRef]

11. Mitrevski, B.S.; Brenna, J.T.; Zhang, Y.; Marriott, P.J. Application of comprehensive two-dimensional gas chromatography to sterols analysis. *J. Chromatogr. A* **2008**, *1214*, 134–142. [CrossRef]

12. Zhang, Y.; Auchu, H.J.T.R.J.; Brenna, J.T. Comprehensive two dimensional gas chromatography fast quadrupole mass spectrometry (GC×GC-qMS) for urinary steroid profiling. mass spectral characteristics with chemical ionization. *Drug Test. Anal.* **2011**, *3*, 857–867. [CrossRef] [PubMed]

13. Barnes, B.B.; Snow, N.H. Analysis of Salvinorin A in plants, water, and urine using solid-phase microextraction-comprehensive two-dimensional gas chromatography–time of flight mass spectrometry. *J. Chromatogr. A* **2012**, *1226*, 110–115. [CrossRef]

14. Vasquez, N.P.; Crosnier de Bellaistre-Bonose, M.; Lévêque, N.; Thioulouse, E.; Doummar, D.; Billette de Villemeur, T.; Rodriguez, D.; Couderc, R.; Robin, S.; Courderot-Masuyer, C.; et al. Advances in the metabolic profiling of acidic compounds in children's urines achieved by comprehensive two-dimensional gas chromatography. *J. Chromatogr. B* **2015**, *1002*, 130–138. [CrossRef] [PubMed]

15. Rocha, S.M.; Caldeira, M.; Carrola, J.; Santos, M.; Cruz, N.; Duarte, I.F. Exploring the human urine metabolomic potentialities by comprehensive two-dimensional gas chromatography coupled to time of flight mass spectrometry. *J. Chromatogr. A* **2012**, *1252*, 155–163. [CrossRef] [PubMed]

16. Remuzzi, G.; Benigni, A.; Remuzzi, A. Mechanisms of progression and regression of renal lesions of chronic nephropathies and diabetes. *J. Clin. Investig.* **2006**, *116*, 288–296. [CrossRef]

17. Liyanage, T.; Ninomiya, T.; Jha, V.; Neal, B.; Patrice, H.M.; Okpechi, I.; Zhao, M.H.; Lu, J.; Garg, A.X.; Knight, J.; et al. Worldwide access to treatment for end-stage kidney disease: A systematic review. *Lancet* **2015**, *385*, 1917–2014. [CrossRef]

18. Lopez-Novoa, J.; Rodrigez-Peria, J.M.A.B.; Ortiz, A.; Martinez-Salgado, C.; Hernandez, L.F.J. Etiopathology of chronic tubular, glomerular and renovascular nephropaties. Clinical implications. *J. Transl. Med.* **2011**, *9*, 13–39. [CrossRef]

19. Couser, W.G. Pathogenesis of glomerular damage in glomerulonephritis. *Nephrol. Dial. Transplant.* **1998**, *13*, 10–15. [CrossRef] [PubMed]

20. Nangaku, M.; Couser, W.G. Mechanisms of immune-deposit formation and the mediation of immune renal injury. *Clin. Exp. Nephrol.* **2005**, *9*, 183–191. [CrossRef]

21. Couser, W.G. Complement inhibitors and glomerulonephritis: Are we there yet? *J. Am. Soc. Nephrol.* **2003**, *14*, 815–818. [CrossRef]

 2020, 25, 3651

Article

An Optimization of Liquid–Liquid Extraction of Urinary Volatile and Semi-Volatile Compounds and Its Application for Gas Chromatography-Mass Spectrometry and Proton Nuclear Magnetic Resonance Spectroscopy

Natalia Drabińska [1,2,*], Piotr Młynarz [3], Ben de Lacy Costello [2,*], Peter Jones [4], Karolina Mielko [3], Justyna Mielnik [3], Raj Persad [5] and Norman Mark Ratcliffe [2]

[1] Institute of Animal Reproduction and Food Research of Polish Academy of Sciences, 10 Tuwima Str., 10-748 Olsztyn, Poland

[2] Institute of Biosensor Technology, University of the West of England, Coldharbour Lane, Frenchay, Bristol BS16 1QY, UK; Norman.Ratcliffe@uwe.ac.uk

[3] Department of Biochemistry, Molecular Biology and Biotechnology, Faculty of Chemistry, Wroclaw University of Science and Technology, 27 Wybrzeże Stanisława Wyspianskiego, 50-370 Wroclaw, Poland; piotr.mlynarz@pwr.edu.pl (P.M.); karolina.mielko@pwr.edu.pl (K.M.); justyna.mielnik@pwr.edu.pl (J.M.)

[4] Indigo Science Ltd., Bristol BS7 9JS, UK; peter.jones@indigoscience.com

[5] Bristol Urological Institute, Southmead Hospital, Bristol BS10 5BN, UK; rajpersad@bristolurology.com

* Correspondence: n.drabinska@pan.olsztyn.pl (N.D.); Ben.DeLacyCostello@uwe.ac.uk (B.d.L.C.); Tel.: +48-89-52-34-641 (N.D.); +44-117-328-2461 (B.d.L.C.)

Academic Editors: Bartolo Gabriele and Alessandra Gentili

Received: 12 June 2020; Accepted: 10 August 2020; Published: 11 August 2020

Abstract: Urinary volatile compounds (VCs) have been recently assessed for disease diagnoses. They belong to very diverse chemical classes, and they are characterized by different volatilities, polarities and concentrations, complicating their analysis via a single analytical procedure. There remains a need for better, lower-cost methods for VC biomarker discovery. Thus, there is a strong need for alternative methods, enabling the detection of a broader range of VCs. Therefore, the main aim of this study was to optimize a simple and reliable liquid–liquid extraction (LLE) procedure for the analysis of VCs in urine using gas chromatography-mass spectrometry (GC-MS), in order to obtain the maximum number of responses. Extraction parameters such as pH, type of solvent and ionic strength were optimized. Moreover, the same extracts were analyzed using Proton Nuclear Magnetic Resonance Spectroscopy (^{1}H-NMR), to evaluate the applicability of a single urine extraction for multiplatform purposes. After the evaluation of experimental conditions, an LLE protocol using 2 mL of urine in the presence of 2 mL of 1 M sulfuric acid and sodium sulphate extracted with dichloromethane was found to be optimal. The optimized method was validated with the external standards and was found to be precise and linear, and allowed for detection of >400 peaks in a single run present in at least 50% of six samples—considerably more than the number of peaks detected by solid-phase microextracton fiber pre-concentration-GC-MS (328 ± 6 vs. 234 ± 4). ^{1}H-NMR spectroscopy of the polar and non-polar extracts extended the range to >40 more (mainly low volatility compounds) metabolites (non-destructively), the majority of which were different from GC-MS. The more peaks detectable, the greater the opportunity of assessing a fingerprint of several compounds to aid biomarker discovery. In summary, we have successfully demonstrated the potential of LLE as a cheap and simple alternative for the analysis of VCs in urine, and for the first time the applicability of a single urine solvent extraction procedure for detecting a wide range of analytes using both GC-MS and ^{1}H-NMR analysis to enhance putative biomarker detection. The proposed method will simplify the transport between laboratories and storage of samples, as compared to intact urine samples.

Keywords: liquid–liquid extraction; volatile compounds; urine; method optimization; GC-MS; ^{1}H-NMR

1. Introduction

Modern metabolomics is now a high-throughput approach for the monitoring of metabolites in biological tissue or fluid in a defined time point. The profile of metabolites may vary during pathological states, hormonal changes, exposure to environmental pollutants, diet, etc., and the changes can be determined in different biological specimens, such as urine, saliva, blood, skin, feces, breath and sweat [1,2]. Urine, because of the non-invasive methods of collection and the richness of metabolites, is a commonly used fluid in metabolite profiling [3,4], potentially giving a large amount of information about the metabolic state of the body. Urine is frequently analyzed using Liquid Chromatography-Mass Spectrometry (LC-MS), Gas Chromatography-Mass Spectrometry (GC-MS) and Nuclear Magnetic Resonance Spectroscopy (NMR) [5–7]. A specific class of metabolomics focused on the profile of volatile compounds (VCs) is termed volatolomics, the applications of which for diagnostic purposes is growing [8–15]. VCs are secreted by cells of the human body, as a result of their metabolism. The changes in the profile of VCs in biological fluids, dependent on the metabolic changes, may reflect the presence of disease. Many studies have suggested that the profile of VCs in urine change in cancer [8–10], nephrological conditions [15], oxidative stress [11], gastrointestinal diseases [12–14] and other disease states.

VCs in urine belong to very diverse chemical classes, such as aldehydes, ketones, organic short chain acids, alcohols, sulfur compounds, etc. They are characterized by different volatilities, polarities and concentrations. These facts complicate the optimization of conditions for VC profiling in a single analytical procedure. To date, many analytical approaches have been used for the analysis of urinary VCs, such as sensor systems [16], Field Asymmetric Ion Mobility Spectrometry (FAIMS) [12], as well as hyphenated techniques based on mass spectrometry, such as High-Pressure Photon Ionization Time-of-Flight-Mass Spectrometry (HPPI-ToF-MS) [17], Proton Transfer Reaction-Mass Spectrometry (PTR-MS) [18], Selected Ion Flow Tube Mass Spectrometry (SIFT-MS) [19], and finally the most ubiquitous technique, GC-MS [10]. The latter is still considered the gold standard in VC analysis.

Direct headspace methods of urine sampling suffer from low sensitivity, which hinders their diagnostic potential for disease diagnoses. An analysis of VCs with GC-MS typically requires a sample preparation step, particularly pre-concentration of analytes. These sample preparation methods for VC analysis in urine are comprised mostly of adsorption techniques, such as solid-phase microextraction (SPME) [8,10,20–22] and thermal desorption with sorbent tubes [23]. These are selective techniques, which are also adversely affected by high water concentrations, limiting the number of compounds detected and reducing their usefulness for non-targeted profiling. Therefore, there is a strong need for alternative methods that enable the detection of a broader range of compounds.

VCs can be extracted from the matrix using more conventional approaches. For example, liquid–liquid extraction (LLE) is a traditional and favored extraction technique in analytical chemistry, because of its simplicity and lack of complicated equipment. LLE is used in a wide range of applications and for the extraction of varied classes of compounds in food chemistry [24,25], environmental analysis [26], drug analysis [27], etc. Even though LLE is so commonly used in analytical chemistry and industry, its application for the extraction of VCs from biological fluids is not as frequent as would have been expected [28]. The extraction of individual VCs has been proposed by Seyler et al. [29], who optimized the method for quantification of six nirosamines in urine. LLE potentially permits relatively high concentrations of a diverse range of VCs to be attained, providing a low-cost simple alternative to more expensive complex extraction technologies, while extracting VC and semi-VCs with a range of polarities.

From the metabolomic viewpoint, it is also important to obtain extracts suitable for analysis utilizing a more diverse range of approaches. The application of different analytical techniques combining GC-MS and ^{1}H-NMR gives the possibility of determining a broader set of metabolites, with varying volatility, from VCs to semi-VCs, including non-polar compounds. The main aim of this study was to optimize a simple and reliable LLE procedure for VCs and semi-VCs analysis in urine using GC-MS. To do so, extraction parameters, such as pH, type of solvent and ionic strength, were optimized by considering the maximum number of deconvoluted peaks detected. Next, the optimized method was validated in term of precision, linearity and sensitivity. Moreover, the same extracts were analysed using ^{1}H-NMR, to evaluate for the first time the applicability of a single urine extract for multiplatform purposes, which should increase the prospects for linking urine metabolites to a particular disease. The particular purpose of the study is to show the enhanced number of (uncharacterized) compounds found using LLE, relative to other methods.

2. Results and Discussion

2.1. Optimization of the Extraction Parameters

In the present study, the type of solvent, acid molarity and ionic strength were selected and evaluated to achieve the optimal condition of LLE, based on the maximum number of GC-MS peaks detected (Table 1). The solvents, acidic pH and amount of salt were preselected based on data from the literature [28–30].

Table 1. Comparison of the factors affecting the extraction efficiency of urine VCs using GC-MS analyses. Values in bold were selected as the most efficient and were used for the optimization of the next parameters.

Factor Analyzed	Number of Peaks Detected	Constant Conditions
Type of solvent (4 mL added)		
DCM *	**205.0 ± 46.1 [a,**]**	
Chloroform	121.0 ± 47.6 [b]	1 M acid, salt addition
Diethyl ether	20.7 ± 4.0 [c]	
Acid molarity (2 mL added)		
0.01 M	137.0 ± 2.7 [b]	
0.1 M	120.0 ± 21.1 [b]	solvent: DCM, salt addition
1 M	**205.0 ± 46.1 [a]**	
Ionic strength		
Salt addition (0.2 g)	**205.0 ± 46.1 [a]**	solvent: DCM, 1 M acid
No salt	156.0 ± 44.7 [a]	

(*) DCM—dichloromethane; (**) Values are a mean number of peaks of three replicates ± SD. Different letters (a, b, c) in a column for each parameter represents significantly different ($p < 0.05$) values (Fisher's Least Significant Difference (LSD), ANOVA) for solvent type and acid molarity and Student t-test for ionic strength.

Dichloromethane (DCM) was found to be the most efficient solvent for VC extraction from urine samples, followed by chloroform (Table 1 and Supplementary Figure S1). DCM is immiscible with water and can dissolve a wide range of organic compounds, hence its extensive use for LLE [31,32]. Diethyl ether is a commonly used solvent for extracting organic compounds from aqueous solutions, however in this study on urine, it was found to be the least effective solvent for LLE, resulting in the detection of only circa 20 peaks. In contrast, Zlatkis et al. [28] noted the presence of 300 compounds in ether extracts (comparison to the efficiency of other solvents was not analyzed), out of which 40 have been identified. However, in their study, a very large amount of urine, 450 mL, was used for extraction with 80 mL of diethyl ether, as compared to 2 mL of urine used in our study. Use of 450 mL urine creates storage issues, and many patients could not produce such an amount. The use of 80 mL of ether for extraction requires more processing time, i.e., more drying agent, evaporation times, and it is undesirable from the health and safety perspective and impractical for certain processing steps such as centrifugation. 2 mL is a realistic volume of urine from a patient sample, which can be collected clinically, therefore the present study was optimized using that volume. In comparison to our methodology, the literature describes VC extraction using absorbents, and smaller numbers (75 and 147 peaks identified, respectively) of VCs were reported, even with larger amounts of urine (using single quadrupole GC-MS) [22,30].

In the present study, the highest molarity of acid (1 M) with pH value near 0 was found to be significantly more efficient for LLE (Table 1). There was no significant difference between 0.1 M (pH value: ~1.5) and 0.01 M (pH value: ~2.5) acid addition. Acidic conditions were previously reported to be more suitable for VC analysis as compared to basic and neutral pH when using SPME [4,14,30]. This is related to the chemical properties of the compounds present in urine. Acidic pH increases the number of compounds in the non-conjugated form [4]. In our study, many of the VCs detected in urine samples contained the carboxylic acid group, therefore the acidic pH facilitates their extraction [21].

Another parameter important to achieving a good extraction is the presence of salt. The results of the experiments carried out with and without salt present are presented in Table 1. The presence of salt changes the nature of the molecular interactions between compounds, causing more ionic activity, and consequently affecting the activity coefficient of metabolites. It was found to have great importance in the SPME method's development, where the presence of salt facilitates the transfer of VCs from the matrix to the headspace [33]. For LLE, salt addition may alter the solubility of certain compounds in the matrix, making them more likely to transfer to the solvent. In the present study, as in LLE, the transfer of VCs to the headspace is not needed, and we did not observe a statistically significant difference in the number of peaks between the presence and absence of salt. However, taking into consideration the slightly increased number of peaks detected (even though not significant) and their size (a summarized peak area of 5.49×10^5 vs. 1.09×10^6 for no salt and salt addition, respectively), we decided to carry out the experiments with the presence of sodium sulphate. It is worth underlining that the urine contains salt, the concentration of which may vary from sample to sample, affecting the results. Therefore, it is important to standardize the urine somehow, e.g., by the analysis of the osmolality of urine samples before metabolomics analyses. In our method, the saturation of urine with salt was achieved, minimizing the differences between samples.

After the evaluation of experimental conditions, the best conditions for the LLE extraction of VCs from urine samples were as follows: 2 mL of 1 M sulfuric acid and 2 mL of urine was added to a 10 mL glass vial containing 0.2 g of sodium sulphate. The vial was vortexed till complete salt dissolution. Then, 4 mL of DCM was added and mixed again for 1 min. After that, the vial was centrifuged for 1 min at 3500 rpm for emulsion separation, and the DCM layer was collected and dried with anhydrous sodium sulphate. 3 mL of the dried DCM extract was quantitatively transferred to a new 5 mL glass vial for evaporation. The last approximately 100 µL of extract was then transferred to a GC vial, and the previously used 5 mL vial was rinsed three times with approximately 100 µL of fresh DCM. After the evaporation to dryness in the heating dry block at 40 °C, the residue was reconstituted in 10 µL of DCM, and 2 µL of the extract was injected into the GC injector port. An example of the chromatogram obtained using the optimized method is presented in Figure 1 (red chromatogram).

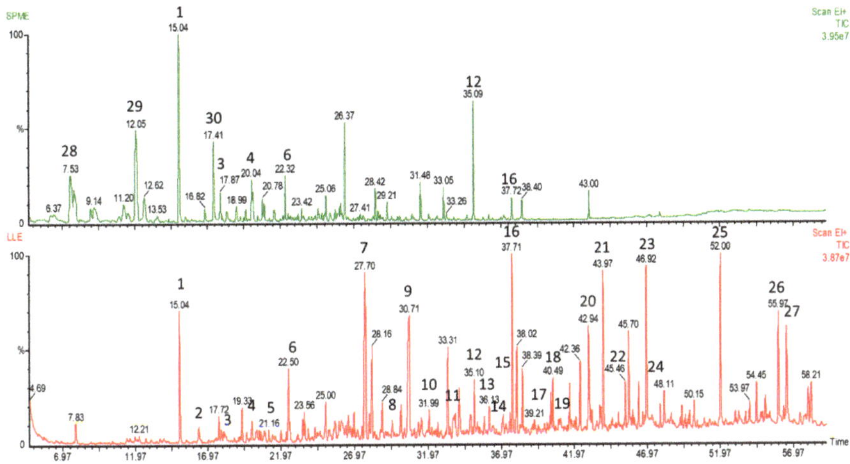

Figure 1. The comparison of total ion mass chromatograms of volatile compounds (VCs) extracted from urine sample using solid-phase microextraction (green chromatogram) and the optimized method (red chromatogram). The tentatively identified compounds: (**1**) Allyl isothiocyanate; (**2**) Tetradecane; (**3**) Acetic acid; (**4**) 2-Butyl-1-octanol; (**5**) Diethyl sulfoxide; (**6**) hexadecane; (**7**) tetrahydro-6-methyl-2*H*-Pyran-2-one; (**8**) 2-Methoxy-phenol; (**9**) Dimethyl sulfone; (**10**) Heptanoic acid; (**11**) 2-Methyl-octanoic acid; (**12**) *p*-Cresol; (**13**) Erucin; (**14**) Nonanoic acid; (**15**) Octenoic acid; (**16**) 2-Methoxy-4-vinylphenol; (**17**) *n*-Decanoic acid; (**18**) Divinyl sulphide; (**19**) 1-Hexadecanol; (**20**) Benzoic acid; (**21**) 7-Methylindole (**22**) Benzeneacetic acid; (**23**) Apocynin; (**24**) Benzamide; (**25**) *n*-Hexadecanoic acid; (**26**) Octadecanoic acid; (**27**) Caffeine; (**28**) 4-heptanone; (**29**) *p*-Cymene; (**30**) 4-Ethenyl-1,2-dimethylbenzene.

2.2. Analytical Performance

The demonstration that the method is of high quality is a crucial step in method development [34]. The validation of the method for non-targeted metabolomics comprises in most cases solely the analysis of precision [2]. Therefore, to check the reliability of the methodology, an external standard method was applied. Seven commercial standards, representing acids and aldehydes with different chain lengths, were selected, and the standard mixture was used for precision, sensitivity and linearity evaluation. The results are presented in Table 2.

Table 2. Validation parameters calculated for a mixture of commercial standards. Compounds are ordered with respect to their increasing retention times.

Retention Time	Compound	Intraday Precision [RSD*%]	Interday Precision [RSD%]	Linear Range [μmol/L]	R^2	LOD** [μmol/L]	LOQ*** [μmol/L]
9.78	heptanal	9	9	8.857–70.853	0.984	4.4	14.8
13.11	octanal	9	19	8.004–64.035	0.986	4.0	13.3
16.15	nonanal	13	17	7.268–58.142	0.976	3.6	12.1
19.84	decanal	6	16	6.642–53.137	0.983	3.3	11.1
29.70	hexanoic acid	15	15	9.997–79.977	0.985	10.0	33.3
32.51	heptanoic acid	13	26	8.814–70.515	0.988	8.8	29.4
35.05	octanoic acid	11	15	3.944–63.102	0.990	3.9	13.1

(*) Relative standard deviation; (**) LOD - Limit of detection; (***) LOQ - Limit of quantification.

The intraday and interday precision ranged from 6.1% to 14.9%, and from 9.1% to 26.4%, respectively. Naz et al. [34] recommended that the Relative Standard Deviation (RSD) values should not exceed 30% in metabolomics studies, therefore the results obtained in our study can be considered satisfactory, especially for the manual injection applied in our study. The correlation coefficients (R^2) of the calibration curves for all selected standards ranged between 0.98 and 0.99, indicating that the method has a highly linear response for the concentration ranges presented in Table 2. The limit of detection (LOD) and limit of quantification (LOQ) values were found to be less than 10 and less than 33.3 μmol/L, respectively, for all the analyzed standards. However, the sensitivity of the method was not the priority of this study.

2.3. Method Application—GC-MS

The applicability of the optimized method was evaluated based on the analysis of six urine samples collected from healthy individuals. The comparison of the samples was not the aim of this study, therefore the urine samples were not normalized. However, for the application of the method in metabolomics studies in the future, the normalization will be necessary [35]. The analysis of urine with more analytical techniques can increase the number of metabolites of different physicochemical properties, and consequently may provide more information about the metabolic state of the body [36]. Therefore, in the present study, we decided to conduct the analysis of the samples using both GC-MS and ^{1}H-NMR spectroscopy, with the aim of detecting greater numbers of peaks corresponding to individual compounds from a single extract.

The GC-MS analysis resulted in a total number of 400 individual deconvoluted peaks representing different compounds, detected in chromatographs, present in at least 50% of samples. The number of VCs detected in each sample is presented in Table 3. A comparison between the optimized method and SPME method is presented in Figure 1. The extraction of VCs from another individual run using both LLE and SPME of a headspace above the urine from the same individual showed a greater number of deconvoluted peaks detected using our optimized method (328 ± 5.66 vs. 234 ± 4.24, for LLE and SPME, respectively). As can be seen in Figure 1, the optimized LLE method results in more peaks at longer retention times, corresponding to heavier molecules, when compared to the SPME method, which is more efficient for small mass VCs. A comparison of the number of peaks detected by our method and those in the literature using SPME fiber pre-concentration technology with urine shows there is a significant improvement using LLE. The extraction using SPME with Divinylbenzene/Carboxen/Polydimethylsiloxane (DVB/CAR/PDMS) fiber and acidic conditions resulted in the detection of 75 VCs in urine in a single run [30]. In another study, a total number of 147 VCs was detected in urine using SPME with CAR/PDMS fiber; however, this number represents a sum of VCs obtained in acidic, basified and neutral pH samples [22]. On the other hand, in the study of Rocha et al. [37], GCxGC-TOFMS analysis allowed for the detection of approximately 700 compounds, of which 294 were tentatively identified, however it resulted from using a very high-cost and complex chromatographic system. Previous attempts at the application of LLE with diethyl ether using 450 mL of urine for VCs analysis resulted in the detection of 300 VCs, 40 of which were identified [28]. The results obtained proved that the optimized method is applicable for the GC-MS profiling of VCs in the urine.

Table 3. The number of deconvoluted peaks detected in urine samples from six apparently healthy individuals by GC-MS method.

Urine Sample	Number of Compounds Detected Using GC-MS
1	336
2	326
3	330
4	337
5	338
6	272

2.4. Method Application—^{1}H-NMR

^{1}H-NMR analysis was performed in both extract parts, because the information obtained is not only complementary but also supportive to some of the metabolites from GC-MS. The ^{1}H-NMR spectra of the polar phase revealed approximately 60 proton signals originating from organic compounds within the urine. The signals are mostly located in the three areas of chemical shifts (δ): (a) 0.77–1.94 ppm characteristic for CH_3 and CH_2 groups; (b) 1.95–4.33 ppm for the aliphatic proton signal of different CH and CH_2 groups, and (c) 6.8–8.07 ppm, distinctly visible peaks which in principle can exhibit protons originating from aromatic compounds. However, as is seen in the spectra (Figure 2), the signals from higher mass molecules are not filtered off completely by the CPMG pulse sequence. The signals in the polar fractions were assigned to the Chenomx references, where over 100 signals were detected, and among them 42 metabolites were found to be present in the urine aqueous phase (Supplementary Data Table S1). Interestingly, there were significant differences between the compounds tentatively identified by both GCMS and ^{1}H-NMR, which proves that simultaneous analysis of the same extracts with these two methods is complementary.

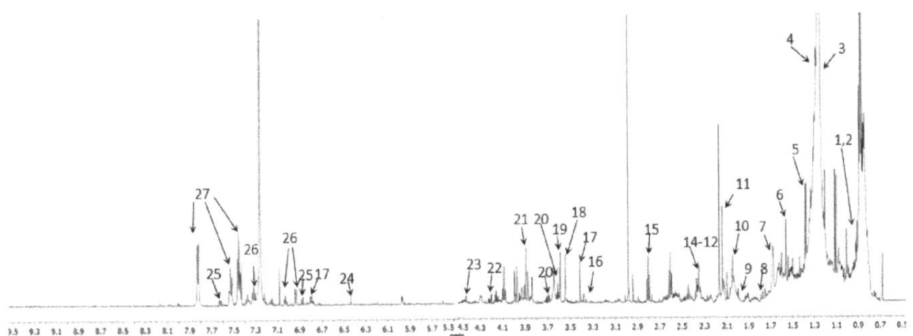

Figure 2. ^{1}H-NMR 600 MHz Carr–Purcell–Meiboom–Gill (CPMG) spectra of urine obtained from non-polar phase sample ($CDCl_3$, T = 300 K); **1**—Cholesterol ester, **2**—Terminal -CH_3, **3**—Acyl chain C4-C7, **4**—(CH_2)$_n$-, **5**—2-hydroxyisobutyric acid, **6**—Saturated C3 acyl chain, **7**—-CO-CH_2-CH_2, **8**—Glycocholic acid, **9**—Acetamide, **10**—Allylic methylene -C=C-CH_2, **11**—O-Acetylcarnitine, **12**—3-hydrixyisovaleric acid, **13**—Pyruvic acid, **14**—Acyl chain C2, **15**—Succinylacetone, **16**—Theophylline, **17**—3,4 Dihydroxybenzeneacetate, **18**—Phenylacetate, **19**—Glycine, **20**—Glycerol, **21**—Glycolic acid, **22**—Sn1+Sn3 -CH_2-O-CO-R, **23**—1,3 dihydroxyacetone, **24**—Fumaric acid, **25**—Xanthurenic acid, **26**—Phenol derivative, **27**—Benzoic acid.

The DCM (LLE) can extract metabolites with a range of polarities (particularly relatively non polar compounds), however it would not be expected to be that efficient in extracting polar compounds, such as ionic compounds [38]. The literature data related to non-polar compound analysis from biological fluids and tissues is very limited, usually including on the ^{1}H NMR spectra a general description of the groups of compounds [39]. In another report, the urine extracts were dissolved in different deuterated solvents (MeOD, DMSO, DMF, MeCN, Acetone, $CDCl_3$ and DCM), with subsequent monitoring of the levels of five metabolites: hippurate, creatinine, lactate, histidine and alanine [38]. Metabolites showed signal variation in the ppm scale depending on the solvent used. However, among the selected metabolites, only hippurate was found to be resolvable in $CDCl_3$.

The study did not avoid some limitations. First, the method was optimized only for GC-MS, not both GC-MS and ^{1}H-NMR. However, the main aim of the study was the optimization of LLE extraction for VC analysis, and the ^{1}H-NMR part was the additional attempt, performed to check if the same extract can be used on multiple platforms. A second limitation is the small number of samples used for applicability testing. It is related to the character of the study, which is method development. The real urine sample analyses were included only to prove that the method is suitable for real sample analysis. Finally, the study does not contain the identification of all the compounds (however the main chemical groups were tentatively identified and mentioned). However, the authors decided not to undertake this exhaustive analysis because the main goal of this study was to detect the highest number of peaks corresponding to individual compounds that could be potential biomarkers.

The identification would be scientifically interesting when defined peaks are identified as biomarkers linked to specific diseases. This will be the target for future studies using the developed LLE method.

3. Materials and Methods

3.1. Chemicals and Preparation of Calibration Solutions

The following analytical standard grade commercial chemicals were used: heptanal, octanal, hexanoic acid from Aldrich Chemicals (Milwaukee, WI, USA), and nonanal, decanal, heptanoic acid and octanoic acid from Acros Organics (Geel, Belgium). Sodium sulphate, DCM, chloroform and diethyl ether were purchased from Fisher Scientific (Hampton, NH, USA) and sulfuric acid was purchased from Aldrich Chemicals (Milwaukee, WI, USA),

The stock solution incorporating standards was prepared by dissolving 1 µL of each standard in MilliQ water (Millipore, Bedford, MA, USA) in a 100 mL measuring flask. The working solutions were prepared by diluting the standard stock solution, in the range 3 to 80 µmol/L.

3.2. Urine Samples

For method development, approx. 50 mL of morning urine sample was collected from one apparently healthy female volunteer, who had an ad hoc omnivore diet. The sample was immediately divided into 2 mL aliquots and stored in the fridge at 4 °C until the analysis which was conducted the same day.

For comparison of the LLE to the SPME method, a urine sample from one apparently healthy male volunteer was collected. The sample was treated as described above.

For method application, urine samples from six healthy individuals have been used. Samples were obtained from Liverpool Bio-Innovation Hub (LBIH) Biobank. The LBIH Biobank has Research Tissue Bank status and is licensed by the Human Tissue Authority (HTA). The collection and storage of biosamples has been ethically approved by the North West 5 Research Ethics Committee. Samples were stored in 2 mL aliquots at −80 °C after collection, and defrosted in the fridge at 4 °C before analysis.

All procedures involving human participants were performed in accordance with the 1964 Helsinki Declaration and its later amendments or comparable ethical standards.

3.3. Extraction Optimization

The highest number of individually defined and chromatographically resolved peaks corresponding to VCs across the entire GC chromatogram was used as a measure of the best extraction performance conditions during method development. The deconvoluted peaks, subtracted from the blank samples, were counted.

Type of solvent (DCM, chloroform, diethyl ether), ionic strength (0 or 0.4 g of anhydrous sodium sulphate) and the molarity of sulfuric acid (0.01, 0.1 and 1 M) were investigated for their effect on extraction efficiency as presented in Figure 3. The extractions were performed by adding 2 mL of acid solution to 2 mL of urine and 0.2 g of anhydrous sodium sulphate, followed by vortexing until the salt dissolved. Next, 4 mL of organic solvent was added and vortexed for 1 min. The layers were separated by centrifugation for 1 min at 3500 rpm using a Beckman Coulter Aliegra X-22R Centrifuge (Brea, CA, USA). The solvent layer was dried with anhydrous sodium sulphate, collected and evaporated to dryness in a dry heating block at 40 °C for approximately 30 min. The dry residue was stored in a freezer at −20 °C until analysis and then dissolved in 10 µL of organic solvent used for the extraction. A 2 µL aliquot of the extract was immediately injected manually into the inlet of a gas chromatograph.

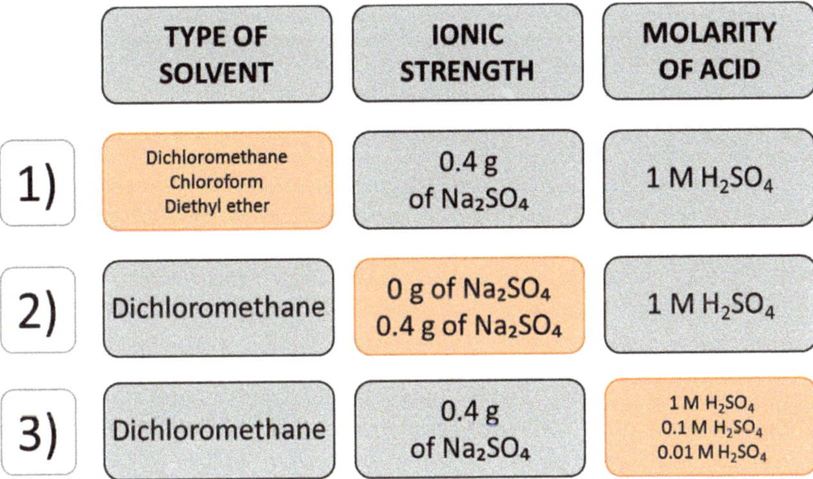

Figure 3. Flow diagram of the one-factor-at-a-time design of extraction optimization. The orange chart refers to conditions which were compared at a time.

3.4. Analytical Performance

The optimized LLE procedure was validated in terms of precision, linearity and sensitivity based on the peak areas of heptanal, octanal, nonanal, decanal, hexanoic acid, heptanoic acid and octanoic acid used as external standards. The standard solutions were extracted using the optimized method. Precision was calculated and expressed as the RSD of six replicates of an aqueous stock standard solution with a concentration of 0.01 μL/mL (interday precision), followed by repeating the interday precision the next day (intraday precision). The linearity was determined by evaluation of the regression curves of the standard peak areas versus the concentration and expressed as the squared determination coefficient R^2. The linear ranges were obtained by creating the calibration curves using six sequential dilutions of the working standard solutions. Sensitivity expressed as LOD and LOQ was calculated based on the signal-to-noise ratio (S/N). LOD was defined as the lowest concentration with a S/N ratio of 3, whereas the LOQ used a S/N ratio of 10.

3.5. SPME

The SPME method was performed according to the method described by Silva et al. [40] with slight modifications. Briefly, 2 mL of urine, 0.2 g of sodium sulphate and 2 mL of 1 M sulfuric acid were added to 20 mL headspace vial. The extraction was conducted manually by inserting the CAR/PDMS fiber into the sample vial and exposing it for 60 min at 50 °C. After the extractions, the fiber was introduced into the GC inlet and the VOCs were thermally desorbed for 10 min at 245 °C in a splitless mode.

3.6. Gas Chromatography-Mass Spectrometry

The analysis of urinary VCs was performed using a Hewlett Packard HP5890 series II GC coupled to an HP5971, single quadrupole Mass Selective Detector (MSD) with an HP Chemstation (Hewlett Packard, Bracknell, UK). Chromatographic separation was performed on a Stablewax DA capillary column, 30 m × 0.25 mm × 0.25 μm (Restek, Benner Circle, Bellefonte, PA, USA). The carrier gas was 99.9995% pure helium (AirProducts, Crewe, UK) with a constant flow rate of 1.2 mL/min. The GC was operated under the following conditions: temperature program, 40 °C with 2 min of hold time, ramping at 4 °C/min to the final temperature of 245 °C and then held at 245 °C for 6 min, giving a total run time of 59.25 min. The injection port was operated in splitless mode (purge off 0.7 min) at 245 °C. After a solvent delay of 4.6 min, mass spectra were acquired in full scan mode with a scan range of *m/z* 35–450 used for data acquisition. The operating conditions for the MS system were as follows: electron ionization mode at an energy of 70 eV; transfer line and ion source temperatures were 280 °C and 180 °C, respectively. Total ion chromatograms (TIC) were analyzed and peaks were integrated automatically using the Turbomass software (PerkinElmer, Inc., Waltham, MA, USA) with an initial detection threshold of 10.0. The analyst reviewed the automated integration and made adjustments and manually integrated the peaks, if it was necessary, keeping the proper judgement. Moreover, the chromatograms were analyzed using the free Automated Mass Spectral Deconvolution and Identification System (AMDIS) software by the National Institute of Standards and Technology (NIST, Gaithersburg, MD, USA). The deconvoluted peaks were tentatively identified, where possible, by comparison of the mass spectra with the NIST/EPA/NIH Mass Spectral Library (version 2.2, 2014, Gaithersburg, MD, USA). For identification, only the components with a match factor >80% were listed.

3.7. ^{1}H-NMR

The urine extracts (organic and polar phases) were dissolved in 0.55 mL deuterochloroform (with an internal standard, TMS) and ^{1}H-NMR spectra of the urine samples were recorded using an Avance II spectrometer (Brucker, Billerica, MA, USA) that was operating at a proton frequency of 600.58 MHz. The ^{1}H-NMR spectra were collected using standard one-dimensional Carr–Purcell–Meiboom–Gill (CPMG) pulse sequence with water presaturation, at 300 K temperature. For each sample, 128–512 consecutive scans (NS) with a 400 µs spin-echo delay were collected; there were 80 loops for the T2 filter, with a 3.5 s relaxation delay and a 2.73 s acquisition time, a time-domain of 64k, and a spectral width of 20.02 ppm. The spectra were processed with a line broadening of 0.3 Hz and were manually phased and baseline corrected using Topspin 1.3 software (Brucker, Billerica, MA, USA). The water spectrum region was removed from the analysis.

3.8. Statistical Analyses

All the analyses were performed in triplicate. Data are presented as a mean ± standard deviation (SD). The data were compared with the one-way analysis of variance (ANOVA) test or Student *t*-test, as appropriate using the Statistica 10.0 software (StatSoft, Tulsa, OK, USA). Fisher's Least Significant Difference (LSD) test was applied to assess significant differences ($p < 0.05$) between variables.

4. Conclusions

In conclusion, a simple, LLE-based sample preparation protocol for the metabolic profiling of urine samples was optimized and validated. The method was reliable and does not require specific instrumentation for sample preparation. We successfully demonstrated for the first time the applicability of single urine solvent extracts for both GC-MS as well as ^{1}H-NMR analysis, for volatile and semi-volatile compound analysis. The possibility of obtaining a wider range of the metabolites can give more information about the health of the individuals, and may facilitate the identification of biomarkers linked to different diseases. The possibility of using both GC-MS and ^{1}H-NMR platforms allowed the detection of metabolites with different volatilities from a single sample. This may be useful when carrying out future studies aimed at identifying metabolites linked to disease. Often, previous studies have had limited scope and concentrate on a limited range of methods. Using these methods, the detection capability could be increased many-fold by simply increasing the extraction volumes used; for instance, 40 mL DCM and 20 mL urine. The benefit of the proposed method is the possibility of storage of extracts in smaller vials, as compared to intact urine samples, and the possibility of transport to different labs for analysis in a dry way, without the special freezing conditions.

Future work will involve the application of the optimized method to look for diagnostic markers of urological cancers. In summary, LLE has been shown to be superior to the SPME fiber pre-concentration methodology for urine analyses, permitting semi-volatiles to be detected and providing a promising alternative methodology to the use of very costly absorptive technologies.

Supplementary Materials: The following are available online, Supplementary Table S1. Tentative assignment of metabolites found to be presented in urine aqueous phase (^{1}H-NMR analyses). Supplementary Figure S1. The comparison of the extraction using dichloromethane (DCM), chloroform and diethyl ether. Supplementary Figure S2 ^{1}H-NMR 600MHz CPMG spectra of urine obtained from polar phase sample (D2O, T=300K). 1—2-hydroxyisovaleric acid, 2—3-methyl-2-oxovaleric acid, 3—sovaleric acid, 4—Valine, 5—3-hydroxyisobutyric acid, 6—Methylsuccinic acid, 7—Fucose, 8—3-hydroxyisovaleric acid, 9—2-hydroxyisobutyric acid, 10—2-Phenylpropionic acid, 11—Alanine, 12—2-aminoadipic acid, 13—Acetate, 14—Acetamide, 15—Acetone, 16—Acetoacetic acid, 17—Succinic acid, 18—Citrate, 19—Saccrosine, 20—Dimethylamine, 21—Trimethylamine, 22—*N,N*-dimethylgycine, 23—N-methylhydantoin, 24—Creatine, 25—Creatinine, 26—Choline, 27—Methanol, 28—Glycine, 29—Glycolate, 30—π-methylhistidine, 31—Trigoneline, 32—Fumaric acid, 33—trans-Aconitic acid, 34—Xanthuretic acid, 35—Carnosine, 36—3-Indoxylsulfate, 37—Imidazole, 38—Hippurate, 39—Oxypurinol, 40—Adenine, 41—Formic acid, 42—1-methylnicotinamide.

Author Contributions: N.D., P.J. and N.M.R. conceived and designed the research; B.d.L.C. secured funding; N.D., P.M., J.M. and K.M. conducted analytical experiments; N.D. analyzed GC-MS data; P.M., J.M. and K.M. analysed ^{1}H NMR data; N.D., N.M.R. and P.M. interpreted results; N.D. wrote the draft of the manuscript; P.M., B.d.L.C., P.J., R.P. and N.M.R. contributed in the further writing of the manuscript. All authors read and approved the final version of the manuscript.

Funding: The study was supported by Project No. GA2543 from the Above and Beyond Charity in Bristol.

Acknowledgments: Natalia Drabińska is a START 2019 scholarship holder financed by the Foundation for Polish Science (FNP).

Conflicts of Interest: The authors declare no conflict of interest.

References

1. De Lacy Costello, B.; Amann, A.; Al-Kateb, H.; Flynn, C.; Filipiak, W.; Khalid, T.; Osborne, D.; Ratcliffe, N.M. A review of the volatiles from the healthy human body. *J. Breath Res.* **2014**, *8*, 014001. [CrossRef] [PubMed]

2. Živković Semren, T.; Brčić Karačonji, I.; Safner, T.; Brajenović, N.; Tariba Lovaković, B.; Pizent, A. Gas chromatographic-mass spectrometric analysis of urinary volatile organic metabolites: Optimization of the HS-SPME procedure and sample storage conditions. *Talanta* **2018**, *176*, 537–543. [CrossRef] [PubMed]

3. Bouatra, S.; Aziat, F.; Mandal, R.; Guo, A.C.; Wilson, M.R.; Knox, C.; Bjorndahl, T.C.; Krishnamurthy, R.; Saleem, F.; Liu, P.; et al. The Human Urine Metabolome. *PLoS ONE* **2013**, *8*, e73076. [CrossRef] [PubMed]

4. Monteiro, M.; Carvalho, M.; Henrique, R.; Jerónimo, C.; Moreira, N.; De Lourdes Bastos, M.; De Pinho, P.G. Analysis of volatile human urinary metabolome by solid-phase microextraction in combination with gas chromatography-mass spectrometry for biomarker discovery: Application in a pilot study to discriminate patients with renal cell carcinoma. *Eur. J. Cancer* **2014**, *50*, 1993–2002. [CrossRef] [PubMed]

5. Jain, A.; Li, X.H.; Chen, W.N. An untargeted fecal and urine metabolomics analysis of the interplay between the gut microbiome, diet and human metabolism in Indian and Chinese adults. *Sci. Rep.* **2019**, *9*, 9191. [CrossRef]

6. Zheng, L.; Wang, J.; Gao, W.; Hu, C.; Wang, S.; Rong, R.; Guo, Y.; Zhu, T.; Zhu, D. GC/MS-based urine metabolomics analysis of renal allograft recipients with acute rejection. *J. Transl. Med.* **2018**, *16*, 202. [CrossRef]

7. Zabek, A.; Paslawski, R.; Paslawska, U.; Wojtowicz, W.; Drozdz, K.; Polakof, S.; Podhorska, M.; Dziegiel, P.; Mlynarz, P.; Szuba, A. The influence of different diets on metabolism and atherosclerosis processes—A porcine model: Blood serum, urine and tissues 1H-NMR metabolomics targeted analysis. *PLoS ONE* **2017**, *12*, e0184798. [CrossRef]

8. Taunk, K.; Taware, R.; More, T.H.; Porto-Figueira, P.; Pereira, J.A.M.; Mohapatra, R.; Soneji, D.; Câmara, J.S.; Nagarajaram, H.A.; Rapole, S. A non-invasive approach to explore the discriminatory potential of the urinary volatilome of invasive ductal carcinoma of the breast. *RSC Adv.* **2018**, *8*, 25040–25050. [CrossRef]

9. Taware, R.; Taunk, K.; Pereira, J.A.M.; Dhakne, R.; Kannan, N.; Soneji, D.; Câmara, J.S.; Nagarajaram, H.A.; Rapole, S. Investigation of urinary volatomic alterations in head and neck cancer: A non-invasive approach towards diagnosis and prognosis. *Metabolomics* **2017**, *13*, 111. [CrossRef]

10. Khalid, T.; Aggio, R.; White, P.; De Lacy Costello, B.; Persad, R.; Al-Kateb, H.; Jones, P.; Probert, C.S.; Ratcliffe, N. Urinary volatile organic compounds for the detection of prostate cancer. *PLoS ONE* **2015**, *10*, e0143283. [CrossRef]

11. Antón, A.P.; Ferreira, A.M.C.; Pinto, C.G.; Cordero, B.M.; Pavón, J.L.P. Headspace generation coupled to gas chromatography-mass spectrometry for the automated determination and quantification of endogenous compounds in urine. Aldehydes as possible markers of oxidative stress. *J. Chromatogr. A* **2014**, *1367*, 9–15. [CrossRef] [PubMed]

12. Arasaradnam, R.P.; Westenbrink, E.; McFarlane, M.J.; Harbord, R.; Chambers, S.; O'Connell, N.; Bailey, C.; Nwokolo, C.U.; Bardhan, K.D.; Savage, R.; et al. Differentiating coeliac disease from irritable bowel syndrome by urinary volatile organic compound analysis—A pilot study. *PLoS ONE* **2014**, *9*, e107312. [CrossRef] [PubMed]

13. Arasaradnam, R.P.; Ouaret, N.; Thomas, M.G.; Quraishi, N.; Heatherington, E.; Nwokolo, C.U.; Bardhan, K.D.; Covington, J.A. A novel tool for noninvasive diagnosis and tracking of patients with inflammatory bowel disease. *Inflamm. Bowel Dis.* **2013**, *19*, 999–1003. [CrossRef] [PubMed]

14. Drabińska, N.; Azeem, H.A.; Krupa-Kozak, U. A targeted metabolomic protocol for quantitative analysis of volatile organic compounds in urine of children with celiac disease. *RSC Adv.* **2018**, *8*, 36534–36541. [CrossRef]

15. Wang, M.; Xie, R.; Jia, X.; Liu, R. Urinary Volatile Organic Compounds as Potential Biomarkers in Idiopathic Membranous Nephropathy. *Med. Princ. Pract.* **2017**, *26*, 375–380. [CrossRef]

16. Zhu, S.; Corsetti, S.; Wang, Q.; Li, C.; Huang, Z.; Nabi, G. Optical sensory arrays for the detection of urinary bladder cancer-related volatile organic compounds. *J. Biophotonics* **2019**, *12*, c201800165. [CrossRef]

17. Wang, Y.; Hua, L.; Jiang, J.; Xie, Y.; Hou, K.; Li, Q.; Wu, C.; Li, H. High-pressure photon ionization time-of-flight mass spectrometry combined with dynamic purge-injection for rapid analysis of volatile metabolites in urine. *Anal. Chim. Acta* **2018**, *1008*, 74–81. [CrossRef]

18. Zou, X.; Lu, Y.; Xia, L.; Zhang, Y.; Li, A.; Wang, H.; Huang, C.; Shen, C.; Chu, Y. Detection of Volatile Organic Compounds in a Drop of Urine by Ultrasonic Nebulization Extraction Proton Transfer Reaction Mass Spectrometry. *Anal. Chem.* **2018**, *90*, 2210–2215. [CrossRef]

19. Batty, C.A.; Cauchi, M.; Hunter, J.O.; Woolner, J.; Baglin, T.; Turner, C. Differences in microbial metabolites in urine headspace of subjects with Immune Thrombocytopenia (ITP) detected by volatile organic compound (VOC) analysis and metabolomics. *Clin. Chim. Acta* **2016**, *461*, 61–68. [CrossRef]

20. Drabinska, N.; Jarocka-Cyrta, E.; Ratcliffe, N.M.; Krupa-Kozak, U. The profile of urinary headspace volatile organic compounds after 12-week intake of oligofructose-enriched inulin by children and adolescents with celiac disease on a gluten-free diet: Results of a pilot, randomized, placebo-controlled clinical trial. *Molecules* **2019**, *24*, 1341. [CrossRef]

21. Aggio, R.B.M.; Mayor, A.; Coyle, S.; Reade, S.; Khalid, T.; Ratcliffe, N.M.; Probert, C.S.J. Freeze-drying: An alternative method for the analysis of volatile organic compounds in the headspace of urine samples using solid phase micro-extraction coupled to gas chromatography—Mass spectrometry. *Chem. Cent. J.* **2016**, *10*, 9. [CrossRef] [PubMed]

22. Smith, S.; Burden, H.; Persad, R.; Whittington, K.; De Lacy Costello, B.; Ratcliffe, N.M.; Probert, C.S. A comparative study of the analysis of human urine headspace using gas chromatography-mass spectrometry. *J. Breath Res.* **2008**, *2*, 037022. [CrossRef] [PubMed]

23. O'Lenick, C.R.; Pleil, J.D.; Stiegel, M.A.; Sobus, J.R.; Wallace, M.A.G. Detection and analysis of endogenous polar volatile organic compounds (PVOCs) in urine for human exposome research. *Biomarkers* **2018**, *24*, 240–248. [CrossRef] [PubMed]

24. Alves, A.A.R.; Barros, E.B.P.; Rezende, C.M. Chapter 78—Method Development and Optimization of Liquid–Liquid Extraction for the Quantitative Analysis of Volatile Compounds from Brazilian Grape Juices. In *Flavour Science Proceedings from XIII Weurman Flavour Research Symposium*; Ferreira, V., Lopez, R.B.T.-F.S., Eds.; Academic Press: San Diego, CA, USA, 2014; pp. 417–421. ISBN 978-0-12-398549-1.

25. Orak, H.H.; Bahrisefit, I.S.; Sabudak, T. Antioxidant Activity of Extracts of Soursop (Annona muricata L.) Leaves, Fruit Pulps, Peels, and Seeds. *Polish J. Food Nutr. Sci.* **2019**, *69*, 359–366. [CrossRef]

26. Soliman, M.A.; Pedersen, J.A.; Suffet, I.M. Rapid gas chromatography–mass spectrometry screening method for human pharmaceuticals, hormones, antioxidants and plasticizers in water. *J. Chromatogr. A* **2004**, *1029*, 223–237. [CrossRef] [PubMed]

27. Kataoka, H. New trends in sample preparation for clinical and pharmaceutical analysis. *TrAC Trends Anal. Chem.* **2003**, *22*, 232–244. [CrossRef]

28. Zlatkis, A.; Liebich, H.M. Profile of Volatile Metabolites in Human Urine. *Clin. Chem.* **1971**, *17*, 592–594. [CrossRef]

29. Seyler, T.H.; Kim, J.G.; Hodgson, J.A.; Cowan, E.A.; Blount, B.C.; Wang, L. Quantitation of Urinary Volatile Nitrosamines from Exposure to Tobacco Smoke*. *J. Anal. Toxicol.* **2013**, *37*, 195–202. [CrossRef]

30. Cozzolino, R.; De Magistris, L.; Saggese, P.; Stocchero, M.; Martignetti, A.; Di Stasio, M.; Malorni, A.; Marotta, R.; Boscaino, F.; Malorni, L. Use of solid-phase microextraction coupled to gas chromatography-mass spectrometry for determination of urinary volatile organic compounds in autistic children compared with healthy controls. *Anal. Bioanal. Chem.* **2014**, *406*, 4649–4662. [CrossRef]

31. Cequier-Sánchez, E.; Rodríguez, C.; Ravelo, Á.G.; Zárate, R. Dichloromethane as a Solvent for Lipid Extraction and Assessment of Lipid Classes and Fatty Acids from Samples of Different Natures. *J. Agric. Food Chem.* **2008**, *56*, 4297–4303. [CrossRef]

32. Ciska, E.; Drabińska, N.; Honke, J.; Narwojsz, A. Boiled Brussels sprouts: A rich source of glucosinolates and the corresponding nitriles. *J. Funct. Foods* **2015**, *19*, 91–99. [CrossRef]

33. Risticevic, S.; Lord, H.; Górecki, T.; Arthur, C.L.; Pawliszyn, J. Protocol for solid-phase microextraction method development. *Nat. Protoc.* **2010**, *5*, 122–139. [CrossRef] [PubMed]

34. Naz, S.; Vallejo, M.; García, A.; Barbas, C. Method validation strategies involved in non-targeted metabolomics. *J. Chromatogr. A* **2014**, *1353*, 99–105. [CrossRef] [PubMed]

35. Kordalewska, M.; Macioszek, S.; Wawrzyniak, R.; Sikorska-Wiśniewska, M.; Śledziński, T.; Chmielewski, M.; Mika, A.; Markuszewski, M.J. Multiplatform metabolomics provides insight into the molecular basis of chronic kidney disease. *J. Chromatogr. B* **2019**, *1117*, 49–57. [CrossRef]

36. Rosen Vollmar, A.K.; Rattray, N.J.W.; Cai, Y.; Santos-Neto, Á.J.; Deziel, N.C.; Jukic, A.M.Z.; Johnson, C.H. Normalizing Untargeted Periconceptional Urinary Metabolomics Data: A Comparison of Approaches. *Metabolites* **2019**, *9*, 198. [CrossRef]

37. Rocha, S.M.; Caldeira, M.; Carrola, J.; Santos, M.; Cruz, N.; Duarte, I.F. Exploring the human urine metabolomic potentialities by comprehensive two-dimensional gas chromatography coupled to time of flight mass spectrometry. *J. Chromatogr. A* **2012**, *1252*, 155–163. [CrossRef]

38. Görling, B.; Bräse, S.; Luy, B. NMR chemical shift ranges of urine metabolites in various organic solvents. *Metabolites* **2016**, *6*, 27. [CrossRef]

39. Fathi, F.; Brun, A.; Rott, K.H.; Cobra, P.F.; Tonelli, M.; Eghbalnia, H.R.; Caviedes-Vidal, F.; Karasov, W.H.; Markley, J.L. NMR-based identification of metabolites in polar and non-polar extracts of avian liver. *Metabolites* **2017**, *7*, 61. [CrossRef]

40. Silva, C.L.; Passos, M.; Cmara, J.S. Investigation of urinary volatile organic metabolites as potential cancer biomarkers by solid-phase microextraction in combination with gas chromatography-mass spectrometry. *Br. J. Cancer* **2011**, *105*, 1894–1904. [CrossRef]

Sample Availability: Not available

Article

The Stool Volatile Metabolome of Pre-Term Babies

Alessandra Frau [1,*,†], Lauren Lett [1,†], Rachael Slater [1], Gregory R. Young [2], Christopher J. Stewart [3], Janet Berrington [3,4], David M. Hughes [5], Nicholas Embleton [4,6] and Chris Probert [1]

1   Institute of Systems, Molecular and Integrative Biology, University of Liverpool, Crown Street, Liverpool L69 3GE, UK; llett@liverpool.ac.uk (L.L.); rsh14@liverpool.ac.uk (R.S.); mdcsjp@liverpool.ac.uk (C.P.)
2   Hub for Biotechnology in the Built Environment, Northumbria University, Newcastle upon Tyne NE1 8ST, UK; gregory.young@northumbria.ac.uk
3   Translational and Clinical Research Institute, Faculty of Medical Sciences, Newcastle University, Newcastle upon Tyne NE1 8ST, UK; Christopher.Stewart@newcastle.ac.uk (C.J.S.); j.e.berrington@newcastle.ac.uk (J.B.)
4   Department of Neonatology, Newcastle upon Tyne Hospitals NHS Foundation Trust, Newcastle upon Tyne NE1 8ST, UK; nicholas.embleton@newcastle.ac.uk
5   Department of Health Data Science, University of Liverpool, Liverpool, Merseyside L69 3GA, UK; dmhughes@liverpool.ac.uk
6   Population Health Sciences Institute, Newcastle University, Newcastle upon Tyne NE1 8ST, UK
*   Correspondence: afrau@liverpool.ac.uk; Tel.: +44-(0)151-795-2703
†   These authors contributed equally to this work.

Abstract: The fecal metabolome in early life has seldom been studied. We investigated its evolution in pre-term babies during their first weeks of life. Multiple (n = 152) stool samples were studied from 51 babies, all <32 weeks gestation. Volatile organic compounds (VOCs) were analyzed by headspace solid phase microextraction gas chromatography mass spectrometry. Data were interpreted using Automated Mass Spectral Deconvolution System (AMDIS) with the National Institute of Standards and Technology (NIST) reference library. Statistical analysis was based on linear mixed modelling, the number of VOCs increased over time; a rise was mainly observed between day 5 and day 10. The shift at day 5 was associated with products of branched-chain fatty acids. Prior to this, the metabolome was dominated by aldehydes and acetic acid. Caesarean delivery showed a modest association with molecules of fungal origin. This study shows how the metabolome changes in early life in pre-term babies. The shift in the metabolome 5 days after delivery coincides with the establishment of enteral feeding and the transition from meconium to feces. Great diversity of metabolites was associated with being fed greater volumes of milk.

Keywords: metabolome; feces; neonates; fermentation; protein; carbohydrate; short chain fatty acid

Citation: Frau, A.; Lett, L.; Slater, R.; Young, G.R.; Stewart, C.J.; Berrington, J.; Hughes, D.M.; Embleton, N.; Probert, C. The Stool Volatile Metabolome of Pre-Term Babies. *Molecules* **2021**, *26*, 3341. https://doi.org/10.3390/molecules26113341

Academic Editors: Natalia Drabińska, Ben de Lacy Costello and Igor Jerković

Received: 30 March 2021
Accepted: 27 May 2021
Published: 2 June 2021

Publisher's Note: MDPI stays neutral with regard to jurisdictional claims in published maps and institutional affiliations.

1. Introduction

The intestinal metabolome is shaped by the interactions between the microbiota and diet. Before birth, mammals ingest amniotic fluid which contains amino acids (notably taurine), some proteins (including growth factors and hormones), phospholipids [1], and, potentially, bacteria [2] and volatile organic compounds, from the mother [3]. Soon after birth, bacteria and other microbes that will eventually form the microbiota begin to colonize the intestine. During the neonatal period, there is a huge switch in the enteral intake from amniotic fluid, to colostrum and then milk, in the majority of babies. Colostrum and milk also contain microbes which may seed to the baby [4,5]. Babies that are born significantly pre-term are cared for in Neonatal Intensive Care Units (NICUs) where they receive expressed colostrum and breast milk, if possible.

It has been proposed that the study of feces from neonates may be useful in the early identification of necrotizing enterocolitis (NEC) [6–8] and late onset sepsis (LOS), to which preterm babies are at risk. There is a paucity of research on the metabolome in early life and we hypothesize that disease signals may be obscured as the metabolome is rapidly changing.

Here, we have analyzed the metabolome of a new cohort of preterm babies, who did not develop NEC or late onset sepsis, and explore factors that might have an impact on the metabolome. The paper describes the 'normal metabolome of the preterm neonate' as a reference document for others interested in the health of the newborn.

2. Results

2.1. Patients Demographics

Fifty-one healthy infants (not affected by NEC or LOS), all <32 weeks gestation at birth and participating in both the Enteral LactoFerrin In Neonates (ELFIN) and mechanisms affecting the gut of preterm infants in enteral feeding trials (MAGPIE) [9] studies, were used in this sub-study. A total of 152 samples were analysed (distribution of age and samples shown in Table 1). Of the 51 infants, 46 were twins and 7 were singletons; their key neonatal features are summarised below.

Table 1. Summary of basic demographic features and sampling from 51 preterm babies.

	Median	**Range**
Gestational age (weeks)	29	23–31 + 6 d
Birthweight (g)	1095	585–1820
Samples per donor	3	2–6

d: days.

2.2. Metabolomic Profile of Stool Samples from Pre-Term Babies

There were 36 volatile organic compounds (VOCs) present in at least 25% of samples (Table 2, Appendix A Table A1). The three short chain fatty acids, acetic acid, propionic acid and butanoic acid, were present in 91%, 53% and 42% of samples, respectively. Aldehydes and alcohols were the largest groups with 6 compounds in each group.

Table 2. Summary of 36 volatile organic compounds (VOCs) found in at least 25% of samples.

Short Chain Fatty Acids	**Branched Chain Fatty Acids**	**Methylated Aldehydes**	**Esters**
Acetic acid	2-methylbutanoic acid	Isovaleraldehyde	Ethyl acetate
Propionic acid	Isovaleric acid	2-methylbutyraldehyde	Propyl acetate
Butanoic acid		Isobutyraldehyde	Ethyl propionate
			Propyl propionate

Aldehydes	**Alcohols**	**Ketones/Diketones**	**Others**
Hexanal	Ethanol	2-heptanone	2-ethylfuran
Heptanal	Propanol	4-heptanone	2-pentylfuran
Octanal	1-pentanol	6-methyl-5-hepten-2-one	D-limonene
Nonanal	1-hexanol	Acetoin	Methoxy-phenyl-oxime
Benzaldehyde	1-octen-3-ol	2,3-butanedione	1,4-xylene
Phenylacetaldehyde	2-ethylhexanol		Ethylbenzene

We then investigated the impact of the infants' postnatal age (all samples were included, n = 152). A mixed effect regression model of VOC number per patient and postnatal age (days) showed a significant (p-value < 0.0001) increase in VOCs during time of 1.0126 compound per day (95% Wald confidence interval 1.009, 1.016). Subsequentially, samples were grouped by age (Table 3). The number of VOCs was limited in the first 5 days of life (Table 3) (Figure 1). Anova analysis (f-ratio = 16.55624, p < 0.00001, post hoc HSD) showed that the number of VOCs was significantly different among the groups: R1 < R2**, R3****, R4**** and R2 < R3*, R4***. R3 and R4 not significantly different. (* 0.05, ** < 0.01, *** < 0.001, **** < 0.0001).

Table 3. Samples collected in each age group.

	Age Range (d)	Number Samples	Median Number of VOCs (Range)
Samples from R1	0–5	18	13 (6–22)
Samples from R2	6–10	44	17 (8–31)
Samples from R3	11–20	56	22.5 (8–30)
Samples from R4	21–70	34	24 (13–31)

VOCs: volatile organic compounds, d: days.

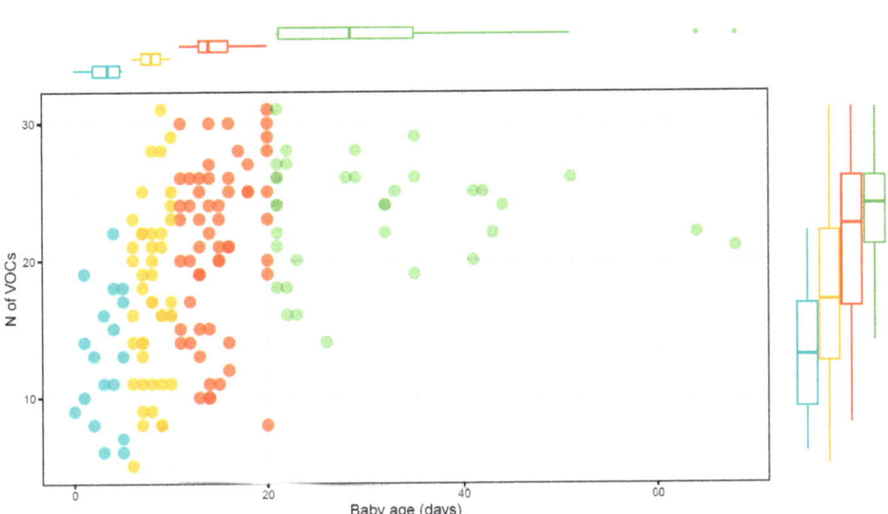

Figure 1. Scatterplot and boxplots to show the number of volatile organic compounds (VOCs) in each of the age groups. Each dot represents a sample (all samples were included, n = 152). (R1 = up to 5 (n = 18), R2 = 6–10 (n = 44), R3 = 11–20 (n = 56), R4 = 21–70 days (n = 34).

Scrutiny of the data showed that some VOCs were present in the majority of babies in each age group. Others started to appear in the second or third groups. Nine VOCs were present in >66% of samples in first group (1–5 days): 4 were aldehydes—hexanal (100%), heptanal (67%), octanal (78%) and nonanal (67%); 2 others were methylated aldehydes, 2-methylbutyraldehyde (67%) and isovaleraldehyde (89%); the remainder were acetic acid (89%), 2-pentylfuran (72%), and 1-octen-3-ol (67%). These 9 VOCs remained common in the later samples. The second group (6-10 days) had 4 further VOCs that were found on >60% of samples: these were 2-methylbutanoic acid (61%), isovaleric acid (70%), 2,3-butanedione (64%) and 6-methyl-5-hepten-2-one (59%). While acetic acid was common in all 4 groups, propionic acid (range 38–64%) and butanoic acid (11–62%) were not.

In Figure 2, we focus on a selection of compounds, showing that some of these, specifically, aldehydes and acetic acid, were present since birth and others (acids, esters, ketones and alcohol), increased after day 5.

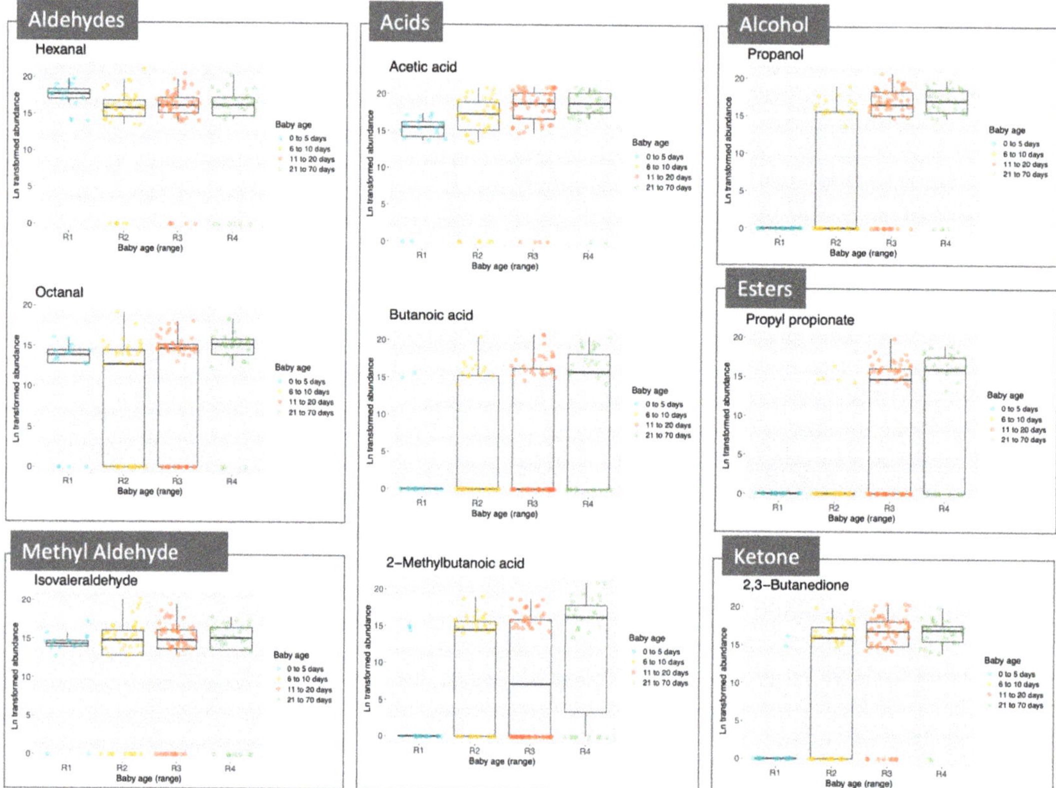

Figure 2. Boxplots for a selection of compounds (abundance/age group). Each boxplot represents a compound, and these are grouped according to the type of molecule (i.e., aldehydes, methyl aldehydes, acids, alcohol, esters, and ketone). All samples were included, n = 152.

Linear mixed-effects (LME) analysis was used to identify compounds that changed over time, results are in Table 4. All of these increased over times (positive slope value). Three other factors were considered in the analysis: batch, gestational age (weeks) and delivery mode. Patient ID was a random effect in the LME analysis. Esters were slightly increased over time in babies with a higher gestational age, meanwhile an alcohol and a ketone show a weak increase in babies born earlier during the pregnancy. Interestingly, the alcohol, 1-octen-3-ol (Table 4 and Figure 3), can be related to fungal metabolism [10,11]. This metabolite and 2-pentylfuran, another compound related to fungal metabolism [12], were also slightly increased in babies born by caesarean section (Table 4 and Figure 3). Most of the compounds that showed significant association with delivery mode were increased in babies born by caesarean section, except for ethyl acetate that was increased in babies born by vaginal delivery.

Table 4. List of volatile organic compounds (VOCs) that were influenced by relevant variables obtained with linear mixed-effects (LME) analysis.

Compound	Postnatal Age (Days)		Gestational Age (Weeks)		Delivery Mode	
	Slope	*p*-Value	Parameter	*p*-Value	Parameter	*p*-Value
Short chain fatty acids						
Butanoic acid	0.28	***	—	—	—	—
Acetic acid	0.08	*	—	—	—	—
Propionic acid	0.15	*	—	—	—	—
Branched chain fatty acids						
Isovaleric acid	0.2	**	—	—	—	—
2-methylbutanoic acid	0.24	***	—	—	—	—
Esters						
Ethyl acetate	0.19	**	0.87	*	−4.07	*
Propyl acetate	0.36	***	1.15	*	—	—
Ethyl propionate	0.27	***	0.88	*	—	—
Propyl propionate	0.36	***	1.02	*	—	—
Aldehydes						
Heptanal	0.11	*	—	—	—	—
Octanal	0.13	*	—	—	3.17	*
Nonanal	0.14	**	—	—	—	—
Benzaldehyde	0.15	**	—	—	—	—
Phenylacetaldehyde	0.2	***	—	—	—	—
Alcohols						
Propanol	0.38	***	—	—	—	—
1-octen-3-ol	—	—	−0.76	*	3.45	**
Ketones/diketones						
2-heptanone	—	—	—	—	3.95	**
4-heptanone	—	—	−1.38	***	—	—
6-methyl-5-hepten-2-one	0.18	***	—	—	—	—
Acetoin	0.18	**	—	—	—	—
2,3-butanedione	0.27	***	—	—	—	—
Others						
2-ethylfuran	—	—	—	—	2.66	*
2-pentylfuran	—	—	—	—	4.11	**
Methoxy phenyl-oxime	0.15	**	—	—	—	—

A positive slope for infant age (days) indicates an increase in the compound over time; a positive value for gestational age indicates that babies born later had more of that compound; a positive value for delivery mode means that babies born through a caesarean section had more of that compound, opposite to a negative slope that refers to a compound being more prevalent in babies born by vaginal delivery. Values that were not significant are not shown (−). Significance codes: − *p* not significant, * *p* < 0.05, ** *p* < 0.01, *** *p* < 0.001. All samples were included, n = 152.

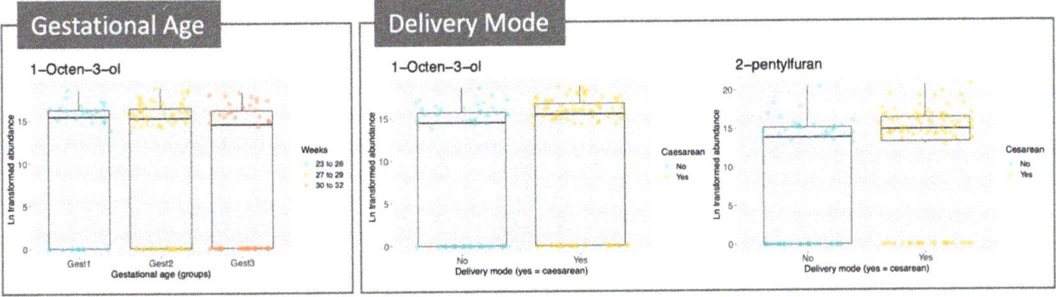

Figure 3. Boxplots for a selection of compounds (abundance/gestational age and delivery mode). Each boxplot represents a compound, and these are grouped according to the variable of interest (gestational age and delivery mode). All samples were included, n = 152.

3. Discussion

This is the largest study of the fecal metabolome in the neonatal period. Samples from the first few days after birth are characterized by the limited range of VOCs and the predominance of acetic acid and aldehydes. We found that acetic acid was found in the majority of these samples, but propionic acid and butanoic acid were not. Studies on the fermentation of taurine have shown that acetic acid is the most common short-chain fatty acid (SCFA) derived from this amino acid [13]: it is plausible that the taurine-rich amniotic fluid is responsible for this pattern of SCFA in the meconium.

The presence of aldehydes was striking. There were four medium-chain aldehydes (C6–C9) and two further branched aldehydes. Aldehydes are a consequence of lipid peroxidation [14,15]. Branched-chain aldehydes arise from amino acids (for example, leucine and isoleucine [16,17]) and are metabolites of lactic acid bacteria, which are abundant in the vagina and are likely to seed to the neonate during delivery.

There was a steady increase in the range (median 13 to 24, ANOVA $p < 0.00001$) of VOCs in faecal samples during the first few weeks of life. The lack of esters was striking. Esters are common in adult faeces and may arise from foods (as flavours in fruit [18]) but may occur by the condensation of fatty acids and alcohols [19].

The previous study of VOCs in preterm new-borns [6] reported 36 samples were obtained from seven babies over 14 days. The same analytical laboratory methods were used although the present study had more consistent stool weights (80.6 mg (range 32.5–100 mg, SD 12.3 mg) than the earlier one (890 mg, range 300–2400 mg, SD 460 mg). The main difference between these two studies was the temporal sampling employed here: the earlier report did not consider the influence of the age of the babies. As a result, no conclusions could be drawn about the evolution of the metabolome. Costello noted that 7 of the 15 most abundant compounds were aldehydes. Acetone and ethanol were also prevalent. 2-ethylhexanol was also common (97%), but it was considered to be a contaminant arising from plasticware: it was found in 61% of samples in the present study, even though samples were collected into glass vials. The three short chain fatty acids are common in the stool of adults (>95%) [19], each had a low prevalence (<10%) in the Costello study.

The paper reports the evolution of the faecal metabolome in the first weeks of life in preterm babies. There is a marked change that occurs in association with the introduction of first milk feeds. The lack of SCFA in the first week of life suggests they are not a requirement for the intestine in utero or early after birth; their appearance when milk is introduced suggests that the faecal microbiota contains bacteria able to ferment carbohydrates and amino acids to synthesize SCFA.

Gestational age and delivery mode were included in our LME model as these factors are known to influence the gut microbiota of infants. A weak increase in fungal metabolites was observed in babies born earlier during the pregnancy and delivered by caesarean section. In full-term infants, mode of delivery is known to influence the microbiota and it has been shown that babies born by caesarean delivery are more susceptible to being colonized by opportunistic pathogen acquired from the hospital environment rather than commensal bacteria that are transmitted by the mother during vaginal delivery [20]. This effect may increase in babies spending a long time in NICU and may explain the increase in signal of fungal volatile (1-octen-3-ol and 2-pentylfuran), as yeasts may colonize the gut in an opportunistic fashion and NICU are a source of yeasts [21]. Similarly, earlier preterm babies showed a weak increase in fungal metabolites. A recent study on interkingdom relationships (bacteria, fungi and archaea) on preterm infants [22] found a defined succession of bacteria genera, however the evolution of the fungal community was less predictable. They found a negative correlation between fungal and bacterial load, and that *Candida* colonization was inhibited by *Staphylococcus*, a pioneer in the establishment of gut microbiota in early life [23].

4. Materials and Methods

4.1. Patients

Patients in this sub-study were part of a large cohort recruited to the MAGPIE study. This study focuses on the children without necrotising enterocolitis or late onset sepsis, who gave a least two stool samples during the first 70 days of life. The overarching study was the ELFIN study. Preterm infants at one of 12 participating NHS hospital trusts (13 separate NICUs) were eligible if they met enrolment criteria for ELFIN which included preterm infants < 32 weeks gestation and <72 h postnatal age. Potential infants meeting the eligibility criteria for MAGPIE were identified and recruited by the local healthcare team. Parents were approached for written informed consent after they had received a verbal and written explanation of MAGPIE. The study protocol was approved by East Midlands—Nottingham 2 Research Ethics Committee 16/EM/0042.

4.2. Extraction of VOCs

Faecal samples collected in glass vials and stored at −80 °C in Newcastle for up to 12 months, before shipping to the Liverpool laboratory, on dry ice, and being stored at −20 °C again. Prior to analysis, samples were weighed, and aliquots transferred to 10 mL glass headspace vials with magnetic septum caps (Sigma-Aldrich, Dorset, UK) in a hood: a mean of 80.6 mg stool (SD 12.3 mg) was used for the analysis. During aliquoting an empty vial remained unsealed in the hood to collect circulating air, later this was then re-sealed in the hood and was stored with the prepared samples. These air samples were analysed alongside the samples to determine whether there were contaminants in the air when the samples were aliquoted.

Volatile organic compound analysis was performed using gas-chromatography mass-spectrometry on a PerkinElmer Clarus 500 GC-MS quadrupole benchtop system (Beaconsfield, UK) and Combi PAL auto-sampler (CTC Analytics, Zwingen, Switzerland). VOCs were extracted using solid phase micro-extraction with a divinylbenzene-carboxen-polydimethylsiloxane (DVB-CAR-PDMS) (Sigma-Aldrich, Dorset, UK) coated fibre, otherwise the protocol and GC-MS conditions were the same as published by Reade et al. (2014) [24]. Samples were heated to 60 °C for 30 min at prior to fibre exposure, the fibre was exposed to the headspace gases at 60 °C for 20 min, then thermally desorbed for 5 min at 220 °C.

The GC column used was a 60 m Zebron ZB-624 (inner diameter 0.25 mm, length 60 m, film thickness 1.4 μm (Phenomenex, Macclesfield, UK). The carrier gas used was 99.996% pure helium (BOC, Sheffield, UK) which was passed through a helium purification system, Excelasorb™ (Supelco, Bellefonte, PA, USA) at 1 mL/min. The initial temperature of the GC oven was set at 40°C and held for 2 min before increasing to 220 °C at a rate of 5 °C/min and held for 4 min with a total run time of 41 min. The MS was operated in electron impact ionization EI + mode, scanning from 10 to 300 m/z with an interscan delay of 0.1 s and a resolution of 1000 at FWHM (Full Width at Half Maximum). Samples were run in two batches, the first batch had 36 samples and the second 116.

4.3. Downstream Data Processing and Analysis

The GC-MS data were processed as CDF files using the Automated Mass Spectral Deconvolution and Identification System software (AMDIS, version 2.73, 2017, Gaithersburg, MD, USA), the NIST mass spectral library ((version 2.0, 2011 purchased from PerkinElmer, Beaconsfield, UK) and the R package Metab [25]. AMDIS and NIST software were used to build a compound library; VOCs were added based on a match criterion of greater than 700, then a probability of a true match (greater than 70%) and finally inspection of fragment patterns. This compound library is then used, with AMDIS, and was applied to deconvolute chromatograms and identifying metabolites. VOCs were named as common names, moreover, the International Union of Pure and Applied Chemistry (IUPAC) [26] names along with PubChem CID number are provided in Appendix A Table A2.

VOCs data were analyzed with R (version 3.6.3, Vienna, Austria) [27] in RStudio (version 1.2.5033, Boston, MA, USA) [28,29]. Firstly, the VOCs table was adjusted as follows: only compounds observed in at least 25% of samples were kept, natural log transformation was performed using the log() function and missing values were imputed to 0. Generalized linear mixed-effects, glmer() function of the lme4 package [30], was used to perform a mixed effect regression model to assess whether there was correlation between the number of VOCs and postnatal baby age (days). Finally, LME model analysis was performed with the lmer() function of the lme4 package [30]. Patients ID was used as a random factor, while baby age (days), GC-MS run batch, gestational age and delivery mode were the fixed factors. ggplot2 [31] package was used to produce the charts.

5. Conclusions

This study shows the evolution of the metabolome in early life in pre-term babies. We observed a clear shift in the metabolome after 5 days from birth that coincides with the establishment of enteral feeding and the transition from meconium to faeces.

Author Contributions: Conceptualization, C.P., N.E., J.B., L.L. and A.F.; methodology, L.L., R.S.; formal analysis, A.F., R.S., C.P., D.M.H., C.J.S.; writing—original draft preparation, A.F. and C.P.; writing—review and editing, G.R.Y., J.B., N.E., C.J.S., L.L., D.M.H., R.S. funding acquisition, N.E. All authors have read and agreed to the published version of the manuscript.

Funding: This research was funded by National Institute for Health Research (NIHR) Efficacy and Mechanistic Evaluation programme (EME) grant number 13/122/02.

Institutional Review Board Statement: The study was conducted according to the guidelines of the Declaration of Helsinki and approved by East Midlands—Nottingham 2 Research Ethics Committee 16/EM/0042.

Informed Consent Statement: Informed consent was obtained from parents of all infants involved in the study.

Data Availability Statement: CDF files, VOCs and metadata tables are available upon reasonable request from the corresponding author.

Conflicts of Interest: The authors declare no conflict of interest.

Sample Availability: Human samples are not available from the authors, in accordance with the Human Tissue Act, these have been used and remainder destroyed.

Appendix A

Table A1. The prevalence of the 36 most abundant volatile in the four age groups.

	Prevalence in Each Age Group (%)				Mean Prevalence (%)
	G1	G2	G3	G4	
Acetic acid	88.9	86.4	92.9	94.1	90.8
Hexanal	100.0	86.4	91.1	82.4	88.8
Nonanal	66.7	68.2	87.5	94.1	80.9
Isovaleraldehyde	88.9	77.3	78.6	76.5	78.9
Heptanal	66.7	63.6	82.1	85.3	75.7
2,3-butanedione	11.1	63.6	85.7	94.1	72.4
Octanal	77.8	54.5	71.4	91.2	71.7
2-methylbutyraldehyde	66.7	63.6	73.2	70.6	69.1
4-heptanone	55.6	70.5	69.6	73.5	69.1
2-pentylfuran	72.2	61.4	69.6	73.5	68.4
Isovaleric acid	16.7	70.5	69.6	70.6	63.8
1-octen-3-ol	66.7	56.8	60.7	76.5	63.8
6-methyl-5-hepten-2-one	33.3	59.1	66.1	82.4	63.8

Table A1. *Cont.*

	Prevalence in Each Age Group (%)				Mean Prevalence (%)
2-ethylhexanol	50.0	59.1	60.7	70.6	61.2
2-heptanone	66.7	59.1	62.5	52.9	59.9
Propanol	0.0	38.6	78.6	88.2	59.9
1-pentanol	61.1	54.5	51.8	64.7	56.6
2-methylbutanoic acid	16.7	61.4	50.0	73.5	54.6
Propionic acid	38.9	40.9	64.3	58.8	53.3
Acetoin	5.6	54.5	60.7	61.8	52.6
Benzaldehyde	38.9	45.5	42.9	76.5	50.7
Phenylacetaldehyde	11.1	38.6	55.4	70.6	48.7
Acetone	16.7	54.5	48.2	41.2	44.7
Butanoic acid	11.1	31.8	48.2	61.8	42.1
Propyl propionate	0.0	11.4	57.1	67.6	39.5
Methoxy-phenyl-oxime	44.4	27.3	37.5	52.9	38.8
Ethyl acetate	0.0	27.3	48.2	55.9	38.2
Propyl acetate	0.0	13.6	53.6	64.7	38.2
Ethanol	0.0	43.2	48.2	35.3	38.2
Isobutyraldehyde	16.7	36.4	37.5	47.1	36.8
Ethyl propionate	5.6	13.6	44.6	58.8	34.2
1,4-xylene	11.1	43.2	41.1	20.6	33.6
1-hexanol	27.8	38.6	33.9	29.4	33.6
D-limonene	0.0	22.7	39.3	47.1	31.6
Ethylbenzene	27.8	29.5	28.6	38.2	30.9
2-ethylfuran	33.3	25.0	25.0	26.5	26.3

Table A2. Names of compounds described as common names, IUPAC names and PubChem CID number.

Common Name	IUPAC Name	CID Number
Acetic acid	Acetic acid	176
Propionic acid	Propanoic acid	1032
Butanoic acid	Butanoic acid	264
2-methylbutanoic acid	2-methylbutanoic acid	8314
Isovaleric acid	3-methylbutanoic acid	10,430
Isovaleraldehyde	3-methylbutanal	11,552
2-methylbutyraldehyde	2-methylbutanal	7284
Isobutyraldehyde	2-methylpropanal	6561
Ethyl acetate	Ethyl acetate	8857
Propyl acetate	Propyl acetate	7997
Ethyl propionate	Ethyl propanoate	7749
Propyl propionate	Propyl propanoate	7803
Hexanal	Hexanal	6184
Heptanal	Heptanal	8130
Octanal	Octanal	454
Nonanal	Nonanal	31,289
Benzaldehyde	Benzaldehyde	240
Phenylacetaldehyde	2-phenylacetaldehyde	998
Ethanol	Ethanol	702
Propanol	Propan-1-ol	1031
1-pentanol	Pentan-1-ol	6276
1-hexanol	Hexan-1-ol	8103
1-octen-3-ol	Oct-1-en-3-ol	18,827

Table A2. *Cont.*

Common Name	IUPAC Name	CID Number
2-ethylhexanol	2-ethylhexan-1-ol	7720
Acetone	Propan-2-one	180
2-heptanone	Heptan-2-one	8051
4-heptanone	Heptan-4-one	31,246
6-methyl-5-hepten-2-one	6-methylhept-5-en-2-one	9862
Acetoin	3-hydroxybutan-2-one	179
2,3-butanedione	Butane-2,3-dione	650
2-ethylfuran	2-ethylfuran	18,554
2-pentylfuran	2-pentylfuran	19,602
D-limonene	(4R)-1-methyl-4-prop-1-en-2-ylcyclohexene	440,917
Methoxy-phenyl-oxime	methyl (Z)-N-hydroxybenzenecarboximidate	9,602,988
1,4-xylene	1,4-xylene	7809
Ethylbenzene	Ethylbenzene	7500

References

1. Underwood, M.A.; Gilbert, W.M.; Sherman, M.P. Amniotic fluid: Not just fetal urine anymore. *J. Perinatol.* **2005**, *25*, 341–348. [CrossRef] [PubMed]
2. Collado, M.C.; Rautava, S.; Aakko, J.; Isolauri, E.; Salminen, S. Human gut colonisation may be initiated in utero by distinct microbial communities in the placenta and amniotic fluid. *Sci. Rep.* **2016**, *6*, 23129. [CrossRef]
3. Minet-Quinard, R.; Ughetto, S.; Gallot, D.; Bouvier, D.; Lemery, D.; Goncalves-Menses, N.; Blanchon, L.; Sapin, V. Volatile organic compounds in amniotic fluid during normal human pregnancy. *Placenta* **2014**, *35*, A19. [CrossRef]
4. Hunt, K.M.; Foster, J.A.; Forney, L.J.; Schütte, U.M.E.; Beck, D.L.; Abdo, Z.; Fox, L.K.; Williams, J.E.; McGuire, M.K.; McGuire, M.A. Characterization of the diversity and temporal stability of bacterial communities in human milk. *PLoS ONE* **2011**, *6*, e21313. [CrossRef]
5. Huurre, A.; Kalliomäki, M.; Rautava, S.; Rinne, M.; Salminen, S.; Isolauri, E. Mode of Delivery—Effects on Gut Microbiota and Humoral Immunity. *Neonatology* **2008**, *93*, 236–240. [CrossRef]
6. De Lacy Costello, B.; Ewen, R.; Ewer, A.K.; Garner, C.E.; Probert, C.S.J.; Ratcliffe, N.M.; Smith, S. An analysis of volatiles in the headspace of the faeces of neonates. *J. Breath Res.* **2008**, *2*, 037023. [CrossRef] [PubMed]
7. Garner, C.E.; Ewer, A.K.; Elasouad, K.; Power, F.; Greenwood, R.; Ratcliffe, N.M.; Costello, B.D.L.; Probert, C.S. Analysis of Faecal Volatile Organic Compounds in Preterm Infants Who Develop Necrotising Enterocolitis: A Pilot Study. *J. Pediatr. Gastroenterol. Nutr.* **2009**, *49*, 559–565. [CrossRef] [PubMed]
8. De Meij, T.G.J.; Van Der Schee, M.P.C.; Berkhout, D.J.C.; Van De Velde, M.E.; Jansen, A.E.; Kramer, B.W.; Van Weissenbruch, M.M.; Van Kaam, A.H.; Andriessen, P.; Van Goudoever, J.B.; et al. Early Detection of Necrotizing Enterocolitis by Fecal Volatile Organic Compounds Analysis. *J. Pediatr.* **2015**, *167*, 562–567.e1. [CrossRef]
9. Embleton, N.D.; Berrington, J.E.; Dorling, J.; Ewer, A.K.; Juszczak, E.; Kirby, J.A.; Lamb, C.A.; Lanyon, C.V.; McGuire, W.; Probert, C.S.; et al. Mechanisms Affecting the Gut of Preterm Infants in Enteral Feeding Trials. *Front. Nutr.* **2017**, *4*, 14. [CrossRef]
10. Combet, E.; Henderson, J.; Eastwood, D.C.; Burton, K.S. Eight-carbon volatiles in mushrooms and fungi: Properties, analysis, and biosynthesis. *Mycoscience* **2006**, *47*, 317–326. [CrossRef]
11. Hung, R.; Lee, S.; Bennett, J.W. Fungal volatile organic compounds and their role in ecosystems. *Appl. Microbiol. Biotechnol.* **2015**, *99*, 3395–3405. [CrossRef] [PubMed]
12. Chambers, S.T.; Syhre, M.; Murdoch, D.R.; McCartin, F.; Epton, M.J. Detection of 2-Pentylfuran in the breath of patients with Aspergillus fumigatus. *Med. Mycol.* **2009**, *47*, 468–476. [CrossRef] [PubMed]
13. Sasaki, K.; Sasaki, D.; Okai, N.; Tanaka, K.; Nomoto, R.; Fukuda, I.; Yoshida, K.; Kondo, A.; Osawa, R. Taurine does not affect the composition, diversity, or metabolism of human colonic microbiota simulated in a single-batch fermentation system. *PLoS ONE* **2017**, *12*, e0180991. [CrossRef]
14. Fritz, K.S.; Petersen, D.R. An overview of the chemistry and biology of reactive aldehydes. *Free Radic. Biol. Med.* **2013**, *59*, 85–91. [CrossRef]
15. Rizzo, W.B. Fatty aldehyde and fatty alcohol metabolism: Review and importance for epidermal structure and function. *Biochim. Biophys. Acta Mol. Cell Biol. Lipids* **2014**, *1841*, 377–389. [CrossRef] [PubMed]
16. Engels, W.J.M.; Alting, A.C.; Arntz, M.M.T.G.; Gruppen, H.; Voragen, A.G.J.; Smit, G.; Visser, S. Partial purification and characterization of two aminotransferases from Lactococcus lactis subsp. cremoris B78 involved in the catabolism of methionine and branched-chain amino acids. *Int. Dairy J.* **2000**, *10*, 443–452. [CrossRef]
17. Smit, B.A.; Engels, W.J.M.; Smit, G. Branched chain aldehydes: Production and breakdown pathways and relevance for flavour in foods. *Appl. Microbiol. Biotechnol.* **2009**, *81*, 987–999. [CrossRef]
18. El Hadi, M.A.M.; Zhang, F.J.; Wu, F.F.; Zhou, C.H.; Tao, J. Advances in fruit aroma volatile research. *Molecules* **2013**, *18*, 8200–8229. [CrossRef]

19. Garner, C.E.; Smith, S.; de Lacy Costello, B.; White, P.; Spencer, R.; Probert, C.S.J.; Ratcliffe, N.M. Volatile organic compounds from feces and their potential for diagnosis of gastrointestinal disease. *FASEB J.* **2007**, *21*, 1675–1688. [CrossRef]
20. Shao, Y.; Forster, S.C.; Tsaliki, E.; Vervier, K.; Strang, A.; Simpson, N.; Kumar, N.; Stares, M.D.; Rodger, A.; Brocklehurst, P.; et al. Stunted microbiota and opportunistic pathogen colonization in caesarean-section birth. *Nature* **2019**, *574*, 117–121. [CrossRef]
21. Heisel, T.; Nyaribo, L.; Sadowsky, M.J.; Gale, C.A. Breastmilk and NICU surfaces are potential sources of fungi for infant mycobiomes. *Fungal Genet. Biol.* **2019**, *128*, 29–35. [CrossRef] [PubMed]
22. Rao, C.; Coyte, K. Multi-kingdom ecological drivers of microbiota assemby in preterm infants. *Nature* **2021**. [CrossRef] [PubMed]
23. Cilieborg, M.S.; Boye, M.; Sangild, P.T. Bacterial colonization and gut development in preterm neonates. *Early Hum. Dev.* **2012**, *88*, S41–S49. [CrossRef]
24. Reade, S.; Mayor, A.; Aggio, R.; Khalid, T.; Pritchard, D.; Ewer, A.; Probert, C. Optimisation of Sample Preparation for Direct SPME-GC-MS Analysis of Murine and Human Faecal Volatile Organic Compounds for Metabolomic Studies. *J. Anal. Bioanal. Tech.* **2014**, *5*. [CrossRef]
25. Aggio, R.; Villas-Bôas, S.G.; Ruggiero, K. Metab: An R package for high-throughput analysis of metabolomics data generated by GC-MS. *Bioinformatics* **2011**, *27*, 2316–2318. [CrossRef] [PubMed]
26. Henri, A.; Favre, W.H.P. *Nomenclature of Organic Chemistry: IUPAC Recommendations and Preferred Names 2013*; Royal Society of Chemistry: Cambridge, UK, 2014.
27. R Core Team. R: A Language And Environment for Statistical Computing. *Vienna, Austria.* 2020. Available online: https://www.R-project.org/ (accessed on 27 May 2021).
28. RStudio Team. *RStudio: Integrated Development for R*; RStudio, PBC: Boston, MA, USA,, 2020; Available online: http://www.rstudio.com/ (accessed on 27 May 2021).
29. Oksanen, J.; Blanchet, F.G.; Kindt, R.; Legendre, P.; Minchin, P.R.; O'Hara, R.B.; Simpson, G.L.; Solymos, P.; Henry, M.; Stevens, M.H.H.; et al. Package 'Vegan': Community Ecology Package. 2015. Available online: https://cran.r-project.org/web/packages/vegan/index.html (accessed on 27 May 2021).
30. Bates, D.; Mächler, M.; Bolker, B.M.; Walker, S.C. Fitting linear mixed-effects models using lme4. *J. Stat. Softw.* **2015**, *67*, 1–48. [CrossRef]
31. Wickham, H. *ggplot2: Elegant Graphics for Data Analysis*; Springer: New York, NY, USA, 2016; ISBN 978-0-387-98141-3.

Article

Using Volatile Organic Compounds to Investigate the Effect of Oral Iron Supplementation on the Human Intestinal Metabolome

Ammar Ahmed [1], Rachael Slater [1], Stephen Lewis [2] and Chris Probert [1,*]

1   The Henry Wellcome Laboratory, Institute of Systems, Molecular and Integrative Biology, University of Liverpool, Liverpool L69 3BX, UK; Ammar.Ahmed@elht.nhs.uk (A.A.); rsh14@liverpool.ac.uk (R.S.)
2   Department of Gastroenterology, University Hospitals Plymouth NHS Trust, Plymouth PL6 8DH, UK; sjl@doctors.org.uk
*   Correspondence: Chris.Probert@liverpool.ac.uk; Tel.: +44-(0)-151-795-4010

Academic Editors: Natalia Drabińska and Ben de Lacy Costello
Received: 6 October 2020; Accepted: 29 October 2020; Published: 3 November 2020

Abstract: Patients with iron deficiency anaemia are treated with oral iron supplementation, which is known to cause gastrointestinal side effects by likely interacting with the gut microbiome. To better study this impact on the microbiome, we investigated oral iron-driven changes in volatile organic compounds (VOCs) in the faecal metabolome. Stool samples from patients with iron deficiency anaemia were collected pre- and post-treatment ($n = 45$ and 32, respectively). Faecal headspace gas analysis was performed by gas chromatography–mass spectrometry and the changes in VOCs determined. We found that the abundance of short-chain fatty acids and esters fell, while aldehydes increased, after treatment. These changes in pre- vs. post-iron VOCs resemble those reported when the gut is inflamed. Our study shows that iron changes the intestinal metabolome, we suggest by altering the structure of the gut microbial community.

Keywords: iron deficiency anaemia; iron supplementation; volatile organic compounds (VOCs); intestinal metabolome; gut microbiome

---

1. Introduction

Iron is an essential element for numerous metabolic processes, including oxygen transport by haemoglobin and myoglobin [1]. Iron deficiency anaemia (IDA) may arise from inadequate dietary iron intake, malabsorption or blood loss, especially from the gastrointestinal tract [2].

Enteral iron absorption is tightly regulated. Its uptake is dependent on Divalent Metal-Transporter-1, which allows the uptake of iron via the enterocytes in the enteral lumen [1,3]. Other key points are the regulation of transferrin, which binds iron in the circulation; ferrireductases, which facilitate absorption by converting iron from the insoluble ferric (Fe^{3+}) form into the soluble ferrous (Fe^{2+}) state [1,3]; and hepcidin, an antimicrobial peptide that controls entry of iron into the plasma by binding to and degrading ferroportin, an iron-exporting protein that is very highly expressed on the basolateral membrane of enterocytes [1–3]. Under physiological conditions, these series of proteins ensure that enough iron is absorbed, without leading to overload.

Oral iron supplements often exceed the absorptive capacity of the small intestine and the surplus enters the colon where it can lead to gastrointestinal side effects [1,2,4,5] by the generation of free radicals, which damage the epithelium [1,2], and by enhancing the growth of some, but not all, enteric bacteria, leading to dysbiosis [6,7]. These adverse effects can be severe and contribute to the poor compliance with iron therapy.

Volatile organic compounds (VOCs) are carbon-based compounds with a high vapour pressure (and low boiling point) that, consequently, readily enter the gaseous phase. They may contribute to the odours that are associated with faeces and other bodily fluids. Faeces represent the end product of enteric metabolism and digestion, and so changes in faecal VOCs can be used to study changes in the intestinal metabolome and, by extension, the microbiome. Our aim was to explore oral iron-driven changes in the enteral metabolome by comparing the VOC profiles of anaemic patients before and after they had received oral iron supplementation.

2. Results

The samples were received as two groups (Table 1).

Table 1. Summary of age and sex of donors in the two groups.

	Unpaired Samples, Group 1			Paired Samples, Group 2		
	Pre	Post	Total Samples	Pre	Post	Total Samples
Male:Female	21:14	8:14	57	4:6	4:6	20
Mean age (y)	71.4	71.1		69.5	NA	

The median number of VOCs from Group 1 was 73.5 and 77 in the pre- and post-treatment samples, respectively; for Group 2, the values were 68 and 71.5, respectively. Comparison of pre- and post-treatment results in Group 1 found four VOCs that changed significantly in abundance: two aldehydes increased (Table 2). Table 3 summarises the fold change in Group 1, in response to the supplement.

Table 2. Volatiles that changed significantly after iron treatment in Group 1.

	p	False Discovery Rate	Trend
Octanal	5.2×10^{-4}	0.004	Increase
Heptanal	9.8×10^{-4}	0.019	Increase
Ethyl hexanoate	4×10^{-4}	0.015	Decrease
2,4-dimethylpentan-3-ol	9.5×10^{-4}	0.019	Decrease

Table 3. Summary of fold change data in Group 1.

VOC That Decreased	Fold Change	VOC That Increased	Fold Change
2,3,5-trimethylpyrazine	15.4	Octanal	7.3
Ethyl 2-phenylacetate	14.9	Heptanal	4.2
Ethyl hexanoate	13.3	2-pentylfuran	4.1
Methyldisulfanylmethane	12.0	Pentanal	3.1
Heptanoic acid	10.7	2-methyltetrazol-5-amine	2.8
4-methylpentanoic acid	8.4	Heptan-2-one	2.7
Methyl pentanoate	8.3	Oct-1-en-3-ol	2.6
Butyl butanoate	7.6		
Ethyl pentanoate	5.9		
Ethyl butanoate	5.6		
Methyl butanoate	4.3		
2,4-dimethylpentan-3-ol	2.9		
Ethenylbenzene	2.7		
1,3-di-tert-butylbenzene	2.6		
Hexanoic acid	2.5		
2-methylbutanoic acid	2.5		
Tetradecane	2.5		
2-methylpropanoic acid	2.4		
5-methyloxolan-2-one	2.4		
2-methylpropanal	2.3		
Ethenyl acetate	2.1		
6,6-dimethyl-2-methylenebicyclo3.1.1heptane	2.1		
Acetic acid	2.1		
1-methyl-3-propan-2-ylbenzene	2.1		

Analysis of the 10 pairs of samples in Group 2 found no compounds that changed to a degree that was statistically significant: this is likely to be a Type 1 error. We looked at the trend in fold change in this cohort (Table 4). There were 27 VOCs that had a greater than two-fold difference between the first and second sample: 6 compounds appeared more and 21 were less abundant; those that increased included pentanal and 2-pentylfuran.

Table 4. Summary of fold change data in Group 2.

VOC That Decreased	Fold Change	VOC That Increased	Fold Change
(1R,5S,6R,7S,10R) 4,10-dimethyl-7-propan-2-yltricyclo(4.4.0.0,5)dec-3-ene	20.1	2-pentylfuran	3.5
3-isopropenyl-1-isopropyl-4-methyl-4-vinylcyclohexene	14.0	methyldisulfanylmethane	3.4
ethyl butanoate	6.4	cyclohexanecarboxylic acid	2.9
butan-1-ol	6.4	hexanal	2.4
4-hydroxy-4-methylpentan-2-one	5.0	pentanal	2.3
4Z-4,11,11-trimethyl-8-methylidenebicyclo(7.2.0)undec-4-ene	4.6	pentane-2,3-dione	2.2
1-methyl-4-propan-2-ylcyclohexa-1,4-diene	4.5		
1-methyl-3-propan-2-ylbenzene	4.4		
(5s)- 2-methyl-5-propan-2-ylcyclohexa-1,3-diene	4.3		
5Z-2,6,10-trimethyl-1,5,9-undecatriene	4.1		
4-methyl-1-propan-2-ylbicyclo(3.1.0)hex-3-ene	4.0		
7-methyl-3-methylideneocta-1,6-diene	4.0		
6,6-dimethyl-2-methylenebicyclo(3.1.1)heptane	3.5		
4,7,7-trimethylbicyclo(4.1.0)hept-4-ene	3.4		
5-methylheptan-2-one	3.3		
4,6,6-trimethylbicyclo(3.1.1)hept-3-ene	3.1		
2-phenylethanol	2.9		
ethenylbenzene	2.7		
ethylbenzene	2.3		
ethanol	2.2		
1,2-xylene	2.2		

Heatmaps (Figure 1) were generated to illustrate the change in VOCs with treatment, for the two groups. In the first, there is a clear cluster of nine VOCs that increased in abundance with treatment, including four aldehydes (C5–C8), two secondary ketones (C4 and C7), 2-pentylfuran and octen-3-ol. In the second, the six VOCs that increased included two aldehydes (C5 and C7), two secondary ketones (C6 and C7) and 2-pentylfuran.

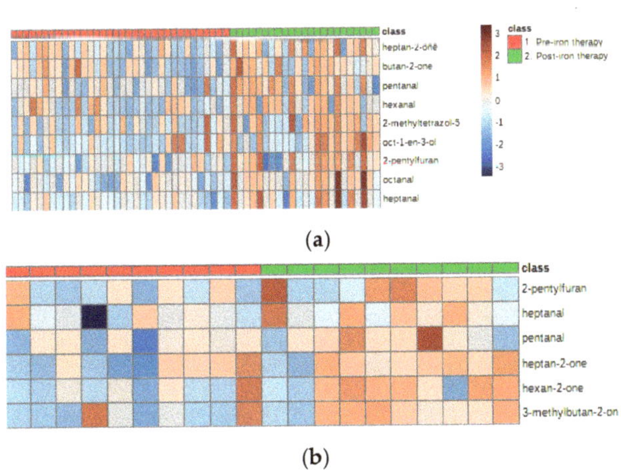

Figure 1. Heatmaps to show the increase in volatile organic compounds (VOCs) with iron therapy in Group 1 (**a**) and Group 2 (**b**).

Box and whisker plots were made to show the change in 2-pentylfuran and pentanal, as these compounds were found to be significantly more abundant in Group 1 and to have greater than two-fold change in abundance in both sets of data in response to treatment (Figure 2). The reduction in esters in Group 1 was of interest as they suggested a change in the metabolism of this class of molecules (Figure 3); however, this was not observed in Group 2.

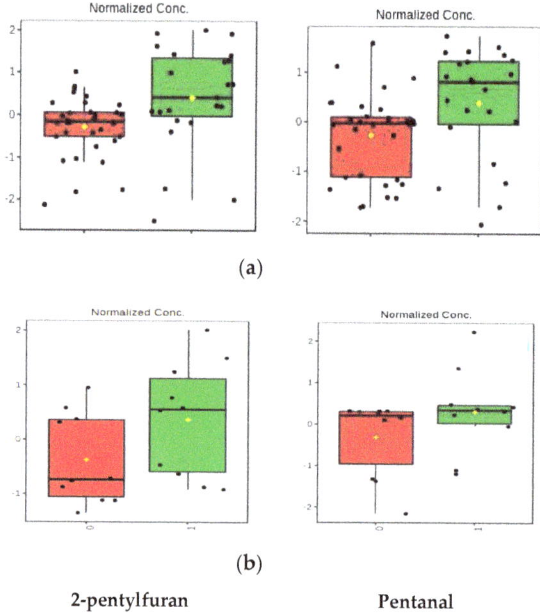

(a)

(b)

2-pentylfuran   Pentanal

Figure 2. Box and whisker plots showing the change in 2-pentylfuran and pentanal in Group 1 (**a**) and Group 2 (**b**).

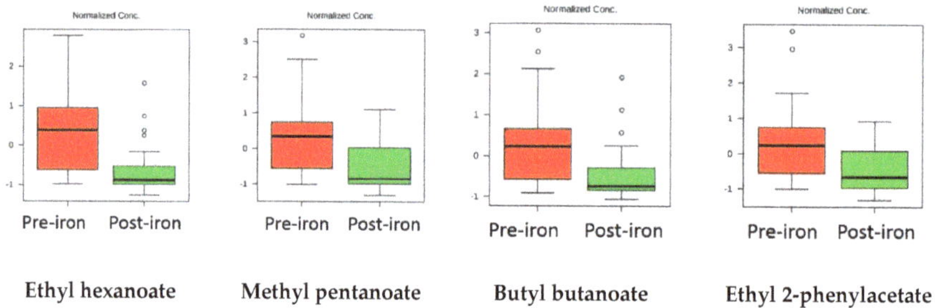

Figure 3. Box and whisker plots show the change in esters in Group 1.

3. Discussion

We investigated the impact of oral iron replacement on the faecal metabolome in two groups of patients. Patients in Group 1 provided unpaired samples of stool, which were taken before and after iron therapy. Patients in Group 2 gave samples twice, enabling a paired analysis. The two groups showed similar results with an increase in faecal aldehydes two months after starting treatment, and there was a reduction in esters.

These changes were more evident, and statistically significant, in Group 1. However, there were similar fold changes in both sets of data (Tables 3 and 4). Several VOCs, most notably aldehydes, had a greater than two-fold increase in response to iron therapy in the follow-up samples (Tables 3 and 4).

Aldehydes may be generated by lipid peroxidation in response to oxidative stress [8,9]. Inflammation may also cause a change in this class [8]. This finding suggests that non-absorbed iron is damaging to the epithelium; however, changes in the microbiome may be responsible for the increase.

Several ketones derived from secondary alcohols were also increased: such ketones may represent oxidative stress but can be generated by the microbiome. The increase in heptan-2-one in the follow-up samples is consistent with previously reported increases of 2-piperidone and 6-methylheptan-2-one in patients with inflammatory bowel disease [8]. Heptan-2-one may play a part in inhibiting enteric *Escherichia coli* [10].

There was an increase in oct-1-en-3-ol, which is strongly associated with fungi. Oct-1-en-3-ol was found be in increased in patients with active Crohn's disease [8]. 2-Pentylfuran is also synthesised by fungi [11,12]. It is plausible that there is a change in the mycobiome in response to iron.

Heptanoic acid was markedly reduced in Group 1, following iron therapy. Recent in vitro models have shown that medium chain fatty acids (MCFAs) are able to bind and form ligands with the gamma class of peroxisome proliferator activated receptors (PPAR-γ). MCFAs are produced by intestinal bacteria and may suppress colitis by activating PPAR-γ in macrophages [13].

The abundance of several short-chain fatty (SCFA) and other carboxylic acids was modestly reduced in the follow-up samples. SCFAs are synthesised by bacterial fermentation of dietary fibre [14]. They are important for intestinal health [14–16]. The reduction in SCFAs suggests that oral iron causes a reduction in SCFA-producing bacteria. Lee et al. showed that oral iron may suppress *Faecalibacterium prausnitzii* and *Ruminococcus bromii* [17]. Mahalhal et al. reported that iron supplements may suppress Firmicutes, which are a major source of SCFAs [18].

Several fatty/carboxylic esters, including methyl pentanoate, ethyl hexanoate, butyl butanoate and ethyl 2-phenylacetate (Figure 3), were decreased in post-iron samples. A decrease in esters was previously reported in inflammatory bowel disease patients [8]. Faecal esters are likely to be derived from the condensation of fatty acids and have been shown to aid the interactions between fatty acids and gut epithelial cells [19]. A decline in both fatty acids and their esters may be a concomitant process driven by the suppression of intestinal bacteria that produce fatty acids. This contrasts with the increased enteric conversion of fatty acids into their esters, which would have been marked by an increase in faecal esters and a decrease in fatty acids. It should be noted that this change was not observed in the paired samples and it may be an artefact of the unpaired samples (Group 1) or a Type 1 error because Group 2 was smaller.

Esters may have an independent anti-inflammatory role in the gut. For example, oral treatment with branched palmitic acid esters of hydroxy stearic acids has demonstrated a reduction in in vivo colonic T cell activation and in expression of proinflammatory cytokines in mice, along with an in vitro reduction in activation of dendritic cells and in the accompanying proliferation of T cells [20].

The patients who took part in the study all had differing underlying causes of IDA, although they were referred to the clinic with suspected gastrointestinal cancer. Future studies in which the analysis of patients is separated according to different IDA pathologies should be considered. Any ongoing gastrointestinal blood lost would have increased liminal iron and reduced the changes observed in this study. Future studies should look at patients taking iron for other indications (post-operatively or in gynaecological clinics).

4. Materials and Methods

4.1. Patient and Stool Sample Selection

Patients with iron deficiency anaemia were recruited from a gastroenterology clinic before the cause of the anaemia was diagnosed. Potential recruited patients gave written informed consent before study procedures took place. Patients were treated with standard ferrous sulphate supplementation, up to 200 mg three times/day. Each participant provided two stool samples: the first before commencing iron therapy and the second two months later. Ethical approval for the study was granted by the UK NHS

Health Research Authority's Research Ethics Service (RES) Committee South West—Central Bristol (REC reference 14/SW/1162).

4.2. Extraction of VOCs

In order to perform metabolomic analysis, patients' samples were sent to the University of Liverpool, where they were stored in freezers at either −20 °C or −80 °C before being processed and run through GC-MS (Perkin Elmer Clarus 500, Beaconsfield, UK) apparatus.

With extensive expert technical assistance in the laboratory, all faecal samples were analysed by way of a quadruple GC-MS benchtop system that was used in conjunction with a CombiPAL autosampler (CTC Analytics, Zwingen, Switzerland). The carrier gas used was helium at a very high purity (BOC, Sheffield, UK).

A local laboratory protocol was followed, which was based on the standardised recommendations of GC-MS method optimisation that were proposed by Reade et al. [21]. Every effort was taken to aliquot at least 500 mg of each faecal sample into a vial that had a magnetic cap and a volume of 10 mL. Although this was not possible for every sample, as some samples had limited faecal quantity, all aliquots had a minimum range of 50–100 mg, which was deemed to be sufficient for GC-MS analysis [21]. Thereafter, the samples were incubated at 60 °C for 30 min. Extraction of VOCs from the vial headspace was itself achieved by utilising solvent-free solid phase micro-extraction (SPME) fibres of the CAR-PDMS 85 μm variety (Sigma-Aldrich, Dorset, UK), which were appropriately pre-conditioned prior to use.

4.3. Downstream Data Processing and Analysis

After thorough evaluation of each chromatogram produced by the GC-MS, additional data inspection and processing were carried out using the Automated Mass Spectral Deconvolution and Identification System (AMDIS version 2.70, https://amdis.software.informer.com/2.7/) software. This was used in tandem with the US National Institute of Standards and Technology's (NIST) Mass Spectral Library (version 2.71, (https://www.perkinelmer.com/uk/product/nist-2011-mass-spectral-library-and-software-n6520220)). By manually analysing chromatograms and mass spectra on AMDIS of over 40 of the first samples that were run on the GC-MS, a project-specific library of compounds was created using the NIST database. In addition to taking into account the highest percentage compound match with NIST spectra, VOCs were only included in the library if they had been identified with a minimum match factor of 800.

At this stage, because these samples were blinded, the library was built by incorporating the VOCs found in both pre-iron and post-iron patients. This library eventually comprised over 300 VOCs, which were all named as per the nomenclature standards set by the International Union of Pure and Applied Chemistry (IUPAC) [22]. Once the library was saturated, the iron status (i.e., pre- or post-) of each sample was categorised, and all the samples were collectively analysed together using AMDIS's batch report function. These data were then further processed and corrected for downstream analysis by using the Metab script package [23], which was utilised within the R (version 3.5, 2018) program.

Subsequently, all statistical analyses were carried out using MetaboAnalyst, a widely used online tool devoted to metabolomic analysis [24]. The principal settings that were utilised included the removal of over 70% of missing metabolite values, though data filtering was not applied. Data were, however, normalised by the median and log-transformed, in addition to being auto-scaled; i.e., mean-centring and then division by the standard deviation of each variable took place. In terms of univariate analysis, fold change analysis and *t*-tests were carried out, with statistical significance being set at a *p*-value less than 0.05. Multivariate analysis centred on principal component analysis (PCA) and partial least squares–discriminant analysis (PLS-DA).

Crucially, in terms of the experimental procedure, VOC extraction was carried out in two batches. Group 1 comprised 35 pre- and 22 post-iron samples that were all from different patients, whereas Group 2 had 10 pre- and 10 post-iron samples each that were all from the same patients (i.e., they were paired samples). Thus, 57 samples were analysed in the first run and a total of 20 samples in the second validation run.

5. Conclusions

This study has shown that oral iron replacement is associated with changes in the faecal metabolome. There is an increase in aldehydes, which may be a result of oxidative stress, and a reduction in esters that may reflect an alteration in the microbiome.

Author Contributions: Conceptualisation by S.L. and C.P.; methodology formulated by S.L., R.S. and C.P.; formal analysis by A.A., R.S. and C.P.; writing—original draft preparation by A.A. and C.P.; writing—review and editing by A.A., R.S., S.L. and C.P. All authors have read and agreed to the published version of the manuscript.

Funding: This research received no external funding.

Conflicts of Interest: The authors declare no conflict of interest.

References

1. Camaschella, C. New insights into iron deficiency and iron deficiency anemia. *Blood Rev.* **2017**, *31*, 225–233. [CrossRef]
2. Camaschella, C. Iron-Deficiency Anemia. *N. Engl. J. Med.* **2015**, *372*, 1832–1843. [CrossRef]
3. Muckenthaler, M.U.; Rivella, S.; Hentze, M.W.; Galy, B. A Red Carpet for Iron Metabolism. *Cell* **2017**, *168*, 344–361. [CrossRef]
4. Tolkien, Z.; Stecher, L.; Mander, A.P.; Pereira, D.I.A.; Powell, J.J. Ferrous Sulfate Supplementation Causes Significant Gastrointestinal Side-Effects in Adults: A Systematic Review and Meta-Analysis. *PLoS ONE* **2015**, *10*, e0117383. [CrossRef]
5. Gereklioglu, C.; Asma, S.; Korur, A.; Erdogan, F.; Kut, A. Medication adherence to oral iron therapy in patients with iron deficiency anemia. *Pak. J. Med. Sci.* **2016**, *32*, 604–607. [CrossRef]
6. Kortman, G.A.; Raffatellu, M.; Swinkels, D.W.; Tjalsma, H. Nutritional iron turned inside out: Intestinal stress from a gut microbial perspective. *FEMS Microbiol. Rev.* **2014**, *38*, 1202–1234. [CrossRef]
7. Polage, C.R. Good and Bad Bacteria Fight for Iron in the Gut. *Sci. Transl. Med.* **2013**, *5*, 199. [CrossRef]
8. Ahmed, I.; Greenwood, R.; Costello, B.; Ratcliffe, N.M.; Probert, C.S. Investigation of faecal volatile organic metabolites as novel diagnostic biomarkers in inflammatory bowel disease. *Aliment. Pharmacol. Ther.* **2016**, *43*, 596–611. [CrossRef]
9. Rezaie, A.; Parker, R.D.; Abdollahi, M. Oxidative Stress and Pathogenesis of Inflammatory Bowel Disease: An Epiphenomenon or the Cause? *Dig. Dis. Sci.* **2007**, *52*, 2015–2021. [CrossRef]
10. Melkina, O.E.; Khmel, I.A.; Plyuta, V.A.; Koksharova, O.A.; Zavilgelsky, G.B. Ketones 2-heptanone, 2-nonanone, and 2-undecanone inhibit DnaK-dependent refolding of heat-inactivated bacterial luciferases in Escherichia coli cells lacking small chaperon IbpB. *Appl. Microbiol. Biotechnol.* **2017**, *101*, 5765–5771. [CrossRef]
11. Bojke, A.; Tkaczuk, C.; Stepnowski, P.; Gołębiowski, M. Comparison of volatile compounds released by entomopathogenic fungi. *Microbiol. Res.* **2018**, *214*, 129–136. [CrossRef]
12. Chambers, S.T.; Bhandari, S.; Scott-Thomas, A.; Syhre, M. Novel diagnostics: Progress toward a breath test for invasiveAspergillus fumigatus. *Med. Mycol.* **2011**, *49*, S54–S61. [CrossRef]
13. Bassaganya-Riera, J.; Viladomiu, M.; Pedragosa, M.; De Simone, C.; Carbo, A.; Shaykhutdinov, R.; Jobin, C.; Arthur, J.C.; Corl, B.A.; Vogel, H.; et al. Probiotic Bacteria Produce Conjugated Linoleic Acid Locally in the Gut That Targets Macrophage PPAR γ to Suppress Colitis. *PLoS ONE* **2012**, *7*, e31238. [CrossRef]
14. Sun, M.; Wu, W.; Liu, Z.; Cong, Y. Microbiota metabolite short chain fatty acids, GPCR, and inflammatory bowel diseases. *J. Gastroenterol.* **2017**, *52*, 1–8. [CrossRef] [PubMed]
15. Singh, B.; Read, S.; Asseman, C.; Malmstrom, V.; Mottet, C.; Stephens, L.A.; Stepankova, R.; Tlaskalova, H.; Powrie, F. Control of intestinal inflammation by regulatory T cells. *Immunol. Rev.* **2001**, *182*, 190–200. [CrossRef]
16. Li, B.; Zheng, S.G. How regulatory T cells sense and adapt to inflammation. *Cell. Mol. Immunol.* **2015**, *12*, 519–520. [CrossRef]

17. Lee, T.; Clavel, T.; Smirnov, K.; Schmidt, A.; Lagkouvardos, I.; Walker, A.; Lucio, M.; Michalke, B.; Schmitt-Kopplin, P.; Fedorak, R.; et al. Oral versus intravenous iron replacement therapy distinctly alters the gut microbiota and metabolome in patients with IBD. *Gut* **2016**, *66*, 863–871. [CrossRef]
18. Mahalhal, A.; Williams, J.M.; Johnson, S.; Ellaby, N.; Duckworth, C.A.; Burkitt, M.D.; Liu, X.; Hold, G.; Campbell, B.J.; Pritchard, D.M.; et al. Oral iron exacerbates colitis and influences the intestinal microbiome. *PLoS ONE* **2018**, *13*, e0202460. [CrossRef]
19. Nepelska, M.; Cultrone, A.; Béguet-Crespel, F.; Le Roux, K.; Doré, J.; Arulampalam, V.; Blottière, H.M. Butyrate Produced by Commensal Bacteria Potentiates Phorbol Esters Induced AP-1 Response in Human Intestinal Epithelial Cells. *PLoS ONE* **2012**, *7*, e52869. [CrossRef]
20. Lee, J.; Moraes-Vieira, P.M.; Castoldi, A.; Aryal, P.; Yee, E.U.; Vickers, C.; Parnas, O.; Donaldson, C.J.; Saghatelian, A.; Kahn, B.B. Branched Fatty Acid Esters of Hydroxy Fatty Acids (FAHFAs) Protect against Colitis by Regulating Gut Innate and Adaptive Immune Responses. *J. Biol. Chem.* **2016**, *291*, 22207–22217. [CrossRef]
21. Reade, S.; Mayor, A.; Aggio, R.; Khalid, T.S.; Pritchard, D.M.; Ewer, A.K.; Probert, C.S. Optimisation of Sample Preparation for Direct SPME-GC-MS Analysis of Murine and Human Faecal Volatile Organic Compounds for Metabolomic Studies. *J. Anal. Bioanal. Tech.* **2014**, *5*. [CrossRef]
22. Favre, H.A.; Powell, W.H. *Nomenclature of Organic Chemistry*; Royal Society of Chemistry (RSC): Cambridge, UK, 2013.
23. Aggio, R.B.M.; Villas–Bôas, S.G.; Ruggiero, K. Metab: An R package for high-throughput analysis of metabolomics data generated by GC-MS. *Bioinformatics* **2011**, *27*, 2316–2318. [CrossRef]
24. Chong, J.; Soufan, O.; Li, C.; Caraus, I.; Li, S.; Bourque, G.; Wishart, D.S.; Xia, J. MetaboAnalyst 4.0: Towards more transparent and integrative metabolomics analysis. *Nucleic Acids Res.* **2018**, *46*, W486–W494. [CrossRef] [PubMed]

Sample Availability: All faecal samples have been destroyed, as required by the Human Tissue Act 2004 (UK).

Publisher's Note: MDPI stays neutral with regard to jurisdictional claims in published maps and institutional affiliations.

MDPI

St. Alban-Anlage 66

4052 Basel

Switzerland

Tel. +41 61 683 77 34

Fax +41 61 302 89 18

www.mdpi.com

Molecules Editorial Office

E-mail: molecules@mdpi.com

www.mdpi.com/journal/molecules